T0320814

INSTANTONS AND LARGE N

An Introduction to Non-Perturbative Methods in Quantum Field Theory

This highly pedagogical textbook for graduate students in particle, theoretical and mathematical physics, explores advanced topics of quantum field theory. Clearly divided into two parts, the first part focuses on instantons with a detailed exposition of instantons in quantum mechanics, supersymmetric quantum mechanics, the large order behavior of perturbation theory, and Yang–Mills theories. The second part moves on to examine the large N expansion in quantum field theory. The organized presentation style, in addition to detailed mathematical derivations, worked examples and applications throughout, enables students to gain practical experience with the tools necessary to start research. The author includes recent developments on the large order behavior of perturbation theory and on large N instantons, and updates existing treatments of classic topics, to ensure that this is a practical and contemporary guide for students developing their understanding of the intricacies of quantum field theory.

MARCOS MARIÑO is Professor of Theoretical Physics and Mathematics at the University of Geneva where his research interests include string theory, mathematical physics and non-perturbative methods in quantum field theory.

INSTANTONS AND LARGE N

An Introduction to Non-Perturbative Methods in Quantum Field Theory

MARCOS MARIÑO

University of Geneva

CAMBRIDGE
UNIVERSITY PRESS

CAMBRIDGE
UNIVERSITY PRESS

University Printing House, Cambridge CB2 8BS, United Kingdom

One Liberty Plaza, 20th Floor, New York, NY 10006, USA

477 Williamstown Road, Port Melbourne, VIC 3207, Australia

314-321, 3rd Floor, Plot 3, Splendor Forum, Jasola District Centre, New Delhi - 110025, India

79 Anson Road, #06-04/06, Singapore 079906

Cambridge University Press is part of the University of Cambridge.

It furthers the University's mission by disseminating knowledge in the pursuit of
education, learning and research at the highest international levels of excellence.

www.cambridge.org
Information on this title: www.cambridge.org/9781107068520

© Cambridge University Press 2015

First published 2015

A catalogue record for this publication is available from the British Library

Library of Congress Cataloging in Publication data
Marino, Marcos, author.
Instantons and large N : an introduction to non-perturbative methods in
quantum field theory / Marcos Mariño (University of Geneva).
pages cm
Includes bibliographical references and index.
ISBN 978-1-107-06852-0
1. Instantons. 2. Quantum field theory. I. Title.
QC793.3.F5M38 2015
530.14′3–dc23
2015017087

ISBN 978-1-107-06852-0 Hardback

Contents

Preface

Quantum Field Theory (QFT) is one of the pillars of modern scientific knowledge, and it is applied in many different areas of physics, from the theory of elementary particles to condensed matter. However, most QFTs cannot be solved exactly and one has to perform some type of approximation in order to extract information from them. The standard approach is a perturbative expansion in a small coupling constant, implemented diagrammatically through Feynman diagrams. This approach has been enormously successful, but it is also insufficient to address many important phenomena which are supposed to be described by QFT.

In this book I give an introduction to two methods in QFT which go beyond the standard perturbative framework: instantons and the large N expansion. Both are quite general, and they have led to many useful insights on the non-perturbative aspects of QFT. The two questions that I will address in this book are the following. What are the kinds of phenomena in a QFT in which we are fundamentally misled if we use conventional perturbation theory? What kind of intuition can we get on these effects by using instantons or large N physics?

It should be said from the very beginning that the main intention of this text is to give a *pedagogical* introduction to these topics. I would like to provide a useful toolbox intended for graduate students and beginners. In line with this, I provide many computational details which are usually skipped in the original literature and in many textbooks.

The treatment of the subject pays tribute to my professional bias: I am a mathematical physicist, not a phenomenologist. Therefore, in this book, after introducing an idea or a technique, I try to illustrate it with a model or example where this idea is implemented in a nice mathematical way, independently of its relevance to measurable physics. This involves typically looking at models in a low number of dimensions (either two, one or even zero). However, I hope that some of the material in this book will be useful for people with a more phenomenological orientation. After all, toy, solvable models should be useful to everybody, and they typically provide useful insights for more realistic applications.

A word about pre-requisites: my ideal reader should have a good knowledge of QFT in the path integral formulation, as well as some working knowledge of non-Abelian gauge theories and their quantization, at the level presented for example in the QFT textbook by Peskin and Schroeder. Of course, in many sections I require less background. Chapters 1 and 3 are devoted to a large extent to advanced topics in Quantum Mechanics. On the other hand, some sections are slightly more difficult. For example, Section 2.5 requires some knowledge of General Relativity. Section 4.5 is probably the most difficult one in the book, and in particular it uses the language of differentiable manifolds (although this is not essential to understand the underlying physics).

I should also comment on the choice of topics, which might strike some experts. There are some topics which I regard as basic building blocks in the theory of instantons and of the large N expansion, and they should be included in any introductory text on the subject. These include instantons in Quantum Mechanics, rudiments of instantons in Yang–Mills theory, and 't Hooft's double-line notation in QCD. The choice of the remaining topics was mostly dictated by my personal taste and/or expertise, and by trying not to duplicate presentations already available in the literature. This has led to choices which might look idiosyncratic. For example, in the discussion on Yang–Mills instantons, it is useful to work out a model with an IR cutoff where semiclassical instanton calculus makes sense. One sensible choice is to introduce a Higgs field, but I have decided instead to consider Yang–Mills theory on a compact manifold. This is not treated in textbooks and it fits better my tastes. Another constraint that shaped my choices is the simple motto "no supersymmetry." Of course, supersymmetry is a wonderful laboratory for many of the ideas presented in this book, but I felt that it should not take too much of a role in an introductory text like this. Therefore, I have only indulged in one supersymmetric excursion, in Chapter 5.

Although this text is pedagogical and elementary, I have included some material which is rarely covered in advanced QFT books, and some practitioners might find it useful. Chapter 3 is devoted to the relation between instantons and the large order behavior of perturbation theory. This is a relatively old idea, but it is not as well known as I think it should be. It paves the way to a proper understanding of exponentially small effects in quantum theories. In addition, I have tried to implement in my presentation of the subject more recent ideas coming from the theory of resurgence of Jean Écalle, which at the time of this writing is being applied in different contexts. Similarly, the last chapter, devoted to instantons in large N theories, puts together observations and examples which are scattered in research papers and have not found their way into any recent textbook on advanced QFT.

Since this text is a pedagogical one, I have tried to give a self-contained presentation, and I provide derivations of almost all the results. Some technical details in long calculations are relegated to the appendices. Some of the results on the heat kernel expansion on curved space, in Appendix B, are listed without a proper derivation, but this does not affect any important results in the book.

The references are listed at the end of each chapter, in the section on bibliographical notes, and they do not pretend to be exhaustive. I have tried to give due credit to seminal contributions, and sometimes to call the attention of readers to unjustly forgotten papers.

This book can be used profitably in advanced courses on Quantum Field Theory, and I have already tested it in graduate courses in Switzerland and Italy. Although I have not included exercises, there are many worked out examples which can be proposed to the students in order to develop their skills. Some parts of the book can be skipped in a one-semester course, for example Sections 1.9, 2.5 or 4.6. Alternatively, these sections could be proposed as advanced study topics.

Some students and colleagues have suggested corrections on preliminary versions of this book. I am particularly indebted to Gerald Dunne, Martin Lüscher and Ricardo Schiappa for their detailed reading of the manuscript and their valuable suggestions.

My deepest gratitude is however to Anna Serra Picamal, who has followed the writing of this book from its inception to its completion, and in addition provided me with the perfect image for its cover. For all this, and for everything else, this book is dedicated to her.

Part I
Instantons

1

Instantons in Quantum Mechanics

1.1 Introduction

In this chapter we start our study of non-perturbative effects by looking at the simplest case: Quantum Mechanics (QM), which can be regarded as a QFT in one dimension. We will focus on effects due to *instantons*, i.e. to non-trivial solutions to the Euclidean equations of motion (EOM). Typically, if g is the coupling constant, these effects go like

$$e^{-A/g}. \tag{1.1.1}$$

Notice that this effect is still small if g is small. However, it is completely invisible in perturbation theory, since it displays an essential singularity at $g = 0$.

Instanton effects are responsible for one of the most important quantum mechanical effects: tunneling through a potential barrier. This effect changes qualitatively the structure of the quantum vacuum. In a potential with a perturbative ground state degeneracy, like the one shown on the left hand side of Fig. 1.1, tunneling effects lift the degeneracy: there is a single ground state, and the energy difference between the ground state and the first excited state is of the form (1.1.1),

$$E_1(g) - E_0(g) \sim e^{-A/g}. \tag{1.1.2}$$

In a potential with an unstable or "false" vacuum, like the one shown on the right hand side of Fig. 1.1, states trapped in the false vacuum will eventually decay due to tunneling effects. This means in particular that the ground state energy associated to this vacuum has a small imaginary part,

$$E_0(g) = \operatorname{Re} E_0(g) + \mathrm{i} \operatorname{Im} E_0(g), \qquad \operatorname{Im} E_0(g) \sim e^{-A/g}, \tag{1.1.3}$$

which also has the dependence on g typical of an instanton effect and is invisible in perturbation theory.

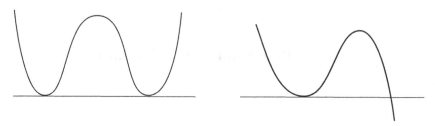

Figure 1.1 Two quantum mechanical potentials where instanton effects change qualitatively our understanding of the vacuum structure.

In this chapter, in order to understand this type of non-perturbative effect in detail, we will focus on observables which vanish in conventional perturbation theory, but have contributions due to instantons. In the case of vacuum decay, this will be the inverse lifetime of the particle; in the case of degenerate vacua, it will be the energy splitting (1.1.2). In our discussion we will focus on one-dimensional problems, where one can find very explicit results for instanton effects.

1.2 Quantum Mechanics as a one-dimensional field theory

In this chapter we will consider quantum systems in one dimension, with a Hamiltonian of the form,

$$H = \frac{1}{2}p^2 + W(q),\tag{1.2.1}$$

where $W(q)$ is the potential. We will set $\hbar = 1$. If this Hamiltonian supports bound states, one basic question to ask is what is the energy of the ground state. This can of course be addressed by elementary methods, like stationary perturbation theory, but we want to formulate the problem in the language of path integrals, so that the intuition gained in this way can be applied to QFTs. The ground state energy of the quantum mechanical system described by (1.2.1) can be extracted from the small temperature behavior of the thermal partition function,

$$Z(\beta) = \operatorname{tr} e^{-\beta H}.\tag{1.2.2}$$

Indeed, if we have a non-degenerate, discrete spectrum with energies

$$E_0 < E_1 < E_2 < \cdots,\tag{1.2.3}$$

the thermal partition function can be written as

$$Z(\beta) = \sum_{n=0}^{\infty} e^{-\beta E_n},\tag{1.2.4}$$

therefore

$$E_0 = - \lim_{\beta \to \infty} \frac{1}{\beta} \log Z(\beta). \tag{1.2.5}$$

On the other hand, the thermal partition function admits a path integral representation in terms of the Euclidean theory, in which we perform a Wick rotation to imaginary time

$$t \to -it, \tag{1.2.6}$$

and, because of the trace in (1.2.2), we have to consider *periodic* trajectories $q(t)$ in imaginary time,

$$q(-\beta/2) = q(\beta/2), \tag{1.2.7}$$

where β is the period of the motion. After Wick rotation, the path integral involves the Euclidean action $S(q)$,

$$S(q) = \int_{-\beta/2}^{\beta/2} dt \left[\frac{1}{2}(\dot{q}(t))^2 + W(q(t)) \right]. \tag{1.2.8}$$

The thermal path integral is then given by

$$Z(\beta) = \int \mathcal{D}[q(t)] e^{-S(q)}, \tag{1.2.9}$$

where the integration is performed over periodic trajectories.

We note that the Euclidean action can be regarded as an action in Lagrangian mechanics,

$$S(q) = \int_{-\beta/2}^{\beta/2} dt \left[\frac{1}{2}(\dot{q}(t))^2 - V(q) \right], \tag{1.2.10}$$

where the potential is

$$V(q) = -W(q), \tag{1.2.11}$$

i.e. it is the inverted potential of the original problem.

It is possible to compute the ground state energy by using Feynman diagrams. We will assume that the potential $W(q)$ is of the form

$$W(q) = \frac{1}{2}q^2 + W_{\text{int}}(q), \tag{1.2.12}$$

where $W_{\text{int}}(q)$ is an interaction term. Then, the path integral defining Z can be computed in perturbation theory by expanding in $W_{\text{int}}(q)$. For concreteness, let us assume that we have a quartic interaction (i.e. an anharmonic, quartic oscillator)

$$W_{\text{int}}(q) = \frac{g}{4}q^4. \tag{1.2.13}$$

At leading order in g, we find,

$$Z(\beta) = Z_G(\beta) \left(1 - \frac{g}{4} \int d\tau \, \langle q(\tau)q(\tau)q(\tau)q(\tau) \rangle_G + \cdots \right). \qquad (1.2.14)$$

Here, $Z_G(\beta)$ is the Euclidean partition function of the theory with the unperturbed Hamiltonian

$$H = \frac{1}{2}p^2 + \frac{1}{2}q^2, \qquad (1.2.15)$$

which is nothing but the thermal partition function of a harmonic oscillator with normalized frequency $\omega = 1$,

$$Z_G(\beta) = \frac{1}{2 \sinh\left(\frac{\beta}{2}\right)}. \qquad (1.2.16)$$

The subscript G indicates that, from the point of view of the path integral, this is a Gaussian theory. The bracket $\langle \cdots \rangle_G$ denotes a normalized vacuum expectation value (vev) in this Gaussian theory, which can be computed by using Wick's theorem. As usual, the calculation can be organized in terms of Feynman diagrams. We will actually work in the limit in which $\beta \to \infty$, since in this limit many features are simpler, for example the form of the propagator, which reads

$$\langle q(\tau)q(\tau') \rangle_G = \int \frac{dp}{2\pi} \frac{e^{ip(\tau-\tau')}}{p^2 + 1} = \frac{e^{-|\tau-\tau'|}}{2}. \qquad (1.2.17)$$

The Feynman rules are illustrated in Fig. 1.2. Since we want to calculate $\log Z(\beta)$, only *connected vacuum diagrams* contribute. In the limit $\beta \to \infty$, the quantity $\log Z(\beta)$ should be given by an overall factor of β, times a β-independent constant, as follows from (1.2.5). Diagrammatically, this is due to the following: the standard

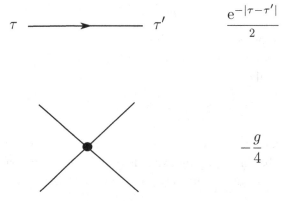

Figure 1.2 Feynman rules for the quantum mechanical quartic oscillator.

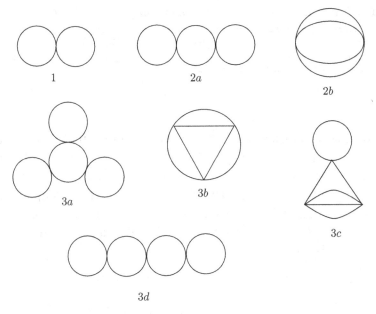

Figure 1.3 Feynman diagrams contributing to the ground state energy of the quartic oscillator up to order g^3.

Feynman rules in position space lead to k integrations, where k is the number of vertices in the diagram. One of these integrations just gives as an overall factor the "volume" of spacetime, and this is the overall factor of β. Therefore, in order to extract $E_0(g)$, we can just perform $k - 1$ integrations over $\mathbb{R}$.

It follows that the ground state energy has the following perturbative expansion:

$$E_0(g) = \frac{1}{2} + \sum_{k=1}^{\infty} a_k \left(\frac{g}{4}\right)^k, \tag{1.2.18}$$

where a_k can be computed diagrammatically as follows. Let $\mathcal{A}_k^{(c)}$ be the set of independent, connected quartic diagrams with k vertices. For $k = 1, 2, 3$, these diagrams are shown in Fig. 1.3. Then,

$$a_k = \sum_{\Gamma \in \mathcal{A}_k^{(c)}} s_\Gamma \mathcal{I}_\Gamma, \tag{1.2.19}$$

where s_Γ is the multiplicity of the graph Γ and $\mathcal{I}_\Gamma$ is the corresponding Feynman integral. Here, the multiplicity is simply the number of contractions which lead to the same topological graph Γ, and we can interpret it as the "number" of graphs

Table 1.1 *Multiplicities of the Feynman diagrams in Fig. 1.3*

Diagram	1	2a	2b	3a	3b	3c	3d
Multiplicity	3	36	12	288	288	576	432

with the topological structure Γ (in the literature one sometimes finds other definitions of the multiplicity, differing typically in the normalization of the coupling constant).

It is now straightforward to calculate $E_0(g)$ to order g^3. The multiplicities of the diagrams shown in Fig. 1.3 are given in Table 1.1. These numbers can be checked by taking into account that the total symmetry factor for a connected diagram with k quartic vertices is given by

$$\frac{1}{k!}\langle (x^4)^k \rangle^{(c)}, \tag{1.2.20}$$

where

$$\langle (x^4)^k \rangle = \frac{\int_{-\infty}^{\infty} dx\, e^{-x^2/2} x^{4k}}{\int_{-\infty}^{\infty} dx\, e^{-x^2/2}}. \tag{1.2.21}$$

is the Gaussian average. By Wick's theorem, this counts all possible pairings among k four-vertices, and we have to take the connected piece. Using that

$$\langle x^{2k} \rangle = (2k-1)!! = \frac{(2k)!}{2^k k!} \tag{1.2.22}$$

we find, for example,

$$\langle x^4 \rangle^{(c)} = \langle x^4 \rangle = 3,$$

$$\frac{1}{2!}\langle (x^4)^2 \rangle^{(c)} = \frac{1}{2}\left(\langle (x^4)^2 \rangle - \langle x^4 \rangle^2 \right) = 48, \tag{1.2.23}$$

in agreement with the results shown in Table 1.1. Putting together the Feynman integrals with the multiplicities, we find, for the different diagrams of Fig. 1.3,

$$1: \quad \frac{3}{4}$$

$$2a: \quad -\frac{36}{16}\int_{-\infty}^{\infty} e^{-2|\tau|}d\tau = -\frac{36}{16},$$

$$2b: \quad -\frac{12}{16}\int_{-\infty}^{\infty} e^{-4|\tau|}d\tau = -\frac{12}{16}\cdot\frac{1}{2},$$

$$3a: \quad \frac{288}{64}\int_{-\infty}^{\infty} e^{-|\tau_1|-|\tau_2|-|\tau_1-\tau_2|}d\tau_1\, d\tau_2 = \frac{288}{64}\cdot\frac{3}{2}, \tag{1.2.24}$$

$$3b: \quad \frac{288}{64} \int_{-\infty}^{\infty} e^{-2|\tau_1|-2|\tau_2|-2|\tau_1-\tau_2|} d\tau_1 \, d\tau_2 = \frac{288}{64} \cdot \frac{3}{8},$$

$$3c: \quad \frac{576}{64} \int_{-\infty}^{\infty} e^{-|\tau_1-\tau_2|-|\tau_1|-3|\tau_2|} d\tau_1 \, d\tau_2 = \frac{576}{64} \cdot \frac{5}{8},$$

$$3d: \quad \frac{432}{64} \int_{-\infty}^{\infty} e^{-2|\tau_1-\tau_2|-2|\tau_2|} d\tau_1 \, d\tau_2 = \frac{432}{64}.$$

This gives,

$$E_0(g) = \frac{1}{2} + \frac{3}{4}\left(\frac{g}{4}\right) - \frac{21}{8}\left(\frac{g}{4}\right)^2 + \frac{333}{16}\left(\frac{g}{4}\right)^3 + \mathcal{O}(g^4). \tag{1.2.25}$$

In Chapter 4 we will be interested in understanding this series in detail, and in particular we will look at the behavior of its coefficients at high order. The method of Feynman diagrams, although it emphasizes the parallelism with field theory, is not the most efficient method to use in order to generate the perturbative series for the ground state. In order to do that, it is better to use the Schrödinger equation

$$\left(-\frac{1}{2}\frac{d^2}{dx^2} + \frac{x^2}{2} + \frac{gx^4}{4}\right)\psi(x) = E_0(g)\psi(x). \tag{1.2.26}$$

We know that, for $g = 0$, the solution to this equation is the ground state of the harmonic oscillator, which is just the Gaussian $e^{-x^2/2}$. We will then write down an ansatz for the solution of the form

$$\psi(x) = e^{-x^2/2} \sum_{n=0}^{\infty} \left(\frac{g}{4}\right)^n B_n(x), \qquad B_0(x) = 1. \tag{1.2.27}$$

Plugging this ansatz into the above equation, and writing the energy as in (1.2.18), we find the following recursive equation for the $B_k(x)$ and the a_k:

$$xB_k'(x) - \frac{1}{2}B_k''(x) + x^4 B_{k-1}(x) = \sum_{p=0}^{k} a_{k-p}B_p(x). \tag{1.2.28}$$

To solve this recursion, we further write

$$B_i(x) = \sum_{j=1}^{2i} x^{2j}(-1)^i B_{i,j}. \tag{1.2.29}$$

By looking at the term of degree zero in (1.2.28), we find that

$$a_k = (-1)^{k+1} B_{k,1}. \tag{1.2.30}$$

The coefficients $B_{i,j}$ satisfy the recursion relation

$$2jB_{i,j} = (j+1)(2j+1)B_{i,j+1} + B_{i-1,j-2} - \sum_{p=1}^{i-1} B_{i-p,1}B_{p,j}. \tag{1.2.31}$$

This recursion can be easily solved to high orders, and one finds for the very first coefficients,

$$a_1 = \frac{3}{4}, \qquad a_2 = -\frac{21}{8}, \qquad a_3 = \frac{333}{16}, \qquad a_4 = -\frac{30885}{128}, \qquad (1.2.32)$$

in agreement with the Feynman diagram calculation (1.2.25).

1.3 Unstable vacua in Quantum Mechanics

Most quantities of interest in a quantum theory will have both perturbative and non-perturbative contributions. For small coupling, perturbative contributions are typically dominant. Therefore, in order to better understand the idiosyncrasies of non-perturbative effects in quantum theory, it is convenient to focus on quantities which vanish in perturbation theory.

A situation where non-perturbative effects dominate the physics is the case of unstable minima in QM. Let us consider a one-dimensional potential $W(q)$ which has a relative minimum at the origin $q = 0$. Near this minimum, the potential is of the form

$$W(q) \approx \frac{1}{2}q^2 + \mathcal{O}(g), \qquad (1.3.1)$$

where g is a coupling constant which gives the strength of the anharmonicity. Examples of such a situation are the cubic potential

$$W(q) = \frac{1}{2}q^2 - gq^3, \qquad (1.3.2)$$

which is depicted in Fig. 1.4 (left), and the inverted quartic potential

$$W(q) = \frac{q^2}{2} + \frac{g}{4}q^4, \qquad g = -\lambda, \quad \lambda > 0, \qquad (1.3.3)$$

which is shown on the right hand side of Fig. 1.4. It is clear that these potentials do not admit bound states, since a particle trapped near the minimum of the potential at $q = 0$ will eventually decay by tunneling through the barrier. However, this is *a priori* not detected by doing conventional stationary perturbation theory in the coupling constant g or λ: in both cases, one finds an infinite power series for the energy of, say, the ground state. In particular, for the inverted quartic oscillator, this series is obtained from (1.2.25) by simply setting $g = -\lambda$:

$$E_0(\lambda) = \frac{1}{2} - \frac{3}{4}\left(\frac{\lambda}{4}\right) - \frac{21}{8}\left(\frac{\lambda}{4}\right)^2 - \cdots. \qquad (1.3.4)$$

This is then a situation where perturbation theory is unable to describe the essential physics of the problem. What are the interesting quantities that can be computed

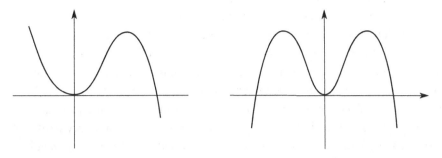

Figure 1.4 Unstable minima in one-dimensional quantum mechanical potentials.

in unstable potentials like the ones shown in Fig. 1.4? It turns out that, although they do not admit bound states, they admit *resonant* states. Resonant states can be defined by considering a scattering-like problem in the above potentials, and requiring that the amplitude of the incoming wavefunction vanish. These boundary conditions (called Gamow–Siegert boundary conditions) select a discrete set of energies in the Schrödinger equation. These energies turn out to be *complex*, with a negative imaginary part, and we can write them as

$$E = \text{Re}\, E - i\frac{\Gamma}{2}, \qquad \Gamma > 0. \tag{1.3.5}$$

The interpretation of resonant states is well known. Since the standard time evolution is given by

$$e^{-iEt} = e^{-it\,\text{Re}\, E} e^{-\Gamma t/2}, \tag{1.3.6}$$

they correspond to unstable states with a lifetime given by $\tau = 1/\Gamma$. The imaginary part of the energy then gives the *decay rate* of the state. Resonant energies can be calculated in many ways, and they typically involve a procedure of analytic continuation which makes it possible to uncover the complex values for the energy. For example, one can use the technique of complex scaling, which is based on analytically continuing the Hamiltonian to a complex-valued function by means of a dilatation operation

$$q \to e^{i\theta} q. \tag{1.3.7}$$

Here θ is a parameter which can be real or complex. When $|\theta|$ is larger than a certain threshold value, the rotated Hamiltonian has eigenfunctions which are square-integrable. The corresponding eigenvalues are the complex resonant energies one is looking for, and they are independent of the value of θ, provided θ is larger than the threshold value. Another possibility for obtaining the complex energies of the resonant states is to make an analytic continuation in the value of the coupling constant. For example, in the case of the quartic potential, one can first consider positive values of the coupling constant $g > 0$. For these values of g the potential admits bound states, whose energies can be computed. Then one can make an analytic continuation to negative values of g. It turns out that, when doing this continuation, the energy levels of the potential develop an imaginary part which corresponds to the resonant energies.

In potentials of the form (1.3.1), the real part of the energy of a resonance has a small g expansion which is precisely the result obtained by stationary perturbation theory. In particular, a resonant state can be regarded as a "perturbed" level of the harmonic oscillator, since as the coupling constant goes to zero one must have

$$\mathrm{Re}\, E \rightarrow N + \frac{1}{2}, \qquad g \rightarrow 0, \tag{1.3.8}$$

where N labels the corresponding level of the harmonic oscillator. The imaginary part of the energy can be estimated with WKB methods and it is *exponentially small* as the coupling constant goes to zero. For example, for the ground state of the quartic potential (i.e. the state which has $N = 0$ in (1.3.8)) one finds

$$\Gamma \approx e^{-A/\lambda}. \tag{1.3.9}$$

In this equation,

$$A = 2 \int_0^{q_+} \sqrt{2W(q)}dq, \tag{1.3.10}$$

and q_+ is the turning point of the potential, satisfying $W(q_+) = 0$. In the previous section we have seen that one can calculate the ground state energy of a quantum mechanical system by looking at the path integral representation of the thermal free energy. It is natural to ask how the imaginary part of the energy of a resonant state can be computed from the path integral point of view. One of the advantages of the path integral method is that it can be easily generalized to field theory, as we will eventually do in Chapter 2.

In order to understand how complex energies appear in the path integral formalism, we will consider a toy model for the quartic potential. As we just mentioned, one way of understanding the appearance of complex energies in this problem is to start with a stable potential with $g > 0$ and then move in the complex plane of the coupling constant until we reach the line $g < 0$. To understand the behavior of the path integral under such analytic continuation, we will perform it first for an ordinary integral, which can be regarded as the reduction of the anharmonic oscillator from one dimension to zero dimensions. This integral is

$$I(g) = \frac{1}{\sqrt{2\pi}} \int_{-\infty}^{+\infty} dz\, e^{-z^2/2 - gz^4/4}. \tag{1.3.11}$$

It is well defined as long as

$$\text{Re}(g) > 0, \tag{1.3.12}$$

but we would like to define it for more general, complex values of g, in particular we would like to define it for negative values of g. To do this, we *rotate the contour of integration* for the z variable, in such a way that

$$\text{Re}(gz^4) > 0 \tag{1.3.13}$$

and the integral is still convergent. Equivalently, we give a phase to z in such a way that

$$\text{Arg } z = -\frac{1}{4}\text{Arg } g. \tag{1.3.14}$$

Once we do this, the integral is no longer real.

In order to define the integral for negative g, we should rotate g towards the negative real axis. This can be done in *two* different ways: clockwise or counterclockwise, as shown in Fig. 1.5. The integration contour for z rotates correspondingly. Since the resulting integration contours are complex conjugate to each other, the two integrals defined in this way are also complex conjugate. For $g \to -|g| + i\epsilon, \epsilon \ll 1$, the integration contour is given by

$$\mathcal{C}_+ : \quad \text{Arg } z = -\frac{\pi}{4}, \tag{1.3.15}$$

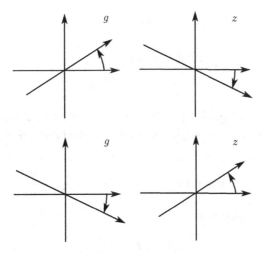

Figure 1.5 We can analytically continue the integral (1.3.11) to negative values of g by rotating the integration contour for z. This can be done in two ways: we can rotate g clockwise, and the integration contour counterclockwise, or the other way around.

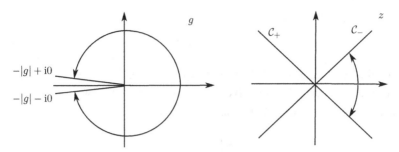

Figure 1.6 The integration contours $C_\pm$ correspond to the negative values of $g = -|g| \pm \mathrm{i}0$.

while for $g \to -|g| - \mathrm{i}\epsilon$, one has

$$C_- : \quad \mathrm{Arg}\, z = \frac{\pi}{4}, \tag{1.3.16}$$

see Fig. 1.6. This means that one can indeed obtain an analytic continuation of the integral $I(g)$ to negative values of g, but the resulting function will have a *branch cut* along the negative real axis. The discontinuity across the cut is given by

$$I(g + \mathrm{i}\epsilon) - I(g - \mathrm{i}\epsilon) = 2\mathrm{i}\, \mathrm{Im}\, I(g) = \frac{1}{\sqrt{2\pi}} \int_{C_+ - C_-} \mathrm{d}z\, \mathrm{e}^{-z^2/2 - gz^4/4}. \tag{1.3.17}$$

The discontinuity (1.3.17) can be computed by saddle-point methods. The saddle points of the integral occur at $z = 0$ or

$$z + gz^3 = 0 \Rightarrow z^2 = -\frac{1}{g}. \tag{1.3.18}$$

Therefore we have two non-trivial saddle points $z_{1,2}$

$$z_{1,2} = \mp \mathrm{e}^{\mathrm{i}(\pi/2 - \phi_g/2)} |g|^{-1/2}, \tag{1.3.19}$$

where ϕ_g is the phase of g. For $g < 0$, they are on the real axis, see Fig. 1.7. The steepest descent trajectories passing through these points are determined by the condition

$$\mathrm{Im}\, f(z) = \mathrm{Im}\, f(z_i), \quad f(z) = \frac{z^2}{2} + \frac{g}{4} z^4. \tag{1.3.20}$$

For $g < 0$ these are hyperbolae

$$x^2 - y^2 = -\frac{1}{g} \tag{1.3.21}$$

passing through the saddle points $z_{1,2}$, see Fig. 1.7. From this figure it is also clear that the contour $C_+ - C_-$ appearing in (1.3.17) can be deformed into the *sum* of the

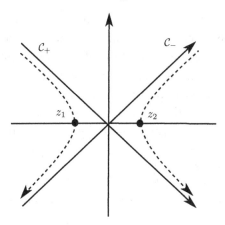

Figure 1.7 The complex plane for the saddle-point calculation of (1.3.11). Here, C^+ and C^- are the rotated contours one needs to consider for $g < 0$. Their sum may be evaluated by the contribution of the saddle point at the origin. Their difference is evaluated by the contribution of the subleading saddle points, here denoted as z_1 and z_2.

steepest descent trajectories passing through $z_{1,2}$, therefore the imaginary part in (1.3.17) is given by

$$\text{Im } I(g) \sim -\frac{1}{\sqrt{2}} \exp\left(\frac{1}{4g}\right), \qquad g \to 0^-. \tag{1.3.22}$$

The overall factor in (1.3.22) is obtained by doing the Gaussian integrations around the two saddle points and adding up the results. Since the integral (1.3.11) is divergent for $g < 0$, the resulting complex function cannot be analytic at $g = 0$. One consequence of this lack of analyticity is that the formal power series expansion of the integral (1.3.11) around $g = 0$,

$$\varphi(g) = \sum_{k=0}^{\infty} a_k g^k, \tag{1.3.23}$$

where

$$a_k = \frac{(-4)^{-k}}{\sqrt{2\pi}} \int_{-\infty}^{\infty} dz \frac{z^{4k}}{k!} e^{-z^2/2} = (-4)^{-k} \frac{(4k-1)!!}{k!}, \tag{1.3.24}$$

has *zero radius of convergence*. Its asymptotic behavior at large k is obtained immediately from Stirling's formula

$$a_k \sim (-4)^k k!. \tag{1.3.25}$$

This factorial divergence is in fact a generic feature of perturbative series in quantum theory, as we will see.

The moral of this simple analysis is that, for negative g, the integral $I(g)$ picks an imaginary part which is given by the contribution of the non-trivial saddle points. By analogy with this integral, we expect that the path integral of the quartic oscillator will have the same behavior. Therefore, we expect an imaginary part in the thermal partition function for negative coupling $g = -\lambda$, and we also expect this imaginary part to be exponentially suppressed at small λ, just as in (1.3.22). The free energy then reads, when expanded formally in powers of Im Z,

$$F(\beta) = -\frac{1}{\beta}\log Z = -\frac{1}{\beta}\log(\text{Re } Z) - \frac{i}{\beta}\frac{\text{Im}Z}{\text{Re }Z} + \cdots. \tag{1.3.26}$$

Therefore, at leading order in the exponentially suppresed factor we have

$$\text{Im } F(\beta) \approx -\frac{1}{\beta}\frac{\text{Im } Z}{\text{Re } Z}, \tag{1.3.27}$$

and

$$\text{Im } E_0(g) = \lim_{\beta\to\infty} \text{Im } F(\beta) \approx -\lim_{\beta\to\infty}\frac{1}{\beta}\frac{\text{Im } Z}{\text{Re } Z}. \tag{1.3.28}$$

Furthermore, as in (1.3.17), we expect that the discontinuity

$$\text{disc } Z(-\lambda) = Z(-\lambda + i\epsilon) - Z(-\lambda - i\epsilon) = 2i \text{ Im } Z(-\lambda) \tag{1.3.29}$$

will be given by the sum of the contributions of the *non-trivial* saddle points of the path integral (1.2.9). We will now calculate these contributions.

1.4 The path integral around an instanton

We will now consider quantum mechanical potentials $W(q)$ which have a relative minimum at $q = 0$. Near this minimum, the potential is of the form (1.3.1). We will also assume that this minimum is unstable, as in Fig. 1.8. We will study the imaginary part of the thermal free energy, $Z(\beta)$, at large β, in order to extract the imaginary part of the ground state energy. To do that, we use the Euclidean path integral (1.2.9). The non-trivial saddle points of this path integral are time dependent, *periodic* solutions of the Euclidean EOM for the *inverted* potential,

$$\ddot{q}_c(t) + V'(q_c) = 0, \tag{1.4.1}$$

having *finite* Euclidean action (otherwise the semiclassical contribution of such saddle points vanishes). Examples of such non-trivial, periodic saddle points are oscillations around the local minima of $V(q)$, as shown in Fig. 1.8. The period of such an oscillation between the turning points q_- and q_+ is given by

$$\beta = 2\int_{q_-}^{q_+}\frac{dq}{\sqrt{2(E - V(q))}}. \tag{1.4.2}$$

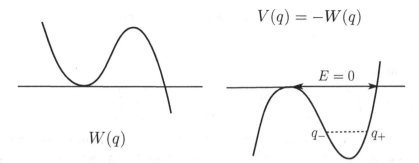

Figure 1.8 A general unstable potential $W(q)$ and the associated inverted potential $V(q)$. A periodic solution with negative energy moves between the turning points $q_\pm$. The trajectory with zero energy, relevant for extracting the imaginary part of the ground state energy, is also shown.

These trajectories satisfy in addition the "energy conservation" constraint

$$\frac{1}{2}\dot{q}^2 + V(q) = E(\beta),\qquad(1.4.3)$$

and the action along such a trajectory is given by

$$S_c \equiv S(q_c(t)) = \mathcal{W}(E) - E\beta,\qquad(1.4.4)$$

where

$$\mathcal{W}(E) = \int_{-\beta/2}^{\beta/2} dt\,(\dot{q}_c(t))^2 = 2\int_{q_-}^{q_+} p(q)dq.\qquad(1.4.5)$$

We will refer to the above solutions of the Euclidean EOM as *instantons*, although in this unstable case some authors prefer to call them *bounces*. Notice that the period (1.4.2) varies between $\beta = \infty$ (corresponding to $E = 0$ in Fig. 1.8) and a minimum critical value β_c corresponding to small oscillations around the minimum q_0 of the potential. This value can be computed as follows. Near the bottom of the inverted potential one has

$$V(q) = V_0 + \frac{1}{2}\omega^2(q - q_0)^2 + \cdots\qquad(1.4.6)$$

where

$$\omega^2 = V''(q_0).\qquad(1.4.7)$$

Let us define ϵ by the equation

$$E = V_0 + \frac{1}{2}\omega^2\epsilon^2.\qquad(1.4.8)$$

Then, at leading order in ϵ, the turning points can be approximated by

$$q_\pm \approx q_0 \pm \epsilon. \tag{1.4.9}$$

We then find,

$$\beta_c = \lim_{\epsilon \to 0} 2 \int_{q_0-\epsilon}^{q_0+\epsilon} \frac{dq}{\sqrt{\omega^2(\epsilon^2 - (q-q_0)^2)}} = \frac{2}{\omega} \int_{-\epsilon}^{\epsilon} \frac{d\zeta}{\sqrt{\epsilon^2 - \zeta^2}} = \frac{2\pi}{\omega}. \tag{1.4.10}$$

For $\beta < \beta_c$ there are no instanton trajectories. In terms of a thermal partition function, this means that for sufficiently high temperatures the instanton degenerates to a solution $q(t) = q_0$. The decay mechanism above the temperature $T_c = 1/\beta_c$ is just due to thermal excitations over the top of the barrier in the original potential $W(q)$.

Example 1.1   In the example of the quartic anharmonic oscillator, the inverted potential is

$$V(q) = -\frac{1}{2}q^2 + \frac{\lambda}{4}q^4, \tag{1.4.11}$$

see Fig. 1.9. We can change variables $q \to \lambda^{-1/2}q$, $E \to E/\lambda$ to set $\lambda = 1$, so that the EOM reads

$$-\ddot{q}(t) + q(t) - q^3(t) = 0. \tag{1.4.12}$$

The inverted potential has minima at $q = \pm 1$ and zeros at $q = \pm\sqrt{2}$. We will focus on the region $q \geq 0$, since results in the region $q < 0$ follow by the symmetry $q \to -q$ of the problem. There is a solution to the EOM (1.4.12), with energy $-1/4 \leq E \leq 0$, and turning points at

$$q_\pm = \sqrt{1 \pm \sqrt{1 + 4E}}, \tag{1.4.13}$$

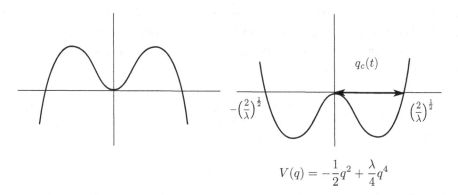

$$q_c(t)$$

$$-\left(\frac{2}{\lambda}\right)^{\frac{1}{2}} \qquad \left(\frac{2}{\lambda}\right)^{\frac{1}{2}}$$

$$V(q) = -\frac{1}{2}q^2 + \frac{\lambda}{4}q^4$$

Figure 1.9 The inverted potential relevant for instanton calculus in the quartic case. The instanton or "bounce" configuration $q_c(t)$ leaves the origin at $t = -\infty$, reaches the zero $(2/\lambda)^{1/2}$ at $t = t_0$, and comes back to the origin at $t = +\infty$.

which can be written in terms of the Jacobi elliptic function dn$(u; k)$:

$$q_c^{t_0}(t) = q_+ \mathrm{dn}\,(u; k)\,, \tag{1.4.14}$$

where

$$u = \frac{q_+}{\sqrt{2}}(t - t_0), \qquad k^2 = 1 - \frac{q_-^2}{q_+^2}. \tag{1.4.15}$$

This solution has a free parameter t_0, which corresponds to the initial point of the trajectory. Since t_0 is defined modulo the period of the motion, we can choose $t_0 \in [-\beta/2, \beta/2]$. The period β can be computed from (1.4.2), and is given by

$$\beta = 2\sqrt{2}\left(\frac{2 - m}{2}\right)^{1/2} K(k), \tag{1.4.16}$$

where $K(k)$ is the complete elliptic integral of the first kind. In terms of k, the energy reads

$$E = -\frac{1 - k^2}{(2 - k^2)^2}. \tag{1.4.17}$$

The value $k = 0$ corresponds to a particle with minimal energy $E = -1/4$, sitting at the bottom of $V(q)$, and the minimum period is $\beta_c = \sqrt{2}/\pi$. Since the frequency of the oscillations around the bottom is $\omega = \sqrt{2}$, this is in accord with the result (1.4.10). The limiting case $k \to 1$ corresponds to a particle with energy $E = 0$ and infinite period $\beta \to \infty$. In this limit the solution (1.4.14) simplifies: the Jacobi function dn$(u; k)$ becomes sech(u), and we find

$$q_c^{t_0}(t) = \frac{\sqrt{2}}{\cosh(t - t_0)}. \tag{1.4.18}$$

This trajectory starts at the origin in the infinite past, arrives at the turning point q_+ at $t = t_0$, and returns to the origin in the infinite future. The trajectory (1.4.18) with $t_0 = 0$ is shown in Fig. 1.10. □

Let us now return to the general case and expand the action around a classical solution $q_c(t)$. We find, after writing

$$q(t) = q_c(t) + r(t), \tag{1.4.19}$$

that, at quadratic order in the fluctuations,

$$S(q) \approx S_c + \frac{1}{2}\int dt_1 dt_2 r(t_1) M(t_1, t_2) r(t_2), \tag{1.4.20}$$

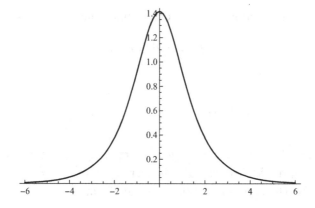

Figure 1.10 The solution (1.4.18) with $t_0 = 0$.

where M is given by

$$M(t_1, t_2) = \frac{\delta^2 S}{\delta q(t_1)\delta q(t_2)}\bigg|_{q(t)=q_c(t)} = \left[-\left(\frac{\mathrm{d}}{\mathrm{d}t_1}\right)^2 - V''(q_c(t_1))\right]\delta(t_1 - t_2).$$

(1.4.21)

This distribution can be regarded as the kernel of an integral operator **M**, which acts on functions as

$$(\mathbf{M}\psi)(t) = \int M(t, t')\psi(t')\mathrm{d}t',$$

(1.4.22)

and is given explicitly by

$$\mathbf{M} = -\frac{\mathrm{d}^2}{\mathrm{d}t^2} - V''(q_c(t)).$$

(1.4.23)

In the quadratic (or one-loop) approximation, the path integral around the configuration $q_c(t)$ is then given by

$$\int \mathcal{D}q(t)\, e^{-S(q)} \approx e^{-S(q_c)} \int \mathcal{D}r(t)\, \exp\left[-\frac{1}{2}\int \mathrm{d}t_1\mathrm{d}t_2\, r(t_1)M(t_1, t_2)r(t_2)\right].$$

(1.4.24)

Since we are integrating over periodic configurations, the boundary conditions for $r(t)$ are

$$r(-\beta/2) = r(\beta/2), \qquad \dot{r}(-\beta/2) = \dot{r}(\beta/2).$$

(1.4.25)

Note that all possible values of the endpoints for $r(t)$ are allowed, since we have to integrate over all possible periodic trajectories, and in particular over all possible endpoints. We now have to perform the Gaussian integration over $r(t)$. In order to do this, we consider a complete set q_n of orthonormal eigenfunctions of $\mathbf{M}$, labelled by $n = 0, 1, \ldots$,

$$\mathbf{M}q_n = \lambda_n q_n, \tag{1.4.26}$$

and satisfying the same periodic boundary conditions as (1.4.25). The eigenvalue problem can be written explicitly as

$$\left[-\frac{d^2}{dt^2} - V''(q_c(t)) \right] q_n(t) = \lambda_n q_n(t), \quad n \geq 0, \tag{1.4.27}$$

and orthonormality means that

$$\int_{-\beta/2}^{\beta/2} dt \, q_n(t) q_m(t) = \delta_{nm}. \tag{1.4.28}$$

Here, we have assumed that the spectrum is discrete. As we will see, in many cases this is not the case, but the formalism we are developing can be easily modified to account for a continuous spectrum. We now expand the fluctuations as

$$r(t) = \sum_{n \geq 0} c_n q_n(t). \tag{1.4.29}$$

This can be regarded as a change of variables from the set of paths $r(t)$ to the coefficients c_n. The measure for $r(t)$ is then defined as the normalized Gaussian measure for the c_n, up to an overall normalization constant $\mathcal{N}$ which is independent of the potential:

$$\mathcal{D}r(t) = \mathcal{N} \prod_{n \geq 0} \frac{dc_n}{\sqrt{2\pi}}, \tag{1.4.30}$$

and we find

$$\int \mathcal{D}r(t) \, \exp\left[-\frac{1}{2} \int dt_1 dt_2 \, r(t_1) M(t_1, t_2) r(t_2) \right]$$
$$= \mathcal{N} \int \prod_{n \geq 0} \frac{dc_n}{\sqrt{2\pi}} \, e^{-\frac{1}{2}\sum_{n \geq 0} \lambda_n c_n^2} = \mathcal{N} \, (\det \mathbf{M})^{-1/2}, \tag{1.4.31}$$

where

$$\det \mathbf{M} = \prod_{n \geq 0} \lambda_n. \tag{1.4.32}$$

This derivation has been purely formal, and in order to calculate the determinant of $\mathbf{M}$ we have to take into account many subtleties.

First of all, if we take a further derivative with respect to t in (1.4.1) we find

$$\frac{d^2}{dt^2} \dot{q}_c(t) + V''(q_c(t)) \dot{q}_c(t) = 0. \tag{1.4.33}$$

Since $q_c(t)$ is periodic, $\dot{q}_c(t)$ is periodic as well, and they satisfy the boundary conditions (1.4.25). Therefore, $\dot{q}_c(t)$ is a *zero mode* of M, i.e. an eigenfunction with zero eigenvalue. It is also a normalizable function, therefore it must be (up to normalization) one of the eigenfunctions $q_n(t)$ of M, say $q_1(t)$. We will then write,

$$q_1(t) = \frac{1}{\|\dot{q}_c\|}\dot{q}_c(t). \tag{1.4.34}$$

We will see in a moment that this eigenfunction is the first excited state of the spectrum of M, and the eigenfunction of the ground state will be denoted by $q_0(t)$. The norm appearing in (1.4.34) is given by

$$\|\dot{q}_c\|^2 = \int_{-\beta/2}^{\beta/2} dt \, (\dot{q}_c(t))^2 = \mathcal{W}(E), \tag{1.4.35}$$

where we used (1.4.5). Note that, in the limit of large β, relevant to extracting the ground state energy, the trajectory $q_c(t)$ has $E = 0$, and we find

$$\mathcal{W}(E) = S_c, \qquad \beta \to \infty. \tag{1.4.36}$$

The origin of the zero mode $\dot{q}_c(t)$ can be explained by time translation invariance. As we made clear in the examples in (1.4.14) and (1.4.18), the solution $q_c(t)$ depends on an arbitrary initial time t_0, and we should write it rather as $q_c^{t_0}(t)$. This trajectory solves the EOM for all values of the parameter t_0,

$$\left.\frac{\delta S}{\delta q(t)}\right|_{q(t)=q_c^{t_0}(t)} = 0. \tag{1.4.37}$$

This is a general fact: when we solve for a non-trivial saddle point we find in general a *family of solutions*. The parameters for such a family are called *moduli* or *collective coordinates*. In the case at hand, we have a single modulus, namely the initial time t_0. If we now take a further derivative of (1.4.37) with respect to t_0, we obtain

$$\int dt_2 \left.\frac{\delta^2 S}{\delta q(t_1)\delta q(t_2)}\right|_{q(t)=q_c^{t_0}(t)} \frac{\delta q_c^{t_0}(t_2)}{\delta t_0} = 0. \tag{1.4.38}$$

The second functional derivative of S is $M(t_1, t_2)$, and

$$\frac{\delta q_c^{t_0}(t_2)}{\delta t_0} = -\dot{q}_c^{t_0}(t_2). \tag{1.4.39}$$

We conclude that

$$\int dt_2 \, M(t_1, t_2)\dot{q}_c^{t_0}(t_2) = 0, \tag{1.4.40}$$

therefore $\dot{q}_c^{t_0}(t)$ is a zero mode of $\mathbf{M}$. Equivalently, we can consider the EOM (1.4.1), which is solved by $q_c^{t_0}(t)$ for any t_0, and take a derivative with respect to t_0. In this way one obtains again (1.4.33).

Let us now address the issue raised by the existence of a zero mode, i.e. the fact that $\lambda_1 = 0$. Naively this leads to a vanishing result for the functional determinant (1.4.32), and a diverging result for (1.4.31). This is due to the fact that the mode c_1 has no damping factor in the Gaussian integral, since $\lambda_1 = 0$, and the infinite answer comes from the integration over c_1. We can now isolate this divergence as

$$\int \prod_n \frac{dc_n}{\sqrt{2\pi}} e^{-\frac{1}{2}\sum_{n\geq 0}\lambda_n c_n^2} = \left(\int \frac{dc_1}{\sqrt{2\pi}}\right) (\det' \mathbf{M})^{-1/2}, \qquad (1.4.41)$$

where

$$\det' \mathbf{M} = \prod_{n\neq 1} \lambda_n \qquad (1.4.42)$$

is the determinant of the operator $\mathbf{M}$ once the zero mode has been removed. However, the integration over c_1 should be treated more carefully, since this variable really stands for the collective coordinate t_0. To understand this, let us recall that c_1 was introduced in the expansion (1.4.29) of an arbitrary, periodic function of t. However, we could also expand such a function as

$$q_c^{t_0}(t) + \sum_{n\neq 1} c_n q_n(t), \qquad (1.4.43)$$

where t_0 is now understood as a parameter or a coordinate in the space of path configurations. Indeed, if we vary c_1 in (1.4.29), we obtain

$$q_1(t)\delta c_1 = \frac{1}{\|\dot{q}_c\|}\dot{q}_c^{t_0}(t)\delta c_1, \qquad (1.4.44)$$

while varying t_0 in (1.4.43) gives

$$-\dot{q}_c^{t_0}(t)\delta t_0. \qquad (1.4.45)$$

Both variations are proportional, therefore t_0 in (1.4.43) parametrizes the same fluctuations as c_1 in (1.4.29). The Jacobian of the change of variables from c_1 to t_0 can be easily computed by comparing both variations,

$$J = \left|\frac{\delta c_1}{\delta t_0}\right| = \|\dot{q}_c\| = (\mathcal{W}(E))^{1/2}. \qquad (1.4.46)$$

Therefore, the integration over c_1 gives

$$\frac{1}{\sqrt{2\pi}}\int dc_1 = \frac{J}{\sqrt{2\pi}}\int_{-\beta/2}^{\beta/2} dt_0 = \frac{\beta\,(\mathcal{W}(E))^{1/2}}{\sqrt{2\pi}}, \qquad (1.4.47)$$

where we have used that the "moduli space" for t_0 is $[-\beta/2, \beta/2]$.

To summarize: instantons come in families parametrized by collective coordinates or "moduli." This leads to zero modes in the quadratic operators that are obtained by looking at fluctuations around a fixed solution. The integration over these zero modes has to be translated into an integration over collective coordinates, and removes the apparent divergences associated to the zero modes in a naive treatment of the path integral.

The second important property of the operator $\mathbf{M}$ is that it has one, and only one, negative mode. To see this, we note that it has an eigenfunction $\dot{q}_c(t)$ with zero eigenvalue. But $\dot{q}_c(t)$ changes sign at one of the turning points, and there should be an eigenfunction with a lower eigenvalue, which has to be negative. In the limit $\beta \rightarrow \infty$, this can also be established by regarding (1.4.23) as a one-dimensional Schrödinger operator. The spectrum of such an operator has the well-known property that the ground state has no nodes, the first excited state has one node, etc. The function $\dot{q}_c(t)$ has one node, so it is the first excited state of $\mathbf{M}$ and the ground state must have negative energy. This is the negative mode of $\mathbf{M}$.

Example 1.2 Let us consider again the quartic oscillator in the limit $\beta \rightarrow \infty$ and with $\lambda = 1$. The operator $\mathbf{M}$ is given in this case by

$$\mathbf{M} = -\frac{d^2}{dt^2} + 1 - \frac{6}{\cosh^2(t - t_0)}. \tag{1.4.48}$$

Using translation invariance we can just set $t_0 = 0$ to study the spectrum. It is easy to see that

$$\mathbf{M}\psi(t) = -3\psi(t), \qquad \psi(t) = \frac{1}{\cosh^2(t)}, \tag{1.4.49}$$

which is the single negative mode of this operator, and $q_0(t) \propto \psi(t)$. □

Since $\lambda_0 < 0$, in calculating the Gaussian integral (1.4.31) we have to make sense of the Gaussian integral with the "wrong" sign involving the mode c_0,

$$\int \frac{dc_0}{\sqrt{2\pi}} e^{\frac{1}{2}|\lambda_0|c_0^2}. \tag{1.4.50}$$

This is done by analytic continuation: we rotate the integration contour of c_0 an angle of $\pi/2$, so that the resulting integral is done along the imaginary axis and is convergent. The result of the integration will be $\pm i|\lambda_0|^{-1/2}$, depending on whether we make the rotation clockwise or counterclockwise. Therefore, the final answer for (1.4.31) is *imaginary*, and there is sign ambiguity due to the analytic continuation. Equivalently, since $\mathbf{M}$ has one and only one negative eigenvalue, det$'\mathbf{M}$ is negative, and in extracting its square root we will obtain an imaginary result. The sign ambiguity corresponds to the choice of branch cut of the square root.

We now put everything together, and obtain

$$2i \operatorname{Im} Z \approx \mathcal{N} e^{-S_c} \frac{\beta \, (\mathcal{W}(E))^{1/2}}{\sqrt{2\pi}} (\det' \mathbf{M})^{-1/2}. \tag{1.4.51}$$

Note that, as we just discussed, the right hand side is imaginary due to the negative mode of $\mathbf{M}$, which is consistent with the fact that we are computing the imaginary part of the partition function. Finally, we should fix the normalization of the path integral measure, $\mathcal{N}$. In order to do this, it is convenient to use the (unperturbed) harmonic oscillator with $\omega = 1$ as a reference point. Its thermal partition function is given in (1.2.16). A path integral evaluation of this partition function along the lines of what we have done gives the well-known result

$$Z_G(\beta) = \mathcal{N} (\det \mathbf{M}_0)^{-1/2}, \tag{1.4.52}$$

where

$$\mathbf{M}_0 = -\frac{d^2}{dt^2} + 1. \tag{1.4.53}$$

We then find,

$$\operatorname{Im} Z(\beta) \approx \frac{1}{2i} Z_G(\beta) \left(\frac{\det' \mathbf{M}}{\det \mathbf{M}_0} \right)^{-1/2} \frac{\beta \, (\mathcal{W}(E))^{1/2}}{\sqrt{2\pi}} e^{-S_c}. \tag{1.4.54}$$

This is the one-loop approximation to the full result. Since we have assumed that our potential is a perturbed quadratic potential, the real part of $Z(\beta)$ is given, at one-loop, by

$$\operatorname{Re} Z \approx Z_G(\beta). \tag{1.4.55}$$

Therefore, after we take the limit $\beta \to \infty$, or $E \to 0$, we find the following one-loop result for the imaginary part of the ground state energy,

$$\operatorname{Im} E_0 \approx \pm \frac{S_c^{1/2}}{2\sqrt{2\pi}} e^{-S_c} \lim_{\beta \to \infty} \left(-\frac{\det' \mathbf{M}}{\det \mathbf{M}_0} \right)^{-1/2}. \tag{1.4.56}$$

Here we used the fact that, for $\beta \to \infty$, $\mathcal{W}(E)$ becomes S_c, as we pointed out in (1.4.36). The choice of branch cut for the square root in (1.4.56) corresponds to the choice of contour rotation in (1.4.50). The formula (1.4.56) gives the imaginary part of the ground state energy for general unstable potentials obtained by perturbing a quadratic potential. Note that the above expression involves the limit as β goes to ∞ of the quotient of the determinants. This can be computed either by first taking the limit of the operators (and then one considers operators on the real line like for example (1.4.48)), or by computing the determinants for arbitrary β and then taking the limit. Below, we will do the computation in both ways.

Although we have discussed two of the subtleties appearing in the computation of the functional determinant, we still have to make sense of the infinite product appearing in (1.4.42). The eigenvalues of a Schrödinger operator of the form (1.4.23), on an interval $[-\beta/2, \beta/2]$, grow like

$$\lambda_n \approx n^2, \qquad n \gg 1, \tag{1.4.57}$$

and the infinite product has to be regularized in an appropriate way. There are various ways to do this. One possibility is to consider quotients of determinants of different operators, as we have done in (1.4.56). In many cases, the divergent parts of the determinants cancel each other, and one is left with a finite piece. Another possibility is to use a regularization prescription which removes the divergences and leads to a physically meaningful finite piece. A particularly useful regularization is the zeta function regularization, which we explain in detail in Appendix B. We will now calculate the relevant determinants in our quantum mechanical problem using these two methods, which of course lead to the same result.

1.5 Calculation of functional determinants I: solvable models

We will first consider a very direct approach to the computation of the determinant of the operators $\mathbf{M}$ in some special cases, starting from a computation of their spectrum. As is well known from QM, the spectrum of Schrödinger operators can only be computed exactly in some special solvable models. Therefore, we have to consider potentials $V(x)$ such that the operator appearing in (1.4.23) belongs to the class of solvable Schrödinger operators. It turns out that the operators one finds in the case of the cubic and the quartic oscillators, in the limit $\beta \to \infty$, are precisely of this type. They belong to a general family of operators called *Pöschl–Teller operators*. These operators are labelled by two parameters ℓ, m, and they have the form

$$\mathbf{M}_{\ell,m} = -\frac{d^2}{dt^2} + m^2 - \frac{\ell(\ell+1)}{\cosh^2(t)}. \tag{1.5.1}$$

They can be regarded as Schrödinger operators in an inverted cosh squared potential, also called the Pöschl–Teller potential. Remarkably, the spectrum of these operators can be determined exactly. This is due to a factorization property first studied by Schrödinger, and which can be substantially clarified in the light of supersymmetric QM, as we will do in Chapter 6. Let us introduce the operators

$$A_\ell = \frac{d}{dt} + \ell \tanh t, \qquad A_\ell^\dagger = -\frac{d}{dt} + \ell \tanh t. \tag{1.5.2}$$

It is immediate to compute that

$$A_\ell^\dagger A_\ell = \mathbf{M}_{\ell,m} + \ell^2 - m^2, \qquad A_\ell A_\ell^\dagger = \mathbf{M}_{\ell-1,m} + \ell^2 - m^2. \tag{1.5.3}$$

Notice that for $\ell = 0$ we recover the free particle. Also, we can obtain the ground state for the full family of potentials just by solving

$$A_\ell \psi_0^{(\ell)}(t) = 0. \tag{1.5.4}$$

This is a first order ODE with solution

$$\psi_0^{(\ell)}(t) \propto \frac{1}{\cosh^\ell(t)}. \tag{1.5.5}$$

The ground state energy for the operator $\mathbf{M}_{\ell,m}$ is simply

$$E_{\ell,m}^{(0)} = m^2 - \ell^2. \tag{1.5.6}$$

The properties above also make it possible to calculate the excited states. To do this, notice that if $\psi^{(\ell-1)}(t)$ is an eigenfunction of $\mathbf{M}_{\ell-1,m}$ with eigenvalue $\mu_{\ell-1}$, then

$$\psi^{(\ell)}(t) = A_\ell^\dagger \psi^{(\ell-1)}(t) \tag{1.5.7}$$

is an eigenfunction of $\mathbf{M}_{\ell,m}$ with the same eigenvalue. Indeed,

$$\begin{aligned}
\mathbf{M}_{\ell,m}\psi^{(\ell)}(t) &= \left(A_\ell^\dagger A_\ell + m^2 - \ell^2\right) A_\ell^\dagger \psi^{(\ell-1)}(t) \\
&= A_\ell^\dagger \left(\mathbf{M}_{\ell-1,m} + \ell^2 - m^2\right) \psi^{(\ell-1)}(t) + \left(m^2 - \ell^2\right) \psi^{(\ell)}(t) \\
&= \mu_{\ell-1}\psi^{(\ell)}(t).
\end{aligned} \tag{1.5.8}$$

We can then construct the spectrum of $\mathbf{M}_{\ell,m}$ by starting with the free particle $\ell = 0$ and applying the operators $A_\ell^\dagger$ (see Fig. 1.11). For $\ell = 0$, the eigenfunctions are just plane waves (scattering states)

$$e^{ikt} \tag{1.5.9}$$

with energies

$$E_{\ell,m}(k) = k^2 + m^2. \tag{1.5.10}$$

Applying $A_1^\dagger$ we obtain the scattering states of the $\ell = 1$ potential

$$\psi_1^{(k)}(t) \propto A_1^\dagger e^{ikt}. \tag{1.5.11}$$

On top of that, we have the ground state (1.5.5) with $\ell = 1$,

$$\psi_1^{(0)} \propto \frac{1}{\cosh(t)}. \tag{1.5.12}$$

To go to $\ell = 2$, we apply $A_2^\dagger$ to the states above, and we obtain the scattering states

$$\psi_2^{(k)}(t) \propto A_2^\dagger A_1^\dagger e^{ikt}, \tag{1.5.13}$$

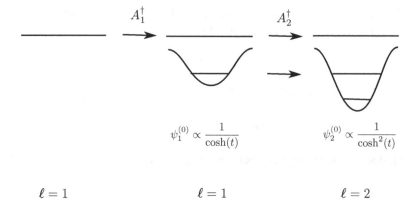

$$\ell = 1 \qquad\qquad \ell = 1 \qquad\qquad \ell = 2$$

Figure 1.11 The recursive solution of the spectrum of the Pöschl–Teller potential.

a bound state

$$\psi_2^{(1)}(t) \propto A_2^\dagger \frac{1}{\cosh(t)} \tag{1.5.14}$$

and the new ground state

$$\psi_2^{(0)}(t) \propto \frac{1}{\cosh^2(t)}. \tag{1.5.15}$$

Proceeding in this way we obtain the full spectrum of the ℓth potential. It consists of scattering states

$$\psi_\ell^{(k)}(t) \propto A_\ell^\dagger \cdots A_1^\dagger e^{ikt}, \tag{1.5.16}$$

with energies

$$E_{\ell,m} = k^2 + m^2, \tag{1.5.17}$$

and ℓ bound states

$$\psi_\ell^{(0)}(t) \propto \frac{1}{\cosh^\ell(t)},$$

$$\psi_\ell^{(j)}(t) \propto A_\ell^\dagger \cdots A_{\ell-j+1}^\dagger \frac{1}{\cosh^{\ell-j}(t)}, \qquad j = 1, \ldots, \ell - 1, \tag{1.5.18}$$

with energies

$$E_{\ell,m}^{(j)} = m^2 - (\ell - j)^2, \qquad j = 0, \ldots, \ell - 1. \tag{1.5.19}$$

Since the spectrum of the operator $\mathbf{M}_{\ell,m}$ is known, we should be able to compute its determinant. However, as we have explained above, this requires a regularization. We will do that by considering the quotient of the determinant of $\mathbf{M}_{\ell,m}$ by a "reference" determinant. As in (1.4.54), it is convenient to take as our reference operator the harmonic oscillator operator $M_{0,m}$. From the point of view of Pöschl–Teller potentials, this corresponds to the free particle. However, since we are working in the limit of infinite volume $\beta \to \infty$, there is a part of the spectrum which is continuum. In our considerations in the previous section we have assumed that the spectrum is discrete, which is actually the case for finite β, so we have to be precise about what we mean by the determinant in this more general case. If an operator M has a discrete spectrum $\{\lambda_n\}$ and a continuum spectrum $\lambda(k)$, the determinant should be understood as

$$\log \det M = \sum_n \log(\lambda_n) + \int dk \, \rho(k) \log(\lambda(k)), \qquad (1.5.20)$$

where $\rho(k)$ is the density of states for the continuum part. This is easily determined by introducing an IR regulator and putting the system in a box. In the case of the Pöschl–Teller potential, the calculation of $\rho(k)$ goes as follows. A scattering state will experience phase shifts. Indeed, as $t \to \pm\infty$, we have

$$A_j^\dagger \approx -\frac{d}{dt} \pm j, \qquad (1.5.21)$$

and the asymptotic form of the scattering states will be

$$\psi_\ell^{(k)}(t) \to \prod_{j=1}^\ell (-ik \pm j) \, e^{ikt}, \qquad t \to \pm\infty. \qquad (1.5.22)$$

Therefore the phase shifts, defined by

$$\psi_\ell^{(k)}(t) \to \exp\left[i\left(kt \pm \frac{\theta(k)}{2} \right) \right], \qquad (1.5.23)$$

are given by

$$\frac{\theta(k)}{2} = -\sum_{j=1}^\ell \tan^{-1}\left(\frac{k}{j} \right) + \frac{\pi}{2}. \qquad (1.5.24)$$

The quantization condition, once we put these scattering states in a box of length β, is just

$$ik\beta + i\theta(k) = 2\pi i n, \qquad (1.5.25)$$

and the density of states is

$$\rho(k) = \frac{dn}{dk} = \rho_{\text{free}}(k) + \rho_\theta(k), \qquad (1.5.26)$$

where

$$\rho_{\text{free}}(k) = \frac{\beta}{2\pi}, \qquad \rho_\theta(k) = \frac{1}{2\pi} \theta'(k). \qquad (1.5.27)$$

In our case

$$\rho_\theta(k) = -\frac{1}{\pi} \sum_{j=1}^{\ell} \frac{j}{k^2 + j^2}. \tag{1.5.28}$$

Let us now compute $\log \det' \mathbf{M}_{\ell,m}$ by using these results. In the expression (1.5.20), the sum over λ_n is over non-zero, discrete eigenvalues, and it is finite. The integration over the continuum part is divergent due to the contribution of $\rho_{\text{free}}(k)$. However, if we subtract $\log \det \mathbf{M}_{0,m}$, this part cancels and we find

$$\log\left(\frac{\det' \mathbf{M}_{\ell,m}}{\det \mathbf{M}_{0,m}}\right) = \sum_{1 \le j \le \ell, \, j \ne m} \log(m^2 - j^2) + \int_{-\infty}^{\infty} dk \, \rho_\theta(k) \log(k^2 + m^2)$$

$$= \sum_{1 \le j \le \ell, \, j \ne m} \log(m^2 - j^2)$$

$$- \frac{1}{\pi} \sum_{j=1}^{\ell} j \int_{-\infty}^{\infty} \frac{dk}{k^2 + j^2} \log(k^2 + m^2). \tag{1.5.29}$$

Since

$$\int_{-\infty}^{\infty} \frac{dk}{k^2 + j^2} \log(k^2 + m^2) = \frac{2\pi}{j} \log(j + m), \tag{1.5.30}$$

the end result is

$$\frac{\det' \mathbf{M}_{\ell,m}}{\det \mathbf{M}_{0,m}} = \frac{\prod_{1 \le j \le \ell, \, j \ne m} (m^2 - j^2)}{\prod_{1 \le j \le \ell} (m + j)^2}, \tag{1.5.31}$$

for $(\ell, m) \ne (1, 1)$. In the case $\ell = m = 1$, the numerator should be taken as 1, since there is no contribution from discrete states, and one has

$$\frac{\det' \mathbf{M}_{1,1}}{\det \mathbf{M}_{0,1}} = \frac{1}{4}. \tag{1.5.32}$$

1.6 Calculation of functional determinants II: the Gelfand–Yaglom method

It turns out that the determinant of the operator appearing in (1.4.56) can be computed without explicit knowledge of the spectrum, for any general potential $V(x)$ admitting periodic orbits, and for finite β. This is due to a remarkable result originally due to Gelfand and Yaglom. In this section we will state the result and apply it to our particular problem. A proof can be found in Appendix B.3.

Let us consider the eigenvalue problem for the following Schrödinger operator in the interval $[-\beta/2, \beta/2]$,

$$\left[-\frac{d^2}{dt^2} + u(t)\right]\psi(t) = \lambda\psi(t). \tag{1.6.1}$$

Let us denote by $\psi_\lambda^{1,2}(t)$ two independent solutions to this ODE, with boundary conditions

$$\begin{aligned}
\psi_\lambda^1(-\beta/2) &= 1, & \dot{\psi}_\lambda^1(-\beta/2) &= 0, \\
\psi_\lambda^2(-\beta/2) &= 0, & \dot{\psi}_\lambda^2(-\beta/2) &= 1.
\end{aligned} \tag{1.6.2}$$

Consider now the fundamental matrix of the ODE,

$$M_\lambda(t) = \begin{pmatrix} \psi_\lambda^1(t) & \psi_\lambda^2(t) \\ \dot{\psi}_\lambda^1(t) & \dot{\psi}_\lambda^2(t) \end{pmatrix}. \tag{1.6.3}$$

Any solution $\psi_\lambda(t)$ to (1.6.1) with initial conditions $\psi_\lambda(-\beta/2)$, $\dot{\psi}_\lambda(-\beta/2)$ can be written as a linear combination of $\psi_\lambda^{1,2}(t)$, as

$$\begin{pmatrix} \psi_\lambda(t) \\ \dot{\psi}_\lambda(t) \end{pmatrix} = M_\lambda(t)\begin{pmatrix} \psi_\lambda(-\beta/2) \\ \dot{\psi}_\lambda(-\beta/2) \end{pmatrix}. \tag{1.6.4}$$

Let us now define the monodromy matrix $T(\lambda)$ as

$$T(\lambda) = M_\lambda(\beta/2). \tag{1.6.5}$$

Notice that

$$\det\left(M_\lambda(t)\right) = W\left[\psi_\lambda^1(t), \psi_\lambda^2(t)\right] \tag{1.6.6}$$

is the Wronskian of the two solutions, so it is constant:

$$\det\left(M_\lambda(t)\right) = 1, \qquad t \in [-\beta/2, \beta/2]. \tag{1.6.7}$$

Let us now consider the operator

$$-\frac{d^2}{dt^2} + u(t) - \lambda \tag{1.6.8}$$

acting on functions satisfying periodic boundary conditions,

$$\psi_\lambda(-\beta/2) = \psi_\lambda(\beta/2), \qquad \dot{\psi}_\lambda(-\beta/2) = \dot{\psi}_\lambda(\beta/2). \tag{1.6.9}$$

If we define the determinant of (1.6.8) by using zeta function regularization, as explained in Appendix B.2, one has the following result

$$\det\left(-\frac{d^2}{dt^2} + u(t) - \lambda\right) = \text{Tr}\left(T(\lambda) - \mathbf{1}\right). \tag{1.6.10}$$

As a check of this statement, note that both sides of (1.6.10), as functions of the complex variable λ, have the same set of zeros. Indeed, the zeros of the left hand side are the eigenvalues of the Schrödinger operator $-d^2/dt^2 + u(t)$, where the corresponding eigenfunctions $\psi_\lambda(t)$ satisfy the periodic boundary conditions (1.6.9). On the other hand, if this is the case, the matrix $T(\lambda)$ has an eigenvector

$$\begin{pmatrix} \psi_\lambda(\beta/2) \\ \dot{\psi}_\lambda(\beta/2) \end{pmatrix} \tag{1.6.11}$$

with eigenvalue 1, due to (1.6.4). But this happens if and only if

$$\mathrm{Tr}\,(T(\lambda) - 1) = 0, \tag{1.6.12}$$

since $\det(T(\lambda)) = 1$ due to (1.6.7). A detailed proof of (1.6.10) is presented in Appendix B.3.

It is easy to verify that (1.6.10) reproduces the standard result for the Euclidean partition function of the harmonic oscillator, i.e. for $u(t) = 0$, $\lambda = -1$. The corresponding operator is $\mathbf{M}_0$, in (1.4.53), and the fundamental matrix (1.6.3) reads in this case,

$$M_{-1}(t) = \begin{pmatrix} \cosh(t + \beta/2) & \sinh(t + \beta/2) \\ \sinh(t + \beta/2) & \cosh(t + \beta/2) \end{pmatrix}. \tag{1.6.13}$$

We then find,

$$\det \mathbf{M}_0 = \mathrm{Tr}\,(T(\lambda) - 1) = 2\,(\cosh \beta - 1) = 4\sinh^2\left(\frac{\beta}{2}\right), \tag{1.6.14}$$

which agrees with (1.2.16). This means that zeta function regularization gives the standard value for the path integral of the harmonic oscillator, without further normalizations. In other words, with this regularization, the constant $\mathcal{N}$ introduced in (1.4.30) and appearing in (1.4.52) can be set to 1.

Let us now use (1.6.10) to compute the determinant $\det'(\mathbf{M})$ appearing in (1.4.56). Let us then suppose that the operator $-d^2/dt^2 + u(t)$ has zero modes. The eigenvalues of (1.6.8) are the eigenvalues of $-d^2/dt^2 + u(t)$, shifted by $-\lambda$. Therefore, we can write

$$\det'\left(-\frac{d^2}{dt^2} + u(t)\right) = -\frac{\partial}{\partial \lambda} \det\left(-\frac{d^2}{dt^2} + u(t) - \lambda\right)\Big|_{\lambda=0}$$

$$= -\frac{\partial}{\partial \lambda} \mathrm{Tr}\,(T(\lambda) - 1)\Big|_{\lambda=0}. \tag{1.6.15}$$

To calculate the quantity in the second line, it is enough to compute $\psi_\lambda^{1,2}(\beta/2)$ to first order in λ. Let us denote by $\psi_0^k(t)$, $k = 1, 2$, the solutions for $\lambda = 0$. Then, the functions

$$\psi_\lambda^k(t) = \psi_0^k(t) - \lambda \mathcal{I}^k(t), \qquad k = 1, 2, \tag{1.6.16}$$

where

$$\mathcal{I}^k(t) = \int_{-\beta/2}^t \left[\psi_0^2(t)\psi_0^1(t') - \psi_0^1(t)\psi_0^2(t') \right] \psi_0^k(t')dt', \qquad k = 1, 2, \quad (1.6.17)$$

solve (1.6.1) with the boundary conditions (1.6.2), at first order in λ. This is easy to check by noting that

$$\dot{\mathcal{I}}^k(t) = \dot{\psi}_0^2(t) \int_{-\beta/2}^t \psi_0^1(t')\psi_0^k(t')dt' - \dot{\psi}_0^1(t) \int_{-\beta/2}^t \psi_0^2(t')\psi_0^k(t')dt',$$

$$\ddot{\mathcal{I}}^k(t) = \psi_0^k(t) + \ddot{\psi}_0^2(t) \int_{-\beta/2}^t \psi_0^1(t')\psi_0^k(t')dt' - \ddot{\psi}_0^1(t) \int_{-\beta/2}^t \psi_0^2(t')\psi_0^k(t')dt',$$

$$(1.6.18)$$

where in the second equation we have used (1.6.7). It follows that

$$\det{}' \left(-\frac{d^2}{dt^2} + u(t) \right) = \psi_0^2(\beta/2) \int_{-\beta/2}^{\beta/2} \left(\psi_0^1(t) \right)^2 dt - \dot{\psi}_0^1(\beta/2) \int_{-\beta/2}^{\beta/2} \left(\psi_0^2(t) \right)^2 dt$$

$$+ \left(\dot{\psi}_0^2(\beta/2) - \psi_0^1(\beta/2) \right) \int_{-\beta/2}^{\beta/2} \psi_0^1(t)\psi_0^2(t)dt. \quad (1.6.19)$$

This equation expresses the determinant of the Schrödinger operator, after removing the vanishing eigenvalue, in terms of two solutions of the zero mode problem

$$\left[-\frac{d^2}{dt^2} + u(t) \right] \psi(t) = 0 \qquad (1.6.20)$$

with the boundary conditions (1.6.2).

Let us now apply this formalism to the operator (1.4.23). We have to find two zero modes. We have already found one of them: it is $\dot{q}_c(t)$. To find the other mode, we notice that the solution $q_c(t)$ depends in addition on the energy E of the trajectory. Let us write it as $q_c(t; E)$ to make this dependence manifest. Since the EOM (1.4.1) is solved for any energy in the available range, we deduce that the function

$$\chi(t) = \frac{\partial q_c(t; E)}{\partial E} \qquad (1.6.21)$$

is also a zero mode. However, this function is *not* periodic, and it does not contribute to the determinant of the operator **M**. From the energy conservation condition (1.4.3), one obtains, by taking a derivative with respect to E,

$$\dot{q}_c(t)\dot{\chi}(t) - \ddot{q}_c(t)\chi(t) = W[\dot{q}_c(t), \chi(t)] = 1. \qquad (1.6.22)$$

We can now use time translation invariance and choose the modulus t_0 entering into the solution in such a way that

$$\ddot{q}_c(-\beta/2) = 0. \qquad (1.6.23)$$

This just means that at $t = -\beta/2$ the particle is at the minimum of the inverted potential. The Wronskian condition (1.6.22) then implies that

$$\dot{q}_c(-\beta/2)\chi(-\beta/2) = 1. \tag{1.6.24}$$

The choice (1.6.23) simplifies the form of the $\psi_0^k(t)$, which are given by

$$\psi_0^1(t) = \frac{\dot{q}_c(t)}{\dot{q}_c(-\beta/2)}, \qquad \psi_0^2(t) = -\chi(-\beta/2)\dot{q}_c(t) + \dot{q}_c(-\beta/2)\chi(t). \tag{1.6.25}$$

It can easily be checked that they satisfy the boundary conditions (1.6.2).

We now use (1.6.19) to evaluate the determinant. Since

$$\psi_0^1(\beta/2) = \dot{\psi}_0^1(\beta/2) = 1, \qquad \dot{\psi}_0^1(\beta/2) = 0, \tag{1.6.26}$$

we conclude that

$$\det'\left(-\frac{d^2}{dt^2} - V''(q_c(t))\right) = \frac{\chi(\beta/2) - \chi(-\beta/2)}{\dot{q}_c(\beta/2)} \int_{-\beta/2}^{\beta/2} (\dot{q}_c(t))^2\, dt. \tag{1.6.27}$$

This can be further simplified as follows. Since $q_c(-\beta/2) = q_c(\beta/2)$ and $\dot{q}_c(-\beta/2) = \dot{q}_c(\beta/2)$, the function $q_c(t)$ can be extended to a periodic function of period β on the whole real axis $t \in \mathbb{R}$. This function is continuous and differentiable, and we will also denote it by $q_c(t; E)$. The period β and the energy E are then related by the periodicity condition

$$q_c(t + \beta; E) = q_c(t; E), \qquad t \in \mathbb{R}. \tag{1.6.28}$$

Taking a derivative with respect to β, we find

$$\frac{\chi(t + \beta) - \chi(t)}{\dot{q}_c(t)} = -\frac{\partial \beta}{\partial E}, \qquad t \in \mathbb{R}, \tag{1.6.29}$$

where the period of the motion, β, is regarded as a function of E. By setting $t = -\beta/2$ in this equation, and taking into account (1.4.5), we conclude that

$$\det'\left(-\frac{d^2}{dt^2} - V''(q_c(t))\right) = -W(E)\left(\frac{\partial E}{\partial \beta}\right)^{-1}. \tag{1.6.30}$$

Notice that, as the energy increases, the period β grows, so that the derivative in the right hand side is positive, and the determinant is negative. As we already know, this is due to the single negative mode appearing in this type of problem.

Example 1.3 Let us consider the quartic oscillator, which we studied in Example 1.1. In this case, the additional zero mode $\chi(t)$ can be explicitly computed as

$$\chi(t) = \frac{dq_+}{dE}\mathrm{dn}(u; k) - \mathrm{sn}(u; k)\mathrm{cn}(u; k)\left\{uk^2\frac{dq_+}{dE} + \frac{q_+ u}{2}\frac{dk^2}{dE} - q_+\frac{E(u, k)}{2(1 - m)}\frac{dk^2}{dE}\right\}$$

$$- \frac{q_+}{2(1 - k^2)}\frac{dk^2}{dE}\mathrm{sn}^2(u; k)\mathrm{dn}(u; k), \tag{1.6.31}$$

where sn and cn are the Jacobi elliptic sine and cosine functions, $E(u, k)$ is the incomplete elliptic integral of the second kind, and u, k were defined in (1.4.15). Notice that this function is not periodic, as advertised. By using that

$$\frac{\partial \beta}{\partial E} = -\frac{2(2 - k^2)^{7/2}}{k^4} \left[\frac{K(k)}{2 - k^2} - \frac{E(k)}{2(1 - k^2)} \right], \tag{1.6.32}$$

as well as standard properties of Jacobi elliptic functions, one can verify that (1.6.29) is indeed satisfied. □

1.7 Decay rates in unstable vacua from instantons

We now have all the necessary technical ingredients to calculate the imaginary part of the ground state energy in unstable vacua, given by (1.4.56). We can calculate the determinant of the relevant operators by using either the special results for solvable potentials, or the general expression (1.6.30). However, this expression was obtained for general β (or, equivalently, E), and to extract the ground state energy we are interested in the limit $\beta \to \infty$, i.e. in the orbits of zero energy. This can be done as follows. First of all, we normalize our potential in such a way that the unstable minimum is at the origin. Periodic orbits of zero energy go from the unstable minimum $q_- = 0$ (which is also one of the turning points) to the other turning point q_+. Clearly, as $E \to 0$, $\beta \to \infty$. Let us consider the derivative of E with respect to β. In the large β limit, we write (1.4.2) as

$$\beta = 2 \int_{q_-}^{q_+} \left\{ \frac{1}{\sqrt{2(E - V(x))}} - \frac{1}{\sqrt{x^2 + 2E}} + \frac{1}{\sqrt{x^2 + 2E}} \right\} dx. \tag{1.7.1}$$

For E small, we have

$$q_- \approx \sqrt{-2E} + \mathcal{O}(E). \tag{1.7.2}$$

The last integral gives

$$2 \log \left(x + \sqrt{x^2 + 2E} \right) \Big|_{q_-}^{q_+} = \log q_+^2 - \log(-E/2) + \mathcal{O}(E), \qquad E \to 0. \tag{1.7.3}$$

The first two terms in (1.7.1) have a smooth limit at $E \to 0$, and they give

$$2 \int_0^{q_+} \left(\frac{1}{\sqrt{2W(x)}} - \frac{1}{x} \right) dx. \tag{1.7.4}$$

It follows that, as $\beta \to \infty$,

$$E(\beta) \approx -2q_+^2 \exp \left[2 \int_0^{q_+} \left(\frac{1}{\sqrt{2W(x)}} - \frac{1}{x} \right) dx \right] e^{-\beta}, \tag{1.7.5}$$

therefore

$$\frac{\partial E}{\partial \beta} \approx 2q_+^2 \exp\left[2\int_0^{q_+}\left(\frac{1}{\sqrt{2W(x)}}-\frac{1}{x}\right)dx\right]e^{-\beta}, \qquad (1.7.6)$$

which is manifestly positive, as anticipated above.

We can now use this limiting expression in (1.6.30). Note that, due to the exponential dependence on β, $\det' \mathbf{M}$ diverges exponentially when $\beta \to \infty$. However, after dividing by the determinant of the Gaussian operator $\mathbf{M}_0$, we obtain a finite expression in this limit, given by

$$\lim_{\beta\to\infty}\frac{\det'\mathbf{M}}{\det\mathbf{M}_0} = -\frac{S_c}{2q_+^2}\exp\left[-2\int_0^{q_+}\left(\frac{1}{\sqrt{2W(x)}}-\frac{1}{x}\right)dx\right]. \qquad (1.7.7)$$

Plugging this expression into (1.4.56) we finally obtain a general formula for the width of an unstable level in QM (at one-loop):

$$\text{Im } E_0 \approx \pm\frac{1}{2\sqrt{\pi}}q_+\exp\left[\int_0^{q_+}dx\left(\frac{1}{\sqrt{2W(x)}}-\frac{1}{x}\right)\right]e^{-S(q_c)}. \qquad (1.7.8)$$

It is an interesting exercise to evaluate (1.7.7) for concrete potentials and check that indeed it agrees with (1.5.31).

Example 1.4 *Anharmonic oscillator.* The action of the instanton is given by

$$S_c = 2\int_0^{\sqrt{2/\lambda}} x\sqrt{1-\frac{\lambda}{2}x^2}\,dx = -\frac{2\left(2-\lambda x^2\right)^{3/2}}{3\sqrt{2\lambda}}\Bigg|_0^{\sqrt{2/\lambda}} = \frac{4}{3\lambda}. \qquad (1.7.9)$$

The integral appearing in the exponent in (1.7.7) is

$$\int_0^{q_+}\left(\frac{1}{\sqrt{2W(x)}}-\frac{1}{x}\right)dx = \int_0^{\sqrt{2/\lambda}}\frac{\sqrt{2}-\sqrt{2-\lambda x^2}}{x\sqrt{2-\lambda x^2}}rdx$$

$$= -\log\left(\sqrt{2}\sqrt{2-\lambda x^2}+2\right)\Bigg|_0^{\sqrt{2/\lambda}} = \log 2. \qquad (1.7.10)$$

The determinant is then given by

$$\frac{\det'\mathbf{M}}{\det\mathbf{M}_0} = -\frac{1}{12}. \qquad (1.7.11)$$

This can also be calculated by using Pöschl–Teller operators: if we compare (1.4.48) to (1.5.1) we see that it is given by $\mathbf{M}_{2,1}$, and formula (1.5.31) gives the same result as (1.7.11). Using finally (1.4.56), we derive an explicit formula for the imaginary part of the ground state energy,

$$\text{Im } E_0 \approx \frac{2}{2\sqrt{\pi}}\cdot\sqrt{\frac{2}{\lambda}}\cdot 2\cdot e^{-4/3\lambda} = \frac{4}{\sqrt{2\pi\lambda}}e^{-4/3\lambda}. \qquad (1.7.12)$$

□

Example 1.5 *Cubic oscillator.* Let us now study the lifetime of a particle in the ground state of the cubic potential

$$W(x) = \frac{1}{2}x^2 - gx^3. \tag{1.7.13}$$

The turning points are $q_- = 0$ and

$$q_+ = \frac{1}{2g}. \tag{1.7.14}$$

The instanton solution is

$$q_c(t) = \frac{1}{2g \cosh^2\left(\frac{t}{2}\right)}, \tag{1.7.15}$$

and the operator $\mathbf{M}$ reads

$$\mathbf{M} = -\frac{d^2}{dt^2} + 1 - \frac{3}{\cosh^2\left(\frac{t}{2}\right)}. \tag{1.7.16}$$

The action of the instanton is

$$S_c = 2\int_0^{1/(2g)} \sqrt{x^2 - 2gx^3}\,dx = \frac{2}{15g^2}. \tag{1.7.17}$$

The integral appearing in the exponent in (1.7.7) is

$$\int_0^{1/(2g)} \frac{x - \sqrt{x^2 - 2gx^3}}{x\sqrt{x^2 - 2gx^3}}\,dx = \log 4, \tag{1.7.18}$$

and we find

$$\frac{\det' \mathbf{M}}{\det \mathbf{M}_0} = -\frac{1}{60}. \tag{1.7.19}$$

Therefore, the general formula (1.4.56) gives,

$$\operatorname{Im} E_0(g) \approx \frac{1}{2\pi^{1/2}} \cdot \frac{1}{2g} \cdot 4 \cdot e^{-2/(15g^2)} = \frac{1}{\sqrt{\pi g^2}} e^{-2/(15g^2)}. \tag{1.7.20}$$

This agrees with the result obtained with the WKB method. The result (1.7.19) can also be obtained with the help of the appropriate Pöschl–Teller operator. After rescaling $t \to 2t$ we find that

$$\mathbf{M} = \frac{1}{4}\mathbf{M}_{3,2}, \qquad \mathbf{M}_0 = \frac{1}{4}\mathbf{M}_{0,2}. \tag{1.7.21}$$

The behavior of the determinant of an operator under rescaling requires a careful analysis. Let us suppose that we rescale

$$\mathbf{M}_{\ell,m} \to \xi\mathbf{M}_{\ell,m}, \qquad \mathbf{M}_{0,m} \to \xi\mathbf{M}_{0,m}. \tag{1.7.22}$$

Then,

$$\frac{\det' \mathbf{M}_{\ell,m}}{\det \mathbf{M}_{0,m}} \to \xi^{N'_{\ell,m} - N_{0,m}} \frac{\det' \mathbf{M}_{\ell,m}}{\det \mathbf{M}_{0,m}}, \tag{1.7.23}$$

where $N'_{\ell,m} - N_{0,m}$ is the number of non-zero modes of $\mathbf{M}_{\ell,m}$ minus the number of modes of $\mathbf{M}_{0,m}$. This can be computed by adapting the procedure in (1.5.29): $\mathbf{M}_{\ell,m}$ has $\ell - 1$ discrete non-zero modes for $m \leq \ell$, plus a continuum. To calculate the difference between the zero modes in the continuum for $\mathbf{M}_{\ell,m}$ and $\mathbf{M}_{0,m}$ we can again use the density of states. We find,

$$N'_{\ell,m} - N_{0,m} = j - 1 + \int_{-\infty}^{\infty} dk\, \rho(k) = j - 1 - \frac{1}{\pi} \sum_{j=1}^{\ell} j \int_{-\infty}^{\infty} \frac{dk}{k^2 + j^2}$$

$$= j - 1 - j = -1. \tag{1.7.24}$$

Therefore, we conclude that after a rescaling, the quotient of determinants behaves as

$$\frac{\det' \mathbf{M}_{\ell,m}}{\det \mathbf{M}_{0,m}} \to \frac{1}{\xi} \frac{\det' \mathbf{M}_{\ell,m}}{\det \mathbf{M}_{0,m}}. \tag{1.7.25}$$

Therefore, in the case at hand $\ell = 3, m = 2$, we have that

$$\frac{\det' \mathbf{M}}{\det \mathbf{M}_0} = 4 \frac{\det' \mathbf{M}_{3,2}}{\det \mathbf{M}_{0,2}}, \tag{1.7.26}$$

which is in agreement with (1.7.19).   □

1.8 Instantons in the double-well potential

The double-well potential, shown in the figure on the left in Fig. 1.1, illustrates one of the most important applications of instantons: their ability to lift perturbation theory degeneracies. Indeed, in perturbation theory, the double-well potential has two different ground states located around the two classical degenerate minima. This implies, in particular, that parity symmetry is spontaneously broken in perturbation theory. However, in the full theory, this cannot be the case: we know from elementary QM that the spectrum of the Schrödinger operator in this bound state problem must be discrete, and that the true vacuum is described by a symmetric wavefunction. This wavefunction corresponds, in the limit of vanishing coupling, to the symmetric combination of the two perturbative vacua. However, in perturbation theory, the energy split between the symmetric and antisymmetric combinations is invisible and goes like $\exp(-1/g)$, a typical instanton effect. This energy split can be computed with the WKB method. In this section we will derive it with instanton techniques in the path integral approach.

Consider the double-well potential with a Hamiltonian of the form (1.2.1) and

$$W(q) = \frac{g}{2}\left(q^2 - \frac{1}{4g}\right)^2, \qquad g > 0. \tag{1.8.1}$$

In perturbation theory one finds two degenerate ground states, located around the minima

$$q = \pm\frac{1}{2\sqrt{g}}. \tag{1.8.2}$$

The frequency of oscillations around these minima has been normalized to be $\omega = 1$, and the ground state energy obtained in stationary perturbation theory is a formal power series of the form

$$E_0(g) = \frac{1}{2} - g - \frac{9}{2}g^2 - \frac{89}{2}g^3 - \cdots. \tag{1.8.3}$$

The Hamiltonian is invariant under the parity symmetry

$$q \to -q, \tag{1.8.4}$$

and thus it commutes with the corresponding parity operator P, whose action on wavefunctions is

$$P\psi(q) = \psi(-q). \tag{1.8.5}$$

Since P commutes with H, we can diagonalize them simultaneously:

$$H\psi_{\epsilon,N}(q) = E_{\epsilon,N}(g)\psi_{\epsilon,N}(q), \qquad P\psi_{\epsilon,N}(q) = \epsilon\,\psi_{\epsilon,N}(q), \tag{1.8.6}$$

where $\epsilon = \pm 1$ is the parity. The quantum number N can be uniquely assigned to a given state by the requirement that, as $g \to 0$,

$$E_{\epsilon,N}(g) = N + 1/2 + \mathcal{O}(g), \tag{1.8.7}$$

i.e. it corresponds to the Nth energy level of the unperturbed harmonic oscillator.

Let us now focus on the ground state energy. As we know, the energy levels $E_{\epsilon,0}(g)$ are degenerate in perturbation theory, but they are split by non-perturbative effects. Therefore, they will have a perturbative contribution, given by (1.8.3), and they will differ by non-perturbative corrections. In order to study non-perturbative effects, it is always convenient to focus on a quantity which is purely non-perturbative, i.e. which vanishes in perturbation theory. In our case, this quantity is clearly

$$E_{+,0} - E_{-,0}. \tag{1.8.8}$$

This is, morally speaking, the analogue of Im E_0 in the case of unstable potentials, which is also purely non-perturbative. We would now like to find a quantity which can be computed in the path integral formalism and which is sensitive to the difference of energies (1.8.8). Clearly, the thermal partition function is not the most appropriate quantity. However, one can consider the "twisted" partition function

$$Z_a(\beta) = \text{Tr}\left(P\,e^{-\beta H}\right), \tag{1.8.9}$$

where P is the parity operator (1.8.5). For large β and small coupling constant one finds that,

$$Z_a(\beta) \approx e^{-\beta E_{+,0}} - e^{-\beta E_{-,0}} \approx -\beta e^{-\beta/2}\left(E_{+,0} - E_{-,0}\right), \tag{1.8.10}$$

so in principle we can use this twisted partition function to extract (1.8.8). In addition, $Z_a(\beta)$ can be written in terms of a path integral with "twisted" boundary conditions,

$$Z_a(\beta) = \int_{q(\beta/2)=P(q(-\beta/2))} \mathcal{D}q(t)\,\exp\left[-S\left(q(t)\right)\right]. \tag{1.8.11}$$

In the case of the double-well potential we are studying, the boundary condition reads

$$q(-\beta/2) = -q(\beta/2). \tag{1.8.12}$$

In the infinite β limit, the leading contributions to the path integral come from paths which are solutions of the Euclidean EOM and have zero energy. In the case of $Z_a(\beta)$, constant solutions of the equation of motion do not satisfy the boundary conditions. Therefore, when we compute $Z_a(\beta)$ in the semiclassical approximation, the relevant Euclidean saddle points are paths which connect the two minima of the potential (1.8.2), like in Fig. 1.12. These are given by

$$q_{\pm}^{t_0}(t) = \pm\frac{1}{2\sqrt{g}}\tanh\left(\frac{t-t_0}{2}\right). \tag{1.8.13}$$

The solutions $q_{\pm}^{t_0}$ are called *(anti)instantons* of center t_0. They are represented in Fig. 1.12 for $g = 1/4$. Notice that both depend on an integration constant or modulus t_0.

The operator $\mathbf{M}$ in (1.4.21) is given in this case by

$$\mathbf{M} = -\frac{d^2}{dt^2} + 1 - \frac{3}{2\cosh^2\left(\frac{t-t_0}{2}\right)}, \tag{1.8.14}$$

which is proportional to the Pöschl–Teller operator $\mathbf{M}_{2,2}$:

$$\mathbf{M} = \frac{1}{4}\mathbf{M}_{2,2}, \tag{1.8.15}$$

after rescaling $t \to t/2$. Notice that this operator has a zero mode but does *not* have a negative mode. This reflects the fact that the quantum mechanical state we

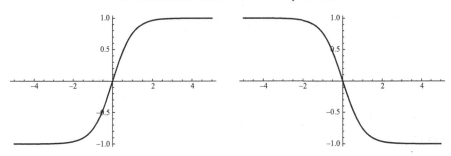

Figure 1.12 Left: an instanton configuration with center at $t_0 = 0$. Right: an anti-instanton configuration with center at $t_0 = 0$. Both are plotted for $g = 1/4$.

are now studying is stable. To evaluate the contribution of these configurations to the path integral, we can repeat the arguments that led to (1.4.54) with minimal modifications. We obtain

$$Z_a(\beta) \approx 2Z_G(\beta) \left[\frac{\det' \mathbf{M}}{\det \mathbf{M}_0}\right]^{-1/2} \frac{\beta S_c^{1/2}}{\sqrt{2\pi}} e^{-S_c}, \qquad (1.8.16)$$

where the extra factor of 2 is due to the fact that the two solutions $q_{\pm}^{t_0}(t)$ give the same contribution. From the relation (1.8.10) we deduce that

$$E_{+,0}(g) - E_{-,0}(g) = -2\frac{S_c^{1/2}}{\sqrt{2\pi}} e^{-S_c} \left[\frac{\det' \mathbf{M}}{\det \mathbf{M}_0}\right]^{-1/2}. \qquad (1.8.17)$$

Let us now compute the quantities involved in this expression. First of all, the classical action, evaluated at the classical trajectories (1.8.13), is

$$S_c = \frac{1}{6g}. \qquad (1.8.18)$$

We see that, at leading order in the effective coupling constant g, and for $\beta \to \infty$, (1.8.17) is proportional to $e^{-1/(6g)}$ and therefore it is non-perturbative. For the quotient of determinants we can use the general result for Pöschl–Teller operators,

$$\frac{\det' \mathbf{M}}{\det \mathbf{M}_0} = 4\frac{\det' \mathbf{M}_{2,2}}{\det \mathbf{M}_{0,2}} = \frac{1}{12}. \qquad (1.8.19)$$

One finally obtains the non-perturbative splitting between the symmetric and the antisymmetric wavefunctions as

$$E_{+,0}(g) - E_{-,0}(g) \approx -\frac{2}{\sqrt{\pi g}} e^{-1/6g}, \qquad (1.8.20)$$

at leading order in g and $e^{-1/(6g)}$. It follows that the true ground state corresponds to the symmetric wavefunction, as expected. The energies of these states can be written as

$$E_{\epsilon,0}(g) = E_0^{(0)}(g) + E_{\epsilon,0}^{(1)}(g), \tag{1.8.21}$$

where $\epsilon = \pm$ and

$$E_0^{(0)}(g) = \frac{1}{2} + \mathcal{O}(g), \qquad E_{\epsilon,0}^{(1)}(g) = -\frac{\epsilon}{\sqrt{\pi g}} e^{-1/6g} \left(1 + \mathcal{O}(g)\right), \tag{1.8.22}$$

and they are due to the perturbative and to the one-instanton contributions, respectively.

1.9 Multi-instantons in the double-well potential

In the previous section we have computed an exponentially small correction to physical energies due to instanton configurations. This correction involves the new small parameter $e^{-A/g}$, besides the usual coupling g, and we should expect that the full answer will be a generalized series involving *both* g and $e^{-A/g}$. This type of series is sometimes called a "trans-series" in the mathematical literature. In particular, we should expect to have corrections of the form $e^{-nA/g}$, where n is a positive integer. In some cases, these corrections are associated to generalized, non-trivial saddle points in the path integral, also called *multi-instantons*. In the case of QM, as we will see in this section, multi-instantons are approximate saddle points, but in other circumstances (like in Yang–Mills theory), multi-instantons are true saddle points. Clearly, in a complete theory of instanton effects, we should incorporate in a systematic way the corrections due to multi-instantons. Although in general this is difficult, in this section we will address this issue in the case of the double-well potential in Quantum Mechanics, where multi-instantons are well understood.

One quick way to detect the presence of multi-instantons in the double-well potential is to consider the partition function defined by

$$Z_\epsilon(\beta) = \frac{1}{2}(Z(\beta) + \epsilon Z_a(\beta)) = \sum_{N=0}^{\infty} \exp\left(-\beta E_{\epsilon,N}\right), \tag{1.9.1}$$

which selects eigenstates with parity ϵ. For large β, only the lowest energy state with parity ϵ survives, and

$$Z_\epsilon(\beta) \approx e^{-\beta(E_0^{(0)} + E_{\epsilon,0}^{(1)})} \approx e^{-\beta/2} \sum_{n=0}^{\infty} \frac{1}{n!} \left(\frac{\epsilon\beta}{\sqrt{\pi g}}\right)^n e^{-n/6g}. \tag{1.9.2}$$

The term with $n = 1$ is the contribution from the one-instanton (or anti-instanton), which we have already analyzed in detail. The terms with $n > 1$ look like contributions to the path integral of configurations with classical action $n/(6g)$, i.e. n times the contribution of a single (anti)instanton. They are due to multi-instantons or n-instantons. In the double-well potential, multi-instantons are not solutions of the classical equation of motion. They can be described as configurations made out of n instantons which are well separated, and they only become true solutions in the limit of infinite separation. Notice that the sum in (1.9.2) can be written as

$$Z_\epsilon(\beta) \approx e^{-\beta/2} \sum_{k=0}^{\infty} \frac{1}{(2k)!} \left(\frac{\beta}{\sqrt{\pi g}} \right)^{2k} e^{-2k/6g}$$

$$+ \epsilon e^{-\beta/2} \sum_{k=0}^{\infty} \frac{1}{(2k+1)!} \left(\frac{\beta}{\sqrt{\pi g}} \right)^{2k+1} e^{-(2k+1)/6g}. \tag{1.9.3}$$

We see that n *even* contributes to $Z(\beta)$, while n *odd* contributes to $Z_a(\beta)$. The reason for this is that a configuration with $n = 2k$ even can be regarded as a chain of k instanton–anti-instanton pairs, which satisfy the boundary condition

$$q(-\beta/2) = q(\beta/2). \tag{1.9.4}$$

Similarly, $n = 2k+1$ can be regarded as a chain of k instanton–anti-instanton pairs, followed by an instanton or an anti-instanton, therefore satisfying the boundary condition

$$q(-\beta/2) = -q(\beta/2). \tag{1.9.5}$$

In the following, we will perform a change of coordinates and write

$$q \rightarrow \frac{q}{\sqrt{g}} - \frac{1}{2\sqrt{g}}. \tag{1.9.6}$$

In the new coordinates, the Euclidean action is

$$S(q) = \frac{1}{g} \int_{-\beta/2}^{\beta/2} dt \left[\frac{1}{2} (\dot{q}(t))^2 - V(q) \right], \tag{1.9.7}$$

where the inverted potential is given by

$$V(q) = -\frac{1}{2} q^2 (1 - q)^2, \tag{1.9.8}$$

and the minima of $W(q) = -V(q)$ occur now at $q = 0$, $q = 1$. The (anti)instanton configuration (1.8.13) becomes, in the new coordinates,

$$q_{\pm}^{t_0} = \frac{1}{1 + e^{\mp(t-t_0)}}. \tag{1.9.9}$$

Since we now know that multi-instanton configurations are expected, let us analyze their effects. We first construct a two-instanton configuration, which is built out of an instanton–anti-instanton pair. This configuration depends on one additional

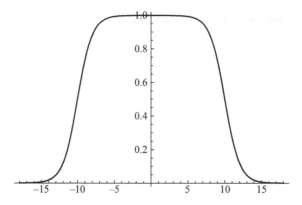

Figure 1.13 A two-instanton configuration of the form (1.9.10), for $\theta = 20$.

parameter θ, which represents the separation between instantons, and it decomposes, in the limit of infinite separation, into an instanton and an anti-instanton. It is given by (see Fig. 1.13)

$$q_c^\theta(t) = q_+^{-\theta/2}(t) + q_-^{\theta/2}(t) - 1 = q_-^{\theta/2}(t) - q_-^{-\theta/2}(t). \qquad (1.9.10)$$

This path is obviously continuous and differentiable. It represents, roughly speaking, an instanton centered at $-\theta/2$, joined to an anti-instanton centered at $\theta/2$. When θ is large, the piece of this path which looks like an (anti)instanton differs from the true (anti)instanton solution only by exponentially small terms, of order $e^{-\theta}$.

Let us now calculate the classical action evaluated on this path, as a function of θ. We first introduce the functions

$$u(t) = q_-^{\theta/2}(t), \qquad v(t) = q_-^{-\theta/2}(t) = u(t + \theta), \qquad (1.9.11)$$

therefore $q_c^\theta = u - v$. The action corresponding to the path (1.9.10) can be written as

$$
\begin{aligned}
g S(q_c^\theta) &= \int_{-\infty}^{\infty} dt \left(\frac{1}{2} \dot{q}_c^2 - V(q_c) \right) \\
&= \int_{-\infty}^{\infty} dt \left(\frac{1}{2} \dot{u}^2 - V(u) + \dot{v}^2 - V(v) \right) \\
&\quad + \int_{-\infty}^{\infty} dt \left(-\dot{u}\,\dot{v} - V(u - v) + V(u) + V(v) \right). \qquad (1.9.12)
\end{aligned}
$$

The first term in the second line is just twice the action of a single instanton, $1/3$. Since q_c is even as a function of t, the last integral is twice the integral for $t > 0$. After integrating by parts the term $\dot{v}\dot{u}$, one finds

$$g S(q_c^\theta) = \frac{1}{3} + 2 \left\{ v(0)\, \dot{u}(0) + \int_0^\infty dt\ (v\,\ddot{u} - V(u-v) + V(u) + V(v)) \right\}.$$

$$(1.9.13)$$

We now want to calculate the expansion of this action in powers of $e^{-\theta/2}$, for large θ. Since, for $t > 0$, v is of order $e^{-\theta}$, this is equivalent to expanding the integrand in powers of v. Taking into account that $V(0) = V'(0) = 0$, $V''(0) = -1$, we find

$$v\,\ddot{u} - V(u-v) + V(u) + V(v) = v\left(\ddot{u} + V'(u)\right) - \frac{v^2}{2}\left(1 + V''(u)\right) + \mathcal{O}(v^3).$$

$$(1.9.14)$$

The first term, which is linear in v, vanishes as a consequence of the equation of motion for u. Since the function v decreases exponentially away from the origin, the main contribution to the integral comes from the neighborhood of $t = 0$, where

$$u = 1 + \mathcal{O}(e^{-\theta/2}),$$

$$(1.9.15)$$

therefore

$$V''(u) \approx V''(1) = -1,$$

$$(1.9.16)$$

and the right hand side of (1.9.14) vanishes at quadratic order in v. On the other hand,

$$v(0) = -\frac{1}{e^{\theta/2} + 1}, \qquad \dot{u}(0) = -\frac{e^{-\theta/2}}{\left(1 + e^{-\theta/2}\right)^2},$$

$$(1.9.17)$$

and

$$v(0)\,\dot{u}(0) \approx -e^{-\theta}$$

$$(1.9.18)$$

at leading order. We conclude that

$$g S(q_c^\theta) = \frac{1}{3} - 2e^{-\theta} + \mathcal{O}\left(e^{-2\theta}\right).$$

$$(1.9.19)$$

Note that, when g is small (and negative), the action favors instanton configurations in which θ is large and the instantons are well separated. It is sometimes useful to regard the instantons as particles in a classical gas confined to a circle (the so-called *instanton gas*), separated by a distance θ, and interacting with a potential $-2e^{-\theta}/g$. The n-instanton contribution corresponds to a gas of n particles.

It is possible to extend the result (1.9.19) to the case in which β is large but finite. To do this, we note that β is a periodic variable, so it parametrizes a circle. The two instantons separated by a distance θ are also separated by a distance $\beta - \theta$, see Fig. 1.14. The symmetry between θ and $\beta - \theta$ then implies

$$g S(q_c^\theta) = \frac{1}{3} - 2e^{-\theta} - 2e^{-(\beta-\theta)} + \cdots.$$

$$(1.9.20)$$

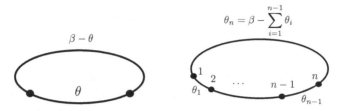

Figure 1.14 Instantons on a circle of length β. In the figure on the left, we have two instantons whose centers are separated by a distance θ in one segment of the circle, and by a distance $\beta - \theta$ in the other segment. In the figure on the right, we have n instantons separated by the distances θ_i.

We now consider an *n-instanton configuration*, i.e. an alternating sequence of instantons and anti-instantons separated by times θ_i, in which there are in total n configurations. For finite β we must have

$$\sum_{i=1}^{n}\theta_i = \beta, \tag{1.9.21}$$

see Fig. 1.14. As noted above, for n even, n-instanton configurations contribute to $Z(\beta)$, while for n odd they contribute to $Z_a(\beta)$. At leading order in θ, we need only consider interactions between nearest neighbor instantons, since other interactions are of higher order in $e^{-\theta}$. The classical action $S(\theta_i)$ is then a direct generalization of (1.9.20):

$$gS(\theta_i) = \frac{n}{6} - 2\sum_{i=1}^{n}e^{-\theta_i} + \mathcal{O}\left(e^{-(\theta_i+\theta_j)}\right). \tag{1.9.22}$$

As we mentioned before, the n-instanton configuration is not a solution to the equation of motion, but it can be seen that the first order variation of the action is exponentially suppressed as the separation between the instantons becomes large. Similarly, at second order we find an operator $\mathbf{M}$ as in (1.4.21). It can be seen that, at leading order in the separation between instantons, the spectrum of $\mathbf{M}$ is just the spectrum appearing in the one-instanton problem, but n-times degenerate, with corrections which are exponentially small in the separation.

We conclude that, although the n-instanton configuration is not a saddle point of the action, at leading order in $e^{-\theta}$ we can expand the functional integral around it as if it were a true saddle point. The determinant of the operator $\mathbf{M}$ is just the determinant obtained in the case $n = 1$, but to the power n. Similarly, since we have n collective time variables, the Jacobian appearing in the n-instanton case is the Jacobian of the one-instanton case to the power n. Therefore, the n-instanton contribution to the partition function (1.9.2), at leading order in $e^{-\theta}$, is given by:

$$Z_\epsilon^{(n)}(\beta) = e^{-\beta/2} \frac{\beta}{n} \left(\epsilon \frac{e^{-1/6g}}{\sqrt{\pi g}} \right)^n \int_{\theta_i \geq 0} \delta \left(\sum_{i=1}^n \theta_i - \beta \right) \exp \left[\frac{2}{g} \sum_{i=1}^n e^{-\theta_i} \right] \prod_{i=1}^n d\theta_i.$$

(1.9.23)

The overall factor β comes from the integration over the collective coordinate given by the "center of mass" of the instanton gas, and the factor $1/n$ arises because the configuration is invariant under a cyclic permutation of the θ_i. Finally, the normalization factor $e^{-\beta/2}$ corresponds to the partition function of the harmonic oscillator, like in (1.4.54).

We will now evaluate (1.9.23) in the approximation in which instanton interactions are neglected. Following the analogy between instantons and a classical interacting gas, this is called the *dilute instanton approximation*. Formally, to suppress the interactions, we should take the limit

$$g \to 0^-,$$

(1.9.24)

since in this case

$$\exp \left[\frac{2}{g} \sum_{i=1}^n e^{-\theta_i} \right] \to 0.$$

(1.9.25)

In fact, as we will see in a moment, the multi-instanton computation is only well defined for $g < 0$, and the dilute instanton approximation corresponds to g negative and small.

When the interaction term is suppressed, the integration over the θ_i is straightforward, since

$$\int_{\theta_i \geq 0} \delta \left(\sum_{i=1}^n \theta_i - \beta \right) \prod_{i=1}^n d\theta_i = \int_{0 \leq \sum_{i=1}^{n-1} \theta_i \leq \beta} \prod_{i=1}^{n-1} d\theta_i = \frac{\beta^{n-1}}{(n-1)!},$$

(1.9.26)

and

$$Z_\epsilon^{(n)}(\beta, g) = e^{-\beta/2} \frac{\beta}{n} \left(\epsilon \frac{e^{-1/6g}}{\sqrt{\pi g}} \right)^n \frac{\beta^{n-1}}{(n-1)!} = \frac{e^{-\beta/2}}{n!} \left(\epsilon \beta \frac{e^{-1/6g}}{\sqrt{\pi g}} \right)^n.$$

(1.9.27)

The sum of the leading order n-instanton contributions leads to an exponential,

$$Z_\epsilon(\beta, g) = e^{-\beta/2} + \sum_{n=1}^\infty Z_\epsilon^{(n)}(\beta, g) \approx e^{-\beta/2} \sum_{n=0}^\infty \frac{1}{n!} \left(\epsilon \beta \frac{e^{-1/6g}}{\sqrt{\pi g}} \right)^n \approx e^{-\beta E_{\epsilon,0}(g)},$$

(1.9.28)

where

$$E_{\epsilon,0}(g) = \frac{1}{2} + \mathcal{O}(g) - \frac{\epsilon}{\sqrt{\pi g}} e^{-1/6g} (1 + \mathcal{O}(g))$$

(1.9.29)

is just the energy (at leading order) of the $N = 0$ level with parity ϵ, and it includes the instanton contribution. This is what we expected, based on (1.9.2).

The dilute instanton approximation for the partition functions gives just the one-instanton contribution to the energies. To go further we have to take into account the interaction between instantons. However, as we pointed out before, when g is positive this interaction is attractive, and the approximation based on well-separated instantons breaks down. Therefore, we have to calculate the instanton contribution when g is small and *negative*. In this case, the interaction between instantons is repulsive and it is consistent to assume that instantons are far away from each other. Once the calculation for negative g is done, we have to perform an analytic continuation to positive g in a consistent way. As we will see, this leads to imaginary parts in the energies, and one needs a more refined treatment of the whole problem based on Borel resummation, which we will address in Chapter 3.

To calculate the integral (1.9.23) we write the δ-function constraint in terms of an integral,

$$\delta \left(\sum_{i=1}^{n} \theta_i - \beta \right) = \frac{1}{2\pi i} \int_{-i\infty}^{i\infty} \exp \left[-s \left(\beta - \sum_{i=1}^{n} \theta_i \right) \right] ds. \tag{1.9.30}$$

Let us now introduce the parameters

$$\lambda(g) = \frac{\epsilon}{\sqrt{\pi g}} e^{-1/6g}, \qquad \mu = -\frac{2}{g}. \tag{1.9.31}$$

λ plays the role of the "fugacity" of the instanton gas, since the n-instanton contribution is multiplied by λ^n. The integral over θ_i now factorizes, and each factor is proportional to the function

$$\mathcal{I}(s, \mu) = \int_0^\infty \exp \left(s\theta - \mu e^{-\theta} \right) d\theta. \tag{1.9.32}$$

We then find

$$Z_\epsilon^{(n)}(\beta) \approx \frac{\beta e^{-\beta/2} \lambda^n}{2\pi in} \int_{-i\infty}^{i\infty} e^{-\beta s} \left[\mathcal{I}(s, \mu) \right]^n ds. \tag{1.9.33}$$

This can be immediately summed for all $n \geq 1$, to obtain

$$Z_\epsilon^{\text{inst}}(\beta) = \sum_{n=1}^\infty Z_\epsilon^{(n)}(\beta) \approx -\frac{\beta}{2\pi i} \int_{-i\infty}^{i\infty} e^{-\beta(s+1/2)} \log \left(1 - \lambda \mathcal{I}(s, \mu) \right) ds. \tag{1.9.34}$$

We want to extract, from this function, the corrections to the energy levels due to the multi-instantons. In order to do this, it is useful to consider two functions which encode information about the spectrum of a Hamiltonian. The first one is the trace of the resolvent operator, which is the Laplace transform of the finite temperature partition function,

$$G(E) = \text{Tr} \frac{1}{H - E} = \int_0^\infty e^{\beta E} Z(\beta) d\beta. \tag{1.9.35}$$

If E_N, $N = 0, 1, \ldots$ are the energy levels of H, we can write

$$G(E) = \sum_{N=0}^{\infty} \frac{1}{E_N - E} \tag{1.9.36}$$

and the poles of $G(E)$ are located at the energy levels of the Hamiltonian H. If the Hamiltonian commutes with the parity operator, as is the case in the double-well potential, one can also consider the trace of the resolvent in a sector of a given parity,

$$G_\epsilon(E) = \text{Tr} \frac{1}{H_\epsilon - E}. \tag{1.9.37}$$

Here, H_ϵ is the Hamiltonian restricted to the subspace of functions of parity ϵ. The second useful function is the spectral or Fredholm determinant of the Hamiltonian H, which is defined by the equation

$$\mathcal{D}(E) = \det(H - E). \tag{1.9.38}$$

It is related to the trace of the resolvent by

$$-\frac{\partial}{\partial E} \log \mathcal{D}(E) = G(E). \tag{1.9.39}$$

We can also consider the spectral determinant restricted to a sector of a given parity,

$$\mathcal{D}_\epsilon(E) = \det(H_\epsilon - E). \tag{1.9.40}$$

Note that, formally,

$$\mathcal{D}(E) \propto \prod_{N \geq 0} \left(1 - \frac{E}{E_N}\right), \tag{1.9.41}$$

and the eigenvalues of the Hamiltonian are then zeros of the spectral determinant. Both the trace of the resolvent and the spectral determinant have to be appropriately regularized, since the sum (1.9.36) and product (1.9.41) over the spectrum appearing in their definitions are not always well defined. For example, in the case of the harmonic oscillator, the trace of the resolvent is given by

$$G_0(E) = \sum_{N=0}^{\infty} \frac{1}{N + 1/2 - E}. \tag{1.9.42}$$

This expression is purely formal, since the infinite sum over N appearing here is divergent. We can however define a regularized trace of the resolvent by using the digamma function $\psi(z)$, which has the expansion

$$\psi(z) = -\gamma_E + \sum_{N=0}^{\infty} \left(\frac{1}{N+1} - \frac{1}{N+z}\right), \tag{1.9.43}$$

where γ_E is Euler's constant. We then set

$$G_0(E) = -\psi\left(\frac{1}{2} - E\right) = \frac{\partial}{\partial E}\log\Gamma\left(\frac{1}{2} - E\right). \tag{1.9.44}$$

This indicates that the spectral determinant of the harmonic oscillator is given, after an appropriate regularization, by

$$\mathcal{D}_0(E) = \frac{1}{\Gamma\left(\frac{1}{2} - E\right)}, \tag{1.9.45}$$

which indeed has zeros at the energy levels $E = N + 1/2$, $N = 0, 1, \ldots$.

Let us now come back to the multi-instantons in the double well, and let us compute the trace of the resolvent associated to the instanton part of the thermal partition function, $Z_\epsilon^{\text{inst}}(\beta)$:

$$G_\epsilon^{\text{inst}}(E) = \int_0^\infty e^{\beta E} Z_\epsilon^{\text{inst}}(\beta)\,d\beta. \tag{1.9.46}$$

This is the instanton contribution to (1.9.37). The calculation of $G_\epsilon^{\text{inst}}(E)$ is easy, since (1.9.34) is essentially an inverse Laplace transform,

$$\begin{aligned}
G_\epsilon^{\text{inst}}(E) &\approx -\frac{1}{2\pi\mathrm{i}}\frac{\partial}{\partial E}\int_{-\mathrm{i}\infty}^{\mathrm{i}\infty} ds\,\log(1 - \lambda\mathcal{I}(s, \mu))\int_0^\infty e^{\beta(E-s-1/2)}d\beta \\
&= -\frac{1}{2\pi\mathrm{i}}\frac{\partial}{\partial E}\int_{-\mathrm{i}\infty}^{\mathrm{i}\infty} ds\,\log(1 - \lambda\mathcal{I}(s, \mu))\frac{1}{s + 1/2 - E} \\
&= -\frac{\partial}{\partial E}\log\phi_\epsilon(E),
\end{aligned} \tag{1.9.47}$$

where

$$\phi_\epsilon(E) = 1 - \lambda\mathcal{I}\left(E - \frac{1}{2}, \mu\right). \tag{1.9.48}$$

In the last line of (1.9.47) we have deformed the integration contour to pick the pole at $s = E - 1/2$. We should now add the zero-instanton contribution, which at leading order in g is simply the trace of the resolvent for the harmonic oscillator, given in (1.9.44). We then find,

$$G_\epsilon(E) \approx G_0(E) - \frac{\partial}{\partial E}\log\phi_\epsilon(E) = -\frac{\partial}{\partial E}\log\Delta_\epsilon(E), \tag{1.9.49}$$

where

$$\Delta_\epsilon(E) = \frac{1}{\Gamma\left(\frac{1}{2} - E\right)} - \lambda\frac{\mathcal{I}(E - \frac{1}{2}, \mu)}{\Gamma\left(\frac{1}{2} - E\right)}. \tag{1.9.50}$$

Note that $\Delta_\epsilon(E)$ gives an approximation at small g to the spectral determinant of the theory in the sector of parity ϵ,

$$\mathcal{D}_\epsilon(E) \approx \Delta_\epsilon(E). \tag{1.9.51}$$

Let us now evaluate the integral (1.9.32). After the change of variables $\mu\, e^{-\theta} = t$ we find

$$\mathcal{I}(s, \mu) = \mu^s \int_0^\mu t^{-1-s} e^{-t} dt = \mu^s \Gamma(-s) - \mu^s \Gamma(-s, \mu), \tag{1.9.52}$$

where

$$\Gamma(z, a) = \int_a^\infty t^{z-1} e^{-t} dt \tag{1.9.53}$$

is the incomplete Gamma function, which behaves at large a as

$$\Gamma(z, a) = a^z e^{-a} \left(\frac{1}{a} + \mathcal{O}\left(\frac{1}{a^2} \right) \right). \tag{1.9.54}$$

The weak coupling limit $g \to 0^-$ corresponds to $\mu \to +\infty$, therefore we conclude that

$$\mathcal{I}(s, \mu) = \mu^s\, \Gamma(-s) + \mathcal{O}(e^{-\mu}). \tag{1.9.55}$$

We thus obtain

$$\Delta_\epsilon(E) \approx \frac{1}{\Gamma(\frac{1}{2} - E)} - \lambda \mu^{E-1/2} \tag{1.9.56}$$

The energies are found by looking at the zeros of $\mathcal{D}_\epsilon(E)$ which, at the order we are working at, are the zeros of $\Delta_\epsilon(E)$. By using Euler's reflection formula for the Gamma function,

$$\Gamma(1-z)\Gamma(z) = \frac{\pi}{\sin(\pi z)}, \tag{1.9.57}$$

we rewrite the equation

$$\Delta_\epsilon(E) = 0 \tag{1.9.58}$$

as

$$\frac{\sin \pi (E - 1/2)}{\pi} = -\frac{\lambda \mu^{E-1/2}}{\Gamma(E + 1/2)}. \tag{1.9.59}$$

We now solve for these zeros as a power series in λ. For $\lambda = 0$, the zeros take place at $1/2 + N$, which are the energies of the harmonic oscillator, therefore

$$E_{\epsilon,N}^{(0)} = \frac{1}{2} + N + \mathcal{O}(\lambda). \tag{1.9.60}$$

Using (1.9.59) we will find a series of the form,

$$E_{\epsilon,N}(g) = \sum_{n=0}^{\infty} E_{\epsilon,N}^{(n)}(g)\lambda^n, \tag{1.9.61}$$

where $E_N^{(n)}(g)$ is the contribution of the nth multi-instanton to the energy. At the order which we are working at, we will be able to calculate this contribution at leading order in g. Let us then write the variable E appearing in (1.9.59) as

$$E = \frac{1}{2} + N + x, \tag{1.9.62}$$

where

$$x = \sum_{n=1}^{\infty} E_{\epsilon,N}^{(n)}(g)\lambda^n \tag{1.9.63}$$

and solves the implicit equation

$$\frac{\sin \pi x}{\pi} + \frac{\hat{\lambda}\, e^{\xi x}}{\Gamma(1 + N + x)} = 0, \tag{1.9.64}$$

and

$$\hat{\lambda} = \left(\frac{2}{g}\right)^N \lambda, \qquad \xi = \log \mu = \log\left(-\frac{2}{g}\right). \tag{1.9.65}$$

This equation can be solved for x as a power series in $\hat{\lambda}$ after expanding in x. One finds,

$$x + \frac{\hat{\lambda}}{N!} + \hat{\lambda} x \frac{\xi - \psi(1 + N)}{N!} + \cdots = 0. \tag{1.9.66}$$

Therefore, at leading order one has

$$x = -\frac{\hat{\lambda}}{N!} + \frac{\xi - \psi(1 + N)}{(N!)^2}\hat{\lambda}^2 + \mathcal{O}(\hat{\lambda}^3), \tag{1.9.67}$$

which means that the one-instanton and two-instanton contributions at one-loop are given by,

$$E_{\epsilon,N}^{(1)}(g) = -\epsilon \frac{1}{N!}\left(\frac{2}{g}\right)^{N+1/2} \frac{e^{-1/6g}}{\sqrt{2\pi}}\,(1 + \mathcal{O}(g))\,,$$

$$E_{\epsilon,N}^{(2)}(g) = \frac{1}{(N!)^2}\left(\frac{2}{g}\right)^{2N+1} \frac{e^{-1/3g}}{2\pi}\left\{\log\left(-\frac{2}{g}\right) - \psi(N + 1) + \mathcal{O}\left(g \log g\right)\right\}. \tag{1.9.68}$$

Notice that a single equation, (1.9.59), gives *all* the multi-instanton contributions to *all* energy levels $E_{\epsilon,N}(g)$ of the double-well potential (albeit at leading order in g).

It is obvious from the form of (1.9.59) that the nth instanton contribution has the form

$$E_{\epsilon,N}^{(n)}(g) = \left(\frac{2}{g}\right)^{n(N+1/2)} \left(-\epsilon \frac{e^{-1/6g}}{\sqrt{2\pi}}\right)^n$$
$$\times \left\{P_n^N\left(\log\left(-\frac{2}{g}\right)\right) + \mathcal{O}\left(g\,(\log g)^{n-1}\right)\right\}, \qquad (1.9.69)$$

where $P_n^N(\xi)$ is a polynomial of degree $n-1$. The results we have obtained are valid for g small and negative. When we consider the analytic continuation of the answer to positive g, we find a paradoxical result: the contributions of the n-instanton correction with $n > 1$ lead to an imaginary part in the energies, due to the presence of the term $\log(-2/g)$. Since the energies are real, these imaginary parts cannot be there. We will solve this problem in Chapter 3, once we have introduced the techniques of Borel resummation.

The equation (1.9.58) can be regarded as a non-perturbative quantization condition which gives the energy levels, taking into account non-perturbative corrections due to instantons. Our calculation of $\Delta_\epsilon(E)$ gives the leading order correction in g, but to all N. It is possible to write down an *exact* quantization condition which incorporates higher order corrections in g, and therefore gives the energy levels as an asymptotic series in the parameters g, $\log(-2/g)$, and the instanton weight $e^{-1/6g}$. This condition, which was first conjectured by Zinn-Justin, has the following form:

$$\left[\Gamma\left(\frac{1}{2} - B(E,g)\right)\right]^{-1} + \frac{\epsilon i}{\sqrt{2\pi}}\left(-\frac{2}{g}\right)^{B(E,g)} \exp\left[-\frac{A(E,g)}{2}\right] = 0. \quad (1.9.70)$$

The functions $A(E,g)$, $B(E,g)$ can be written as formal power series in their arguments,

$$B(E,g) = E + \sum_{k=1}^{\infty} g^k b_{k+1}(E),$$

$$(1.9.71)$$

$$A(E,g) = \frac{1}{3g} + \sum_{k=1}^{\infty} g^k a_{k+1}(E).$$

At leading order in g the quantization condition is precisely (1.9.58), where $\Delta_\epsilon(E)$ is given in (1.9.56). The corrections in g can be computed with the WKB method, and one finds,

$$B(E,g) = E + g\left(3E^2 + \frac{1}{4}\right) + g^2\left(35E^3 + \frac{25}{4}E\right) + \mathcal{O}(g^3),$$

$$A(E,g) = \frac{1}{3g} + g\left(17E^2 + \frac{19}{12}\right) + g^2\left(227E^3 + \frac{187}{4}E\right) + \mathcal{O}(g^3). \quad (1.9.72)$$

By solving the exact quantization condition one finds an expansion for the energy levels of the form

$$E_{\epsilon,N}(g) = \sum_{l=0}^{\infty} E_{N,l}^{(0)} g^l$$

$$+ \sum_{n=1}^{\infty} \left(\frac{2}{g}\right)^{nN} \left(-\epsilon \frac{e^{-1/(6g)}}{\sqrt{\pi g}}\right)^n \sum_{k=0}^{n-1} \left[\log\left(-\frac{2}{g}\right)\right]^k \sum_{l=0}^{\infty} e_{N,nkl} g^l.$$

$$(1.9.73)$$

The first series on the right hand side is purely perturbative in g, and it can be obtained from standard stationary perturbation theory. The second series is a sum over multi-instanton contributions. Notice that, if we neglect the non-perturbative corrections in (1.9.70), only the first term survives, and one finds,

$$B(E, g) = N + \frac{1}{2}. \qquad (1.9.74)$$

This perturbative quantization condition leads to the perturbative series on the right hand side of (1.9.73), i.e. to the answer from perturbation theory. It can also be obtained by using the generalization of the Bohr–Sommerfeld quantization condition, due to Dunham, which takes into account all the perturbative $\hbar$ corrections. The function $B(E, g)$ then has a purely perturbative origin, while the function $A(E, g)$ incorporates the tunneling effects due to instantons, which are non-perturbative in g (and in $\hbar$).

The analysis of the double-well potential that we have performed in this section can be generalized to other potentials with two degenerate minima. One can set up an instanton calculation, in which tunneling effects lift the degeneracy, and a multi-instanton calculation leads to an exact quantization condition which takes these effects into account in a systematic way. A similar situation occurs in the case of periodic potentials, which we analyze now in some detail.

1.10 Instantons in periodic potentials

In previous sections we have analyzed in detail the double-well potential, which has two degenerate minima. We saw that instanton effects lift the degeneracy between the energy levels. In the case of a *periodic* potential, life is even more complicated, since in perturbation theory we have an infinite number of degenerate ground states. We also know from elementary QM that the perturbative picture is misleading: Floquet–Bloch theory tells us that energy eigenstates organize themselves into bands, labelled by a continuous parameter φ. We will now see how this result can be re-interpreted from the point of view of instantons.

Let us start by reviewing some basic aspects of Floquet–Bloch theory, in the case of periodic potentials in one dimension. Let us assume that our Hamiltonian is of the form (1.2.1) where $W(q)$ is a periodic potential with period T:

$$W(q + T) = W(q). \tag{1.10.1}$$

Let us introduce the unitary operator $\mathcal{T}$ which translates wavefunctions by one period of the potential,

$$\mathcal{T}\psi(q) \equiv \psi(q + T). \tag{1.10.2}$$

Due to the periodicity of the potential, this operator commutes with the Hamiltonian. Therefore, $\mathcal{T}$ and H can be diagonalized simultaneously, and we have

$$
\begin{aligned}
\mathcal{T}\psi_{N,\varphi} &= e^{i\varphi}\psi_{N,\varphi}, \quad 0 \le \varphi < 2\pi, \\
H\psi_{N,\varphi} &= E_N(\varphi)\psi_{N,\varphi}, \quad N = 0, 1, 2, \ldots.
\end{aligned}
\tag{1.10.3}
$$

In this equation, the $E_N(\varphi)$ form the discrete spectrum of energies of H, and $\psi_{N,\varphi}(q)$ are the corresponding eigenfunctions. Note that we are decomposing the initial Hilbert space $\mathcal{H}$ into a sum of spaces $\mathcal{H}_\varphi$, each one labelled by a fixed value of the continuous parameter φ. This parameter is restricted to the interval 2π, since values of φ which differ by an integer multiple of 2π give redundant descriptions. We can also regard it as a periodic parameter, with period 2π, hence $E_N(\varphi)$ must be a periodic function of φ. As in Floquet–Bloch theory, we write the wavefunction as

$$\psi_{N,\varphi}(q) = e^{i\varphi q/T} u_{N,\varphi}(q). \tag{1.10.4}$$

It follows from the first equation in (1.10.3) that $u_{N,\varphi}(q)$ is a periodic function of period T. Therefore, the coordinate q can be restricted to one period. Equivalently, it can be regarded as an angular variable parametrizing a circle of length T, and the parameter φ/T can be regarded as a wave vector. We will normalize $u_{N,\varphi}(q)$ as

$$\int_0^T |u_{N,\varphi}(q)|^2 dq = 1. \tag{1.10.5}$$

Let us now incorporate the results of Floquet–Bloch theory in the framework of path integrals. We introduce the partition function restricted to fixed φ, $\mathcal{H}_\varphi$:

$$Z(\beta, \varphi) = \sum_{N \ge 0} e^{-\beta E_N(\varphi)}. \tag{1.10.6}$$

Since this is a periodic function of φ, it has a Fourier decomposition of the form

$$Z(\beta, \varphi) = \sum_{\ell=-\infty}^{\infty} e^{i\ell\varphi} Z_\ell(\beta), \tag{1.10.7}$$

where the Fourier coefficients are given by,

$$Z_\ell(\beta) = \frac{1}{2\pi} \int_0^{2\pi} Z(\beta, \varphi) \, e^{-i\ell\varphi} d\varphi. \tag{1.10.8}$$

They can be interpreted as partition functions with twisted boundary conditions in the full Hilbert space $\mathcal{H}$:

$$Z_\ell(\beta) = \mathrm{Tr}\left(T^\ell e^{-\beta H}\right). \tag{1.10.9}$$

Indeed, we have

$$\mathrm{Tr}\left(T^\ell e^{-\beta H}\right) = \int_0^T dq \, \langle q| T^\ell e^{-\beta H} |q\rangle = \int_0^T dq \, \langle q + \ell T| e^{-\beta H} |q\rangle. \tag{1.10.10}$$

We now introduce the resolution of identity in $\mathcal{H}$. Since we have a decomposition in sectors $\mathcal{H}_\varphi$, we should use

$$1 = \int_0^{2\pi} \frac{d\varphi}{2\pi} \sum_{N \geq 0} |\psi_{N,\varphi}\rangle \langle \psi_{N,\varphi}|. \tag{1.10.11}$$

We then find,

$$\mathrm{Tr}\left(T^\ell e^{-\beta H}\right) = \frac{1}{2\pi} \int_0^{2\pi} d\varphi \sum_{N \geq 0} e^{-\beta E_N(\varphi)} \int_0^T dq \, \psi_{N,\varphi}^*(q + \ell T)\psi_{N,\varphi}(q)$$

$$= \frac{1}{2\pi} \int_0^{2\pi} e^{-i\ell\varphi} \sum_{N \geq 0} e^{-\beta E_N(\varphi)} \, d\varphi, \tag{1.10.12}$$

where we have used (1.10.4) and (1.10.5) to calculate $\psi_{N,\varphi}^*(q + \ell T)$ and its integral. The last term is precisely the right hand side of (1.10.8).

The path integral representation of $Z_\ell(\beta)$ is

$$Z_\ell(\beta) = \int_{q(\beta/2) = q(-\beta/2) + \ell T} [dq(t)] \exp\left[-S(q)\right], \tag{1.10.13}$$

where $S(q)$ is as usual the Euclidean action. The function q can be regarded as a map from a circle of length β to a circle of length T, and we can write it as

$$q : \mathbb{S}_\beta^1 \to \mathbb{S}_T^1. \tag{1.10.14}$$

Maps from one circle to another circle fall into different *topological sectors*. Each sector is characterized by a single integer ℓ called the *winding number*, which indicates how many times the circle $\mathbb{S}_\beta^1$ winds around the circle $\mathbb{S}_T^1$. Mathematically, these maps are classified by the homotopy group $\pi_1(\mathbb{S}_T^1)$, as we will explain in some more detail in Chapter 4. For example, if we parametrize the circle $\mathbb{S}_\beta^1$ by an angle $-\beta/2 \le \theta < \beta/2$, the map

$$q(\theta) = \frac{\ell T}{\beta} \theta \tag{1.10.15}$$

where $q(\theta)$ is interpreted as an angle parametrizing $\mathbb{S}_T^1$, is a map of the winding number ℓ. In the path integral representing $Z_\ell(\beta)$ we restrict the integration domain to maps with a fixed winding number ℓ. Therefore, the partition function (1.10.7) can be regarded as a path integral in which we take into account all possible sectors, and we weight each of them with a phase $e^{i\ell\varphi}$, for a fixed value of φ. We can write the argument of the phase as

$$i\ell\varphi = \frac{i\varphi}{T}(q(\beta/2) - q(-\beta/2)) = \frac{i\varphi}{T}\int_{-\beta/2}^{\beta/2} \dot{q}(t)\,dt. \tag{1.10.16}$$

Therefore, the phase can be incorporated into the path integral by adding the term

$$\frac{i\varphi}{T}\int_{-\beta/2}^{\beta/2} \dot{q}(t)\,dt \tag{1.10.17}$$

to the Euclidean action. This term is an integral of a local functional of the map $q(t)$, but it is rather special since it only takes into account the topological class of the map. It is our first example of a *topological charge*, which we will find again in the more complicated context of Yang–Mills theory. The sum (1.10.7) can then be written as

$$Z(\beta, \varphi) = \int [dq(t)] \exp\left[-S(q) + \frac{i\varphi}{T}\int_{-\beta/2}^{\beta/2} \dot{q}(t)\,dt\right], \tag{1.10.18}$$

where the integration is now over all possible periodic boundary conditions on the circle: $q(\beta/2) = q(-\beta/2) \pmod T$.

The perturbative analysis of a periodic potential goes as follows. A periodic potential has an infinite number of classical vacua, of the form $q_* + \ell T$, where q_* satisfies $W'(q_*) = 0$. Let us assume that, near this vacuum, the Hamiltonian has the form

$$H \approx \frac{p^2}{2} + \frac{1}{2}(q - q_*)^2 + \mathcal{O}(g), \tag{1.10.19}$$

where g is a coupling constant. Then, in perturbation theory, each classical vacuum leads to a quantum vacuum, and we have an infinitely degenerate ground state with energy

$$E_0(g) \approx \frac{1}{2} + \mathcal{O}(g). \tag{1.10.20}$$

This degeneracy is lifted by instanton effects, as in the double-well potential, and the true vacuum is an infinite linear combination of wavefunctions located around the vacuum. If we denote by

$$u_0(q) \approx e^{-(q-q_*)^2/2} \qquad (1.10.21)$$

the perturbative wavefunction located around $q = q_*$, the true wavefunction is, for small g, a linear combination of the localized wavefunctions,

$$\psi_{0,\varphi}(q) \approx \sum_{\ell=-\infty}^{\infty} e^{i\ell\varphi} u_0(q - \ell T), \qquad (1.10.22)$$

which satisfies the first equation of (1.10.3). This is completely analogous to the fact that, in the double-well problem, eigenfunctions of the Hamiltonian could be classified according to their parity. The fact that the potential is periodic leads to the appearance of the angle or quasi-momentum φ.

In order to compute the energies $E_N(\varphi)$ with instanton methods, we notice that it is periodic in φ, so it can be written as a Fourier series,

$$E_N(\varphi) = \sum_{\ell=-\infty}^{\infty} E_{N,\ell} e^{i\ell\varphi} = E_{0,0} + 2\cos\varphi E_{0,1} + \cdots, \qquad (1.10.23)$$

where we have used that, since these energies are real, we must have $E_{N,\ell} = E_{N,-\ell}$. From now on we will focus on the ground state energy $E_0(\varphi)$. At large β, only this energy contributes to the partition function $Z(\beta, \varphi)$, and we find

$$Z(\beta, \varphi) \approx e^{-\beta E_0(\varphi)}, \qquad \beta \gg 1. \qquad (1.10.24)$$

Using now (1.10.18), we can compute this energy in terms of a path integral over all possible paths. These paths are classified by a topological number, ℓ. In the case of $\ell = 0$, we consider paths which are periodic and return to the original point. The saddle-point approximation to $Z_0(\beta)$ is dominated by the constant configuration $q(t) = q_*$, the minimum of the potential. However, the saddle that dominates $Z_\ell(\beta)$ is a path $q(t)$ that starts in the infinite past at q_* and goes in the infinite future to the point $q_* + \ell T$. This is an instanton configuration and we have that

$$Z_1(\beta) \approx e^{-\ell A}, \qquad (1.10.25)$$

where A is the classical action evaluated in the instanton configuration with $\ell = 1$. In typical examples, this is exponentially suppressed in the coupling constant. It follows that the Fourier expansion (1.10.7) can be regarded as an expansion in the small parameter e^{-A}. If we keep just the first instanton correction, we find

$$Z(\beta, \varphi) = Z_0(\beta) + 2\cos\varphi Z_1(\beta) + \cdots, \qquad (1.10.26)$$

and we have used that $Z_1(\beta) = Z_{-1}(\beta)$, again by reality. By comparing the previous equations, we find that

$$E_{0,0} = - \lim_{\beta \to \infty} \frac{1}{\beta} \log Z_0(\beta), \qquad (1.10.27)$$

which is the perturbative ground state energy, while

$$E_{0,1} = - \lim_{\beta \to \infty} \frac{1}{\beta} \frac{Z_1(\beta)}{Z_0(\beta)} \qquad (1.10.28)$$

and is due to the one-instanton configuration.

As a concrete application of instanton techniques in the case of periodic potentials, we will consider a Hamiltonian of the form (1.2.1) with

$$W(q) = \frac{1}{8g} \sin^2 \left(2\sqrt{g}q \right). \qquad (1.10.29)$$

The period of this potential is

$$T = \frac{\pi}{2\sqrt{g}}. \qquad (1.10.30)$$

There is a local minimum at $q = 0$, and one can compute the energy levels as a power series in g, by doing stationary perturbation theory around the harmonic oscillator centered at $q = 0$. For example, for the ground state energy one finds,

$$E_0(g) = \frac{1}{2} - \frac{g}{2} - \frac{g^2}{2} - \cdots . \qquad (1.10.31)$$

As we explained before, in perturbation theory all the vacua at $q = n\pi / (2\sqrt{g})$, $n \in \mathbb{Z}$ are degenerate, and they lead to the same ground state energy $E_0(g) = E_{0,0}(g)$. This degeneracy is lifted by instanton effects. We will now compute the first correction to the ground state energy due to instantons, $E_{0,1}$, at one-loop. We have to evaluate $Z_1(\beta)$ around the instanton configuration satisfying

$$q_c(-\beta/2) = 0, \qquad q_c(\beta/2) = \frac{\pi}{2\sqrt{g}}, \qquad (1.10.32)$$

and in the saddle-point approximation. In the limit $\beta \to \infty$, and for the potential at hand, the relevant solution to the Euclidean EOM (1.4.1) is

$$q_c(t) = \frac{1}{\sqrt{g}} \arctan \left(e^{t-t_0} \right), \qquad (1.10.33)$$

which has Euclidean action

$$S_c = \frac{1}{2g}. \qquad (1.10.34)$$

The fluctuation operator around this configuration is given by

$$\mathbf{M} = -\frac{d^2}{dt^2} + 1 - \frac{2}{\cosh^2(t - t_0)}, \tag{1.10.35}$$

and it coincides with the Pöschl–Teller operator $\mathbf{M}_{1,1}$. The argument leading to (1.8.16) gives in this case,

$$Z_1(\beta) \approx Z_G(\beta) \left[\frac{\det' \mathbf{M}}{\det \mathbf{M}_0}\right]^{-1/2} \frac{\beta S_c^{1/2}}{\sqrt{2\pi}} e^{-S_c}, \tag{1.10.36}$$

and by using (1.5.32) we find

$$\frac{Z_1(\beta)}{Z_0(\beta)} \approx \frac{1}{\sqrt{\pi g}} e^{-1/2g}, \tag{1.10.37}$$

therefore

$$E_{0,1}(g) \approx -\frac{1}{\sqrt{\pi g}} e^{-1/2g}. \tag{1.10.38}$$

As in the case of the double-well potential, it is possible to study multi-instanton effects in the periodic potential (1.10.29), as well as their interactions. Eventually, one finds an exact quantization condition for the energy levels first put forward by Zinn-Justin, and very similar to the condition (1.9.73). This condition again involves two functions, $B_{\mathrm{p}}(E, g)$ and $A_{\mathrm{p}}(E, g)$, which contain perturbative and non-perturbative information, respectively. The very first orders of these functions, as expansions in g, are given by:

$$B_{\mathrm{p}}(E, g) = E + \left(E^2 + \frac{1}{4}\right) g + \left(3 E^3 + \frac{5}{4} E\right) g^2 + \mathcal{O}(g^3),$$

$$A_{\mathrm{p}}(E, g) = \frac{1}{g} + \left(3 E^2 + \frac{3}{4}\right) g + \left(11 E^3 + \frac{23}{4} E\right) g^2 + \mathcal{O}(g^3), \tag{1.10.39}$$

and the exact quantization condition reads

$$\left(\frac{2}{g}\right)^{-B_{\mathrm{p}}(E,g)} \frac{e^{A_{\mathrm{p}}(E,g)/2}}{\Gamma\left(\frac{1}{2} - B_{\mathrm{p}}(E, g)\right)}$$

$$+ \left(-\frac{2}{g}\right)^{B_{\mathrm{p}}(E,g)} \frac{e^{-A_{\mathrm{p}}(E,g)/2}}{\Gamma\left(\frac{1}{2} + B_{\mathrm{p}}(E, g)\right)} = \frac{2 \cos \varphi}{\sqrt{2\pi}}. \tag{1.10.40}$$

1.11 Bibliographical notes

An excellent and comprehensive reference on instantons in QM is given in the last chapters of the monograph [199], as well as in the series of papers [114, 115, 200, 201]. Our sections on multi-instantons and the periodic potential follow these references clósely. A more introductory but insightful reference on instantons in QM can be found in the classic lecture by Coleman, "The uses of instantons," in [50]. The application of instanton methods to QM was initiated by Polyakov in [155].

The classic textbook [86] includes a detailed derivation of the thermal partition function in terms of path integrals. The Feynman diagram approach to the stationary perturbation theory of the quartic oscillator can be found in the paper of Bender and Wu [29], where they also derive the recursion relation (1.2.28). The paper [114] extends the Feynman diagram approach to non-trivial instanton solutions. The energies of resonances for the cubic oscillator are studied in [42, 196] and reviewed in [43]. A modern textbook on Quantum Mechanics including a detailed treatment of resonances is [120]. A careful explanation of the connection between instantons and discontinuities of the partition function across branch cuts can be found in [53, 199]. The role of instantons in the thermal partition function is discussed in [3, 103]. Our conventions for elliptic functions and elliptic integrals follow [8].

Useful references on determinants of Pöschl–Teller operators are [78, 113]. The calculation of the density of states for the continuum part of the spectrum of Pöschl–Teller potentials is similar to the calculation in [119] for two-dimensional solitons. The Gelfand–Yaglom method for computing determinants of one-dimensional Schrödinger operators is explained for example in [78, 172]. The approach in Section 1.6 builds upon (and sometimes corrects) [50, 104, 140, 199]. In particular, [140] discusses in detail the example of the quartic oscillator. Formulae for the one-loop approximations to the quantum propagator appear in many papers, for example [39, 63], and go back to the work by Van Vleck, Maslov and Gutzwiller. A good reference, which also includes a detailed discussion of the resolvent and the spectral determinant, is the second part of the book [59]. The formula for the lifetime of the ground state in a general unstable potential was first derived in [39].

The analysis of multi-instantons in the double-well and in periodic potentials is based on the work of J. Zinn-Justin [198], which is extended and studied further in [200, 201]. A WKB derivation of the multi-instanton expansion in the double-well potential can be found in [11]. The exact quantization conditions of Zinn-Justin have been studied further in [81], which found a simple relationship between the functions $A(E, g)$ and $B(E, g)$. They have also been derived mathematically in [66], using the theory of resurgence.

2

Unstable vacua in Quantum Field Theory

2.1 Introduction

After our detailed study of instanton effects in QM, it is time to more on to QFT. In the last chapter we have seen how one can use the formalism of path integrals to calculate the decay rate of an unstable vacuum in QM. Since this formalism generalizes easily to field theory, we can now analyze various aspects of unstable vacua in QFT, and in particular calculate the lifetime of an unstable or "false" vacuum. We will focus on the simplest example, namely, unstable vacua in scalar QFT.

2.2 Instantons in scalar Quantum Field Theory

We consider a self-interacting scalar field theory in d dimensions. A point in Euclidean space will be denoted by $x = (\tau, \mathbf{x})$, where $\mathbf{x}$ is a $(d-1)$-dimensional vector and τ is the Euclidean time. The Euclidean action has the form

$$S(\phi) = \int \left(\frac{1}{2}(\partial_\mu \phi)^2 + U(\phi) - U(\phi_+) \right) d^d x, \tag{2.2.1}$$

where $U(\phi)$ is the scalar potential of the theory. We will assume that $U(\phi)$ has two non-degenerate minima: a *false* vacuum at ϕ_+ which is quantum mechanically unstable, and a true vacuum at ϕ_-, so that $U(\phi_-) < U(\phi_+)$. We will also assume for simplicity that $U(\phi_+) = 0$.

Example 2.1 A typical example of the above situation is the theory in $d = 4$ with potential

$$U(\phi) = \frac{1}{2}\phi^2 - \frac{1}{2}\phi^3 + \frac{\alpha}{8}\phi^4, \tag{2.2.2}$$

where

$$0 < \alpha < 1. \tag{2.2.3}$$

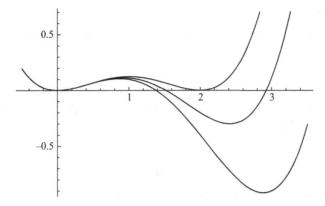

Figure 2.1 The potential (2.2.2) for $\alpha = 0.8, 0.9, 1$, from bottom to top.

This potential is represented in Fig. 2.1 for different values of α. It has a relative, "false" minimum at $\phi_+ = 0$, a true minimum at

$$\phi_- = \frac{3}{2\alpha} + \frac{\sqrt{9 - 8\alpha}}{2\alpha}, \qquad (2.2.4)$$

and a local maximum at

$$\phi = \frac{3}{2\alpha} - \frac{\sqrt{9 - 8\alpha}}{2\alpha}. \qquad (2.2.5)$$

Notice that, for $\alpha = 1$, the two minima are degenerate, and $\phi_- = 2$. $\qquad \square$

As in the quantum mechanical case, we want to compute the imaginary part of the ground state energy, in order to derive the lifetime of the vacuum. The steps we will follow are very similar to what we did in QM. First, we have to look for instantons, which are again defined as non-trivial solutions of the Euclidean EOM

$$\left(-\nabla^2 - \frac{d^2}{d\tau^2} \right) \phi + U'(\phi) = 0, \qquad (2.2.6)$$

where ∇ is the gradient in $d - 1$ spatial dimensions. We also have to impose appropriate boundary conditions. As in the tunneling problem in QM, we want to start from the false vacuum in the infinite past, and come back to it in the infinite future. Therefore,

$$\phi(\tau, \mathbf{x}) \to \phi_+, \qquad \tau \to \pm\infty. \qquad (2.2.7)$$

In order to have a finite action for the instanton, it has to approach the vacuum value at spatial infinity. Hence we have

$$\phi(\tau, \mathbf{x}) \to \phi_+, \qquad \|\mathbf{x}\| \to \infty. \qquad (2.2.8)$$

Since the EOM is invariant under the symmetry group $O(d)$, it is reasonable to look for solutions which are $O(d)$ symmetric, i.e. that only depend on

$$r = \sqrt{\mathbf{x}^2 + \tau^2}. \tag{2.2.9}$$

We also expect that the configuration which minimizes the action is $O(d)$ symmetric, and indeed this turns out to be the case. The equation of motion for the instanton $\phi_c(r)$ reduces to

$$\frac{\mathrm{d}^2\phi_c}{\mathrm{d}r^2} + \frac{d-1}{r}\frac{\mathrm{d}\phi_c}{\mathrm{d}r} = U'(\phi_c). \tag{2.2.10}$$

The boundary conditions (2.2.7) and (2.2.8) become

$$\lim_{r\to\infty} \phi_c = \phi_+. \tag{2.2.11}$$

We also want ϕ_c to be regular at the origin. In view of the second term in the left hand side of (2.2.10), this requires

$$\left.\frac{\mathrm{d}\phi_c}{\mathrm{d}r}\right|_{r=0} = 0. \tag{2.2.12}$$

Analytic solutions to (2.2.10) are not available for non-trivial potentials, but one can show that solutions indeed exist. Numerical solutions for ϕ_c and the potential (2.2.2), for different values of α and for $d = 4$, are shown in Fig. 2.2. Note that, when $\alpha \approx 1$, the field ϕ_c is essentially a step function: it is almost constant and equal to ϕ_- when $r < R$, where R is a characteristic scale which depends on α. For $r > R$ it goes rapidly to zero. We can interpret this solution as a "bubble" of true vacuum ϕ_-, and of radius R, in the middle of the false vacuum ϕ_+. We will later explain this structure in the context of the so-called "thin-wall approximation."

We now give a particularly useful form for the action evaluated at the instanton. We will compute

$$S(\phi_c, \lambda) = S(\phi_c(\lambda x)) = \int \left(\frac{1}{2}(\partial_\mu\phi_c(\lambda x))^2 + U(\phi_c(\lambda x))\right)\mathrm{d}^d x. \tag{2.2.13}$$

If we change variables

$$x \to \lambda x \tag{2.2.14}$$

we find

$$S(\phi_c, \lambda) = \lambda^{2-d}\int \mathrm{d}^d x \frac{1}{2}(\partial_\mu\phi_c(x))^2 + \lambda^{-d}\int \mathrm{d}^d x\, U(\phi_c(x)). \tag{2.2.15}$$

Since $\phi_c(x)$ satisfies the EOM, the action is stationary under variations of λ at $\lambda = 1$:

$$\left.\frac{\mathrm{d}S(\phi_c, \lambda)}{\mathrm{d}\lambda}\right|_{\lambda=1} = (2-d)\int \frac{1}{2}(\partial_\mu\phi_c(x))^2\mathrm{d}^d x - d\int U(\phi_c(x))\mathrm{d}^d x = 0. \tag{2.2.16}$$

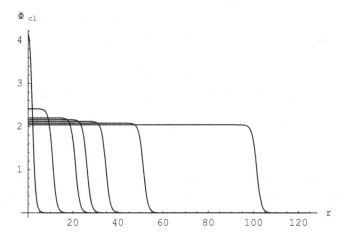

Figure 2.2 In this figure from [79], solutions to (2.2.10) for the potential (2.2.2) are plotted, for various values of $\alpha = 0.5, 0.9, 0.95, 0.96, 0.97, 0.98, 0.99$ (from left to right). (Reprinted with permission from Dunne, G. V. and Min, H., *Phys. Rev.*, D **72**, 125004. Copyright © 2005 American Physical Society.)

Therefore,

$$\int U(\phi_c(x))\mathrm{d}^d x = \frac{2-d}{d} \int \frac{1}{2}(\partial_\mu \phi_c(x))^2 \mathrm{d}^d x, \qquad (2.2.17)$$

and

$$S_c \equiv S(\phi_c) = \frac{1}{d} \int (\partial_\mu \phi_c(x))^2 \mathrm{d}^d x, \qquad (2.2.18)$$

which is manifestly positive.

As in QM, we have to understand the structure of zero modes. There are now d zero modes, corresponding to translation invariance of the instanton in d dimensions: given a solution of the EOM $\phi_c(x_\mu)$, the translated function $\phi_c(x_\mu - x_\mu^{(0)})$ is also a solution. As in (1.4.33), the zero modes are then obtained by taking derivatives with respect to the coordinates. The corresponding functions are

$$\phi_\mu = \partial_\mu \phi_c, \qquad (2.2.19)$$

with norm

$$\int \phi_\mu \phi_\nu \mathrm{d}^d x = \frac{1}{d}\delta_{\mu\nu} \int \mathrm{d}^d x (\partial_\mu \phi_c(x))^2 = \delta_{\mu\nu} S_c. \qquad (2.2.20)$$

In the first equality we have used $O(d)$ invariance of the solution. The normalized zero modes are then

$$\phi_\mu^{(0)} = \frac{1}{S_c^{1/2}}\partial_\mu \phi_c. \qquad (2.2.21)$$

Let us denote by $c_\mu^{(0)}$ the collective coordinates corresponding to the zero modes, which are defined as in (1.4.29). By analogy with (1.4.44), (1.4.45) we have

$$-\partial_\mu \phi \, \delta x_\mu^{(0)} = \phi_\mu^{(0)} \delta c_\mu^{(0)} \qquad (2.2.22)$$

and, as in (1.4.47), the zero modes in the path integral give a factor

$$\frac{1}{(2\pi)^{d/2}} \int \prod_{\mu=1}^{d} dc_\mu^{(0)} = \frac{S_c^{d/2}}{(2\pi)^{d/2}} \int \prod_{\mu=1}^{d} dx_\mu^{(0)} = \frac{S_c^{d/2} V \beta}{(2\pi)^{d/2}}, \qquad (2.2.23)$$

where V is the volume of $(d-1)$-dimensional space and β is the total time. The remaining steps proceed also in analogy with what we did in QM. At one-loop we must analyze the modes of the operator

$$\frac{\delta^2 S}{\delta\phi(\tau, \mathbf{x}\phi(\tau', \mathbf{x}')}\bigg|_{\phi_c} = \mathbf{M}_{\phi_c} \delta(\mathbf{x} - \mathbf{x}') \delta(\tau - \tau'), \qquad (2.2.24)$$

where

$$\mathbf{M}_{\phi_c} = -\frac{d^2}{d\tau^2} - \nabla^2 + U''(\phi_c). \qquad (2.2.25)$$

This operator plays the role of $\mathbf{M}$ in the QM case, while $\mathbf{M}_0$ corresponds to the same operator evaluated at the perturbative, false vacuum at ϕ_+:

$$\mathbf{M}_{\phi_+} = -\frac{d^2}{d\tau^2} - \nabla^2 + U''(\phi_+). \qquad (2.2.26)$$

As in QM, this operator should have one, and only one, negative mode reflecting the instability. It was shown by Coleman, Glaser and Martin that this is the case under quite generic circumstances. We can then write, in analogy with the corresponding formula (1.4.56) in QM,

$$\frac{1}{V} \operatorname{Im} E = \frac{1}{2} \left(\frac{S_c}{2\pi}\right)^{d/2} \left| \frac{\det'\left(-d^2/d\tau^2 - \nabla^2 + U''(\phi_c)\right)}{\det\left(-d^2/d\tau^2 - \nabla^2 + U''(\phi_+)\right)} \right|^{-1/2} e^{-S_c}. \qquad (2.2.27)$$

This is the final formula for the decay rate at one-loop in a scalar QFT.

The only subtlety here, which does not appear in QM, is the issue of renormalization. The above functional determinants involve a one-loop calculation in scalar field theory, and they will have divergences which have to be removed by adding counterterms. The action is then written as

$$S = S_R + \sum_{n=1}^{\infty} \hbar^n S^{(n)}, \qquad (2.2.28)$$

where S_R is the renormalized action, and $S^{(n)}$ includes the counterterms at n loops (we have included $\hbar$ factors explicitly). We then perform the calculation above starting with the renormalized action, and then we incorporate the effects of loop corrections. In particular, the instanton ϕ_c is now computed for S_R, i.e. we solve for

$$\left.\frac{\delta S_R}{\delta \phi}\right|_{\phi_c} = 0. \tag{2.2.29}$$

Let us then take into account the counterterms, which are needed to remove the divergences. Once this is done, it might happen that

$$S(\phi_+) \neq 0. \tag{2.2.30}$$

Since we measure the action of the instanton with respect to the action of the perturbative vacuum at ϕ_+, we should change

$$e^{-S(\phi)} \to e^{-(S(\phi)-S(\phi_+))}. \tag{2.2.31}$$

Once we take into account the effect of loop corrections, the instanton, which is a stationary point of the action, will also receive quantum corrections

$$\phi_c \to \phi_c + \hbar \phi^{(1)} + \cdots, \tag{2.2.32}$$

where $\phi^{(1)}$ is induced by the first order correction to the action. We then have

$$\begin{aligned}
S(\phi_c) \to S_c &\equiv S(\phi_c + \hbar \phi^{(1)} + \cdots) \\
&= S_R(\phi_c + \hbar \phi^{(1)} + \cdots) + \hbar S^{(1)}(\phi_c + \hbar \phi^{(1)} + \cdots) + \cdots \\
&= S_R(\phi_c) + \hbar S^{(1)}(\phi_c) + \cdots
\end{aligned} \tag{2.2.33}$$

where we have used (2.2.29). We then find

$$\begin{aligned}
\frac{1}{V} \text{Im } E &= \frac{1}{2} \left(\frac{S_c}{2\pi}\right)^{d/2} \left|\frac{\det' \left(-d^2/d\tau^2 - \nabla^2 + U''(\phi_c)\right)}{\det \left(-d^2/d\tau^2 - \nabla^2 + U''(\phi_+)\right)}\right|^{-1/2} e^{-S_c + S(\phi_+)} \\
&\approx \frac{1}{2} \left(\frac{S_R(\phi_c)}{2\pi}\right)^{d/2} \left|\frac{\det' \left(-d^2/d\tau^2 - \nabla^2 + U''(\phi_c)\right)}{\det \left(-d^2/d\tau^2 - \nabla^2 + U''(\phi_+)\right)}\right|^{-1/2} \\
&\quad \times e^{-S_R(\phi_c) - \hbar S^{(1)}(\phi_c) + \hbar S^{(1)}(\phi_+)},
\end{aligned} \tag{2.2.34}$$

where we have used that $S_R(\phi_+) = 0$. This is our final, UV finite expression, since the divergences of the one-loop determinants are taken care of by the one-loop counterterms of the action. In physical terms, what we have calculated is the probability per unit time for the formation of a tiny bubble of true vacuum in a given unit volume of space (we assumed that bubbles do not interact, so that this probability is proportional to the volume).

As in the QM example, the only non-trivial piece in the expression for the decay rate (2.2.34) is the functional determinant. In general, this can only be calculated numerically. Examples of such calculations can be found in the references at the end of this chapter.

2.3 The fate of the false vacuum

What happens after the quantum bubble has materialized? This is very similar to what happens to a particle after it has crossed a potential barrier. Such a particle materializes at the point where the potential energy is zero, which is the turning point of the trajectory (see for example Fig. 1.8). It has zero kinetic energy at that point. Starting from those conditions it propagates in the potential, and in a semiclassical approximation we can describe this process with classical mechanics.

Something similar happens with the bubble. After materializing past the barrier at the time $t = 0$ it will evolve with initial conditions

$$\phi(t = 0, \mathbf{x}) = \phi_c(\tau = 0, \mathbf{x}),$$
$$\partial_t \phi(t = 0, \mathbf{x}) = 0. \tag{2.3.1}$$

The last condition is the analogue of having zero kinetic energy at the turning point. The evolution will be governed by the wave equation

$$(\nabla^2 - \partial_t^2)\phi = U'(\phi). \tag{2.3.2}$$

We can solve this equation easily by rotating the instanton solution back to Minkowski space. Take the $O(d)$ invariant instanton $\phi_c(r)$ and define

$$\phi(t, \mathbf{x}) = \phi_c\left(r = (\mathbf{x}^2 - t^2)^{1/2}\right). \tag{2.3.3}$$

This solves the equation above with the same initial conditions (2.3.1). The first condition is obvious. Since

$$\partial_t \phi = \frac{d\phi_c}{dr} \partial_t r = -\frac{t}{r} \frac{d\phi_c}{dr} \tag{2.3.4}$$

vanishes at $t = 0$, the second condition is also satisfied.

What is then the evolution of the solution described by (2.3.3)? Let us assume for simplicity that the instanton is of the form depicted in Fig. 2.2 for $\alpha \approx 1$. Then, at $t = 0$ we have a bubble of true vacuum at the origin, of radius R. The boundary of the bubble will expand in time according to the hyperboloid

$$\mathbf{x}^2 = t^2 + R^2. \tag{2.3.5}$$

Notice that this is a Lorentz invariant evolution, i.e. it has $O(d - 1, 1)$ symmetry inherited from the $O(d)$ invariance of the bounce. We then have a bubble of true vacuum which materializes with radius R and then starts growing, since its boundary expands following (2.3.5). When $t \gg R$, the bubble expands almost at the speed of light.

2.4 The thin-wall approximation

In the study of the decay of the unstable vacuum in scalar field theory presented above, it is hard to obtain analytic solutions, even for simple potentials. It is then useful to have some approximation in which we can write down analytic expressions which capture the physics of the process. A popular approximation scheme is the so-called *thin-wall approximation*, in which the difference of potential energies between the two vacua ϕ_+ and ϕ_- is small:

$$\epsilon = U(\phi_+) - U(\phi_-) = -U(\phi_-) \ll 1. \tag{2.4.1}$$

Let us write the potential in this regime as

$$U(\phi) = U_0(\phi) + \mathcal{O}(\epsilon). \tag{2.4.2}$$

Here, $U_0(\phi)$ is a potential in which $\phi_\pm$ are two degenerate minima,

$$U_0(\phi_+) = U_0(\phi_-) = 0. \tag{2.4.3}$$

In this approximation, the EOM (2.2.10) becomes

$$\frac{d^2\phi}{dr^2} = U_0''(\phi). \tag{2.4.4}$$

In writing this equation, we have neglected the correction to the potential (2.4.2), and also the term involving the first derivative of ϕ. We will see *a posteriori* that this is indeed justified.

The equation (2.4.4) is identical to the EOM in a one-dimensional potential $U_0(\phi)$ with two degenerate minima, in which r plays the role of time (compare with (1.4.1)). The solution must have $\phi(r) \to \phi_+$ as $r \to \infty$, therefore it satisfies $\phi(r) \to \phi_-$ as $r \to -\infty$. In our problem we restrict the solution to positive values of $r \geq 0$, but as we will see $\phi(0)$ will be near ϕ_-. By the conservation of energy for this one-dimensional motion, we have

$$\frac{1}{2}(\phi')^2 - U_0(\phi) = 0, \tag{2.4.5}$$

where the constant of integration on the right hand side is fixed by the condition that ϕ goes to ϕ_+ at infinity, which is a minimum of $U_0(\phi)$ with $U_0(\phi_+) = 0$. The equation (2.4.5) is a first order ODE whose solution is determined by a single integration constant. This can be chosen to be the value of r, denoted $\bar{r}$, at which ϕ takes the average value of its two extreme values,

$$\phi(\bar{r}) = \frac{1}{2}(\phi_- + \phi_+). \tag{2.4.6}$$

We now have a family of approximate solutions to the EOM, parametrized by $\bar{r}$. We will assume that $\bar{r}$ is large compared to the typical scale in which ϕ varies significantly. If this is the case, the function ϕ takes values close to ϕ_- for $r < \bar{r}$, it interpolates rapidly between ϕ_- and ϕ_+ near $r = \bar{r}$, and then stays near ϕ_+ for $r > \bar{r}$. Therefore, $\bar{r}$ represents the radius of the bubble, and ϕ looks almost like a step function. This is the reason for the name "thin-wall" approximation: the instanton solution looks like a big bubble of true vacuum ϕ_- of radius $\bar{r}$, centered at the origin, and separated by a thin wall from the false vacuum ϕ_+ that extends to infinity.

To determine $\bar{r}$, we will compute the Euclidean action as a function of $\bar{r}$ (in the thin-wall approximation) and find the value which extremizes it. We will work in four dimensions, $d = 4$. The action is given by

$$S(\phi) = 2\pi^2 \int \left(\frac{1}{2}(\phi')^2 + U(\phi) \right) r^3 dr, \tag{2.4.7}$$

where the factor $2\pi^2$ is the volume of a three-sphere of unit radius and comes from the integration over angles. The integration over r splits into three natural regions: the region inside the bubble, the region along the wall of the bubble, and the region outside the bubble. Notice that, both inside and outside the wall, ϕ is almost a constant, and takes the values of the vacua, ϕ_- and ϕ_+, respectively. Therefore, for the region inside the bubble we find

$$S_{\text{inside}} \approx 2\pi^2 \int_0^{\bar{r}} U(\phi_-) r^3 dr \approx -\frac{\pi^2}{2} \epsilon \bar{r}^4. \tag{2.4.8}$$

In the region outside the bubble, the integral vanishes, while in the region along the wall of the bubble we can set $r = \bar{r}$ and approximate

$$S_{\text{wall}} \approx 2\pi^2 \bar{r}^3 \int \left(\frac{1}{2}(\phi')^2 + U_0(\phi) - U_0(\phi_+) \right) dr = 2\pi^2 \bar{r}^3 S_1 \tag{2.4.9}$$

where

$$S_1 = \int_{\phi_-}^{\phi_+} [2 \, (U_0(\phi))]^{1/2} \, d\phi. \tag{2.4.10}$$

We conclude that

$$S \approx -\frac{\pi^2}{2} \epsilon \bar{r}^4 + 2\pi^2 \bar{r}^3 S_1, \tag{2.4.11}$$

which is extremized for

$$\bar{r} = \frac{3 S_1}{\epsilon}. \tag{2.4.12}$$

This is indeed large for small ϵ, which is needed for the consistency of the whole scheme. The action is then given by

$$S \approx \frac{27\pi^2 S_1^4}{2\epsilon^3}. \tag{2.4.13}$$

This gives an analytic expression for the exponent of the decay probability, in the thin-wall approximation $\epsilon \ll 1$.

We can finally justify the neglect of the term ϕ'/r in the original EOM (2.2.10): inside and outside the bubble, ϕ is almost constant and ϕ' vanishes. Along the wall of the bubble ϕ' is not small, but $r \sim \bar{r}$ is big and ϕ'/r is small again.

Example 2.2 In the potential (2.2.2), the thin-wall approximation corresponds to $\alpha \to 1$, since

$$\epsilon = U(\phi_+) - U(\phi_-) = 2(1-\alpha) + \mathcal{O}\left((1-\alpha)^2\right). \tag{2.4.14}$$

The potential $U_0(\phi)$ is in this case

$$U_0(\phi) = \frac{1}{8}\phi^2 (\phi - 2)^2, \tag{2.4.15}$$

and it has two degenerate vacua at $\phi_+ = 0$ and $\phi_- = 2$. We can see in the numerical solutions plotted in Fig. 2.2 that, as α becomes closer and closer to 1, the solution has the features discussed above: it becomes closer and closer to a step function interpolating between the true vacuum at $\phi_- \approx 2$ and the false vacuum at $\phi_+ = 0$. Notice that the potential (2.4.15) is identical, up to a trivial rescaling, to the potential appearing in the double-well problem (1.9.8), and the solution to (2.4.4) is

$$\phi(r) = \frac{2}{1 + e^{r - \bar{r}}}. \tag{2.4.16}$$

It is easy to compute that

$$\bar{r} \approx \frac{1}{1-\alpha}, \qquad S \approx \frac{\pi^2}{3(1-\alpha)^3}. \tag{2.4.17}$$

Note that this value for $\bar{r}$ as α becomes close to 1 agrees quite well with the numerical solution of the full EOM displayed in Fig. 2.2. □

2.5 Instability of the Kaluza–Klein vacuum

The same techniques we have used to discuss unstable vacua in scalar field theories can be used to analyze other theories. A particularly striking application of these ideas is the non-perturbative semiclassical instability of the Kaluza–Klein vacuum, discovered by Witten. In this case, the field is the Riemannian metric of a five-dimensional manifold. The classical theory of such a field is General Relativity in five dimensions. The Kaluza–Klein vacuum is a solution of the vacuum EOM (i.e. Einstein's equations $R_{\mu\nu} = 0$ in five dimensions). In this vacuum, the geometry of the five-manifold is of the form

$$X_5 = M_4 \times \mathbb{S}^1, \tag{2.5.1}$$

where M_4 is Minkowski space and $\mathbb{S}^1$ is a circle of radius R. The corresponding metric is

$$ds^2 = dx^2 + dy^2 + dz^2 - dt^2 + d\phi^2, \tag{2.5.2}$$

where we used for simplicity the signature convention $(-, +, +, +)$ for the four-dimensional Minkowski spacetime. The first four terms correspond to the standard Minkowski metric in $\mathbb{R}^4$, while in the last term ϕ is a coordinate for $\mathbb{S}^1$, and it is a periodic variable with period $2\pi R$.

We would like to ask whether the above classical vacuum is quantum mechanically stable or not. A complete analysis of this problem is only possible in a full quantum theory of gravity in five dimensions, but it has been argued that the perturbative stability of this vacuum can be addressed with a one-loop computation. This computation shows that the Kaluza–Klein vacuum is unstable (in the absence of matter). We can also ask whether there are, on top of the perturbative instabilities, non-perturbative instabilities due to instantons. It is clear from the analysis in previous sections that semiclassical, non-perturbative stability can be determined with purely classical data: one has to look for an instanton solution to the classical, Euclidean EOM, which mediates an instability. Such a solution must approach asymptotically the vacuum we want to analyze, and it must have one negative mode.

To start our analysis of the possible non-perturbative instabilities, we have to consider the Euclidean theory. We will then perform a rotation of the metric (2.5.2) to Euclidean space,

$$ds^2 = dx^2 + dy^2 + dz^2 + d\tau^2 + d\phi^2. \tag{2.5.3}$$

Using polar coordinates for $\mathbb{R}^4$ we have

$$ds^2 = dr^2 + r^2 d\Omega_3^2 + d\phi^2, \tag{2.5.4}$$

where $d\Omega_d^2$ is the standard metric on the d-dimensional sphere $\mathbb{S}^d$ of unit radius, and

$$r = \sqrt{x^2 + y^2 + z^2 + \tau^2}. \tag{2.5.5}$$

Is there an instanton solution which mediates an instability? Let us consider the metric

$$ds^2 = \frac{dr^2}{1 - \alpha/r^2} + r^2 d\Omega^2 + \left(1 - \frac{\alpha}{r^2}\right) d\phi^2. \tag{2.5.6}$$

This metric solves the Euclidean Einstein equations in the vacuum, in five dimensions (it is in fact the Euclidean section of the five-dimensional Schwarzschild solution). At spatial infinity, $r \to \infty$, it becomes (2.5.3). The metric (2.5.6) seems to be singular at $r = \alpha$, but this singularity can be avoided by the periodicity of ϕ. To understand why this is the case, let us consider the flat metric in polar coordinates

$$ds^2 = d\rho^2 + \rho^2 d\varphi^2. \tag{2.5.7}$$

This has an apparent singularity at $\rho = 0$, but it is just due to the choice of coordinates. In order to see that the singularity in (2.5.6) is of the same type, we will find a new coordinate in such a way that the part of (2.5.6) involving r, ϕ looks like (2.5.7) near $\rho = 0$. As a consequence, the singularity at $r = \alpha$ will be removed by making ϕ periodic with an appropriate period. Let us set

$$\rho = c \left(1 - \frac{\alpha}{r^2} \right)^\beta, \tag{2.5.8}$$

where c, β are constants to be determined by our requirements. We have

$$r^2 = \frac{\alpha}{1 - (\rho/c)^{1/\beta}}, \qquad dr = \frac{r^3}{2c\alpha\beta} \left(1 - \frac{\alpha}{r^2} \right)^{1-\beta}, \tag{2.5.9}$$

and we deduce

$$\frac{dr^2}{1 - \alpha/r^2} = \frac{\alpha}{4c^2\beta^2} \left(\frac{\rho}{c} \right)^{(1-2\beta)/\beta} \frac{d\rho^2}{\left(1 - (\rho/c)^{1/\beta} \right)^3}. \tag{2.5.10}$$

We want this to look like $d\rho^2$ near $\rho = 0$, therefore we should have

$$\beta = \frac{1}{2}, \qquad c = \alpha^{1/2}. \tag{2.5.11}$$

We can now analyze the periodic part,

$$\left(1 - \frac{\alpha}{r^2} \right) d\phi^2 = \frac{\rho^2}{\alpha} d\phi^2, \tag{2.5.12}$$

where we used that $\beta = 1/2$. Since this should be identified with $\rho^2 d\varphi^2$ in (2.5.7), we conclude that ϕ is periodic with period

$$2\pi \sqrt{\alpha}. \tag{2.5.13}$$

In the original Kaluza–Klein metric, ϕ has period $2\pi R$, where R is the radius of the fifth compact dimension, therefore

$$\alpha = R^2. \tag{2.5.14}$$

We conclude that the metric (2.5.6)

$$ds^2 = \frac{dr^2}{1 - (R/r)^2} + r^2 d\Omega_3^2 + \left(1 - \left(\frac{R}{r} \right)^2 \right) d\phi^2 \tag{2.5.15}$$

is an instanton solution. Notice that the radial coordinate ρ starts at $\rho = 0$, but in view of (2.5.8) this means that

$$r \geq R. \tag{2.5.16}$$

In order to see what is the instability associated to this solution, we recall the analysis of instantons in scalar field theory. After continuation to Minkowski space, the instanton solution represents a bubble of true vacuum. Therefore, in analogy with scalar field theory, we want to rotate the metric (2.5.15) to Minkowski signature. If we did this in the flat case, we would require

$$dr^2 + r^2 d\Omega_3^2 \rightarrow dx^2 + x^2 d\Omega_2^2 - dt^2, \tag{2.5.17}$$

where

$$x^2 = x_1^2 + x_2^2 + x_3^2 \tag{2.5.18}$$

is the squared radius of the space-like coordinates in Minkowski space. To achieve this, we pick a polar angle θ and we write

$$d\Omega_3^2 = d\theta^2 + \sin^2 \theta \, d\Omega_2^2. \tag{2.5.19}$$

We then define a new angle ψ by

$$\theta = \frac{\pi}{2} + i\psi, \tag{2.5.20}$$

and we introduce the coordinates

$$x = r \cosh \psi, \qquad t = r \sinh \psi, \tag{2.5.21}$$

so that

$$r = \sqrt{x^2 - t^2}, \qquad dr = \frac{1}{r}(x dx - t dt),$$
$$\psi = \tanh^{-1} \frac{t}{x}, \qquad d\psi = \frac{1}{r^2}(x dt - t dx). \tag{2.5.22}$$

The line element of the three-sphere becomes

$$d\Omega_3^2 = -d\psi^2 + \cosh^2 \psi \, d\Omega_2^2, \tag{2.5.23}$$

and the four-dimensional Euclidean space metric becomes,

$$dr^2 + r^2 d\Omega_3^2 = dr^2 + r^2 \left(-d\psi^2 + \cosh^2 \psi \, d\Omega_2^2\right)$$
$$= dx^2 - dt^2 + x^2 d\Omega_2^2, \tag{2.5.24}$$

as we wanted. Notice, however, that in this parametrization $x^2 - t^2 = r^2 > 0$ is always positive. Therefore, to be precise, the above continuation (2.5.24) describes the exterior of the light cone in Minkowski space.

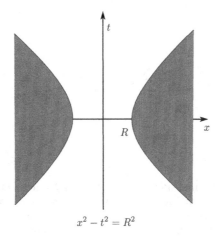

Figure 2.3 Restricted to the x–t plane, the space described by the metric (2.5.25) is the exterior of the hyperboloid $x^2 - t^2 = R^2$.

Now, for the bounce solution, the same continuation and change of variables can be performed, and we obtain the metric

$$ds^2 = \frac{dr^2}{1 - (R/r)^2} + r^2 \left(-d\psi^2 + \cosh^2 \psi \, d\Omega_2^2\right) + \left(1 - \left(\frac{R}{r}\right)^2\right) d\phi^2. \quad (2.5.25)$$

Here, we have as well that $r^2 = x^2 - t^2$, but due to (2.5.16) it starts at $r = R$. Therefore, the omitted part of space here is the full hyperboloid interior bounded by

$$x^2 - t^2 = R^2, \quad (2.5.26)$$

see Fig. 2.3. One might think that this space would have a sharp boundary, but the presence of the fifth dimension makes it smooth. Indeed, the five-dimensional circle now has radius

$$R(1 - R^2/r^2)^{1/2} \quad (2.5.27)$$

which becomes zero as r goes to R. For example, if we restrict the four dimensions to the x axis, the complement of the interval $[-R, R]$ is smoothed out by the fifth dimension into two discs, see Fig. 2.4.

We can now interpret this solution: the instability is generated by the nucleation of a *hole* of radius R in three-dimensional space. From the point of view of a four-dimensional Minkowski observer, this is a hole of nothing which forms at $t = 0$ and then expands in time as

$$x^2 = R^2 + t^2. \quad (2.5.28)$$

Therefore, the Kaluza–Klein vacuum decays into literally *nothing*.

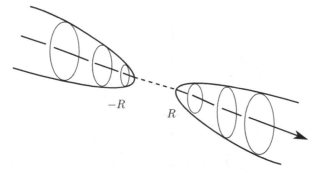

Figure 2.4 The fifth dimension in (2.5.25) is a circle fibered over the four-dimensional space, with radius (2.5.27). As we approach the boundary $r^2 = R^2$, the circle shrinks to zero size and the total space is smooth. When restricted to the x axis, as in the figure, the fifth dimension smooths out the complement of the interval into two discs.

2.6 Bibliographical notes

The analysis of vacuum decay in scalar field theory from the point of view of instantons was pioneered in the papers of Coleman [49] and Callan and Coleman [44] (see also [170]). It is reviewed in various textbooks, for example [50, 199]. In [52] it is shown that an $O(4)$ invariant solution to the EOM of the scalar field theory leads to a minimal action. A detailed calculation of the functional determinant for some simple scalar QFTs can be found in [17, 79]. The thin-wall approximation was introduced by Coleman in [49] and further discussed in [44, 51]. The thin-wall approximation for the quartic potential is also discussed in [17]. Perturbative instabilities of the Kaluza–Klein vacuum are discussed in [15]. The non-perturbative instability of the Kaluza–Klein vacuum presented in this chapter was discovered and studied by Witten in [194]. The idea of removing singularities in Wick-rotated metrics by making some variables periodic was first discussed in [93].

3

Large order behavior and Borel summability

3.1 Introduction

Very often, when doing computations in quantum theory, we have to rely on approximation schemes based on the existence of a small parameter, for example a coupling constant g. The answer obtained in this way is typically a formal power series in g, of the form

$$\varphi(g) = \sum_{k \geq 0} a_k g^k. \tag{3.1.1}$$

An example of this situation is the quantum anharmonic oscillator with Hamiltonian

$$H = \frac{p^2}{2} + \frac{q^2}{2} + \frac{g}{4} q^4, \tag{3.1.2}$$

where $\varphi(g)$ is the series for the energy of the ground state, as obtained in stationary perturbation theory. The very first terms of this series were computed in (1.2.25). There are various questions that we can ask about this kind of series.

1. *Large order behavior.* Is the series (3.1.1) convergent? Note that this depends on the behavior of the coefficients a_k as k grows large.
2. *Summability.* In the case when the above series has zero radius of convergence, is there a way to make sense of the perturbative series?
3. *Non-perturbative effects.* In some situations, on top of the perturbative contribution to the quantity we are computing, there are also contributions which are non-perturbative, and of the form $\exp(-A/g)$. How do we incorporate these corrections in the computation?

These questions are of primary importance in quantum theory, and particularly in QFT, where perturbation theory and Feynman diagrams play a very important role in our physical intuition and in our concrete calculations. It turns out that

77

most of the formal power series appearing in quantum theories, including the series
(1.2.25) for the ground state energy of the anharmonic oscillator, have zero radius
of convergence. This is not just an unfortunate fact of life, but has a deep physical
significance, as pointed out by Dyson in 1952. Dyson's argument goes as follows.
Let us take for concreteness the case of the series (1.2.25). If this series had a finite
radius of convergence, the resulting analytic function around $g = 0$ would describe
the physics of the problem for both positive and negative values of the coupling
constant g, provided this is sufficiently small. However, if the coupling is negative,
the physics is completely different, no matter how small g is in absolute value: the
potential is unstable, and a particle sitting at its bottom will eventually decay by
tunnel effect. Therefore, we should *not* expect a non-zero radius of convergence
for this series.

Dyson's argument indicates that tunneling effects should be deeply related to the
lack of convergence of perturbation theory. In the case of the quartic anharmonic
oscillator, there should be a connection between the imaginary part of $E(g)$ that
gives the decay rate in the unstable potential with $g < 0$, and the large order
behavior of perturbation theory for positive g. This connection was established
quantitatively by Bender and Wu. This leads to a beautiful and important relation
between the instantons of the theory (which compute tunneling amplitudes) and
the perturbative sector.

In this chapter we will analyze in detail all these questions, and in particular
we will explain how to make sense (in some circumstances) of the divergent series
appearing in quantum theory. We will illustrate the results mostly with the quantum
mechanical examples studied in Chapter 1, but the principles at work here are also
important in QFT.

3.2 Asymptotic expansions and Borel resummation

Dyson's argument suggests that the perturbative series for the ground state energy
of the quartic oscillator has zero radius of convergence. It is easy to see that this
is the case by calculating the coefficients of the series: a simple numerical experiment indicates that they grow factorially. It is also possible to establish this growth
through an analysis of the calculation of the coefficients via Feynman diagrams,
which we presented in Chapter 1. As we pointed out in the introduction, this behavior is the norm, rather than the exception: most perturbative series in quantum
theories are not convergent, but asymptotic. In order to understand this behavior in
detail, we need some mathematical tools. Let us start by the formal definition of
asymptotic series. Let us consider a formal power series of the form,

$$\varphi(z) = \sum_{n=0}^{\infty} a_n z^n. \tag{3.2.1}$$

We stress that such a formal power series is not a function, but rather is a convenient way of collecting the coefficients a_n. We say that this formal power series is asymptotic to the function $f(z)$, in the sense of Poincaré, if, for every N, the remainder after $N + 1$ terms of the series is much smaller than the last retained term as $z \to 0$. More precisely,

$$\lim_{z \to 0} z^{-N} \left(f(z) - \sum_{n=0}^{N} a_n z^n \right) = 0, \tag{3.2.2}$$

for all $N > 0$. In an asymptotic series, the remainder does not necessarily go to zero as $N \to \infty$ for a fixed z, in contrast to what happens in convergent series. Note that analytic functions might have asymptotic expansions. For example, the Stirling series for the Gamma function

$$\left(\frac{z}{2\pi} \right)^{1/2} \left(\frac{z}{e} \right)^{-z} \Gamma(z) = 1 + \frac{1}{12z} + \frac{1}{288z^2} + \cdots \tag{3.2.3}$$

is an asymptotic series for $|z| \to \infty$.

In practice, asymptotic expansions are characterized by the fact that, as we vary N, the partial sums

$$\varphi_N(z) = \sum_{n=0}^{N} a_n z^n \tag{3.2.4}$$

will first approach the true value $f(z)$, and then, for N sufficiently large, they will diverge. A natural question is then to find the partial sum which gives the best possible estimate of $f(z)$. To do this, one has to find the N that truncates the asymptotic expansion in an optimal way. This procedure is called *optimal truncation*. Usually, the way to find the optimal value of N is to retain terms up to the smallest term in the series, and discard all terms of higher degree. Let us assume (as it is the case in all interesting examples) that the coefficients a_n in (3.2.1) grow factorially at large n,

$$a_n \sim A^{-n} n!, \qquad n \gg 1. \tag{3.2.5}$$

The smallest term in the series, for a fixed $|z|$, is obtained by minimizing N in

$$\left| a_N z^N \right| = c N! \left| \frac{z}{A} \right|^N. \tag{3.2.6}$$

By using the Stirling approximation, we can rewrite this as

$$c \exp \left\{ N \left(\log N - 1 - \log \left| \frac{A}{z} \right| \right) \right\}. \tag{3.2.7}$$

The above function has a saddle at large N given by

$$N_* = \left| \frac{A}{z} \right|. \tag{3.2.8}$$

If $|z|$ is small, the optimal truncation can be performed at large values of N, but as $|z|$ increases, fewer terms of the series can be used. We can now estimate the error made in the optimal truncation by evaluating the next term in the asymptotics,

$$\epsilon(z) = C_{N_*+1} |z|^{N_*+1} \sim e^{-|A/z|}. \tag{3.2.9}$$

Therefore, the maximal "resolution" we can expect when we reconstruct a function $f(z)$ from an asymptotic expansion is of order $\epsilon(z)$. This type of ambiguity is sometimes called a *non-perturbative ambiguity*, since it is not seen in perturbation theory. Indeed, the exponential term

$$e^{-A/z} \tag{3.2.10}$$

is not analytic at $z = 0$, and therefore it does not contribute to the perturbative series. We conclude that, in general, an asymptotic expansion does not determine the function $f(z)$ uniquely, and some additional information is required. Note that the absolute value of A gives the "strength" of the non-perturbative ambiguity.

It is instructive to see optimal truncation at work in a simple example. Let us consider the quartic integral $I(g)$ defined by (1.3.11), which is well defined as long as $\text{Re}(g) > 0$. In (1.3.23) we obtained a formal power series

$$\varphi(g) = \sum_{k=0}^{\infty} a_k g^k, \tag{3.2.11}$$

where the coefficients a_k are given in (1.3.24). As we pointed out in Chapter 1, this series has a zero radius of convergence, but it can easily be shown that it provides an asymptotic expansion of the integral $I(g)$. The asymptotic behavior of the coefficients at large k is given by (1.3.25), therefore $|A| = 1/4$. In Fig. 3.1 we plot the difference

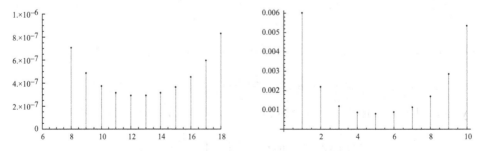

Figure 3.1 We illustrate the method of optimal truncation for the quartic integral (1.3.11) by plotting the difference (3.2.12) between the integral and the partial sum of order N of its asymptotic expansion, as a function of N, for $g = 0.02$ (left) and $g = 0.05$ (right).

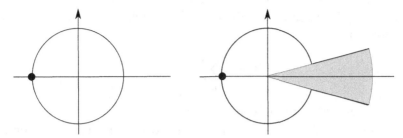

Figure 3.2 The Borel transform is analytic in a neighborhood of $\zeta = 0$, of radius $\rho = |A|$. Typically we encounter a singularity on the circle $|\zeta| = |A|$, but we can continue the transform analytically to a wider region.

$$|I(g) - \varphi_N(g)| \qquad (3.2.12)$$

as a function of N, for two values of g. The optimal values are seen to be $N_* = 12$ and $N_* = 5$, in agreement with the estimate (3.2.8). Optimal truncation gives a reasonable approximation to the original function for some values of the coupling constant, but it typically becomes a poor approximation for other values. In addition, in optimal truncation only a finite number of terms in the asymptotic expansion are actually used, and the remaining terms cannot be exploited to improve the approximation, as one does with convergent series. In fact, we can do better than optimal truncation and take into account the information contained in *all* the terms of the series. The way to do that is Borel resummation, which we now explain.

Let us consider a series (3.2.1), where the coefficients a_n behave like (3.2.5) when n is large (such series are sometimes called *Gevrey-1*). The *Borel transform* of φ, which we will denote by $\widehat{\varphi}(\zeta)$, is defined as the series

$$\widehat{\varphi}(\zeta) = \sum_{n=0}^{\infty} \frac{a_n}{n!} \zeta^n. \qquad (3.2.13)$$

Notice that, due (3.2.5), the series $\widehat{\varphi}(\zeta)$ has a finite radius of convergence $\rho = |A|$ and it defines an analytic function in the circle $|\zeta| < |A|$, see Fig. 3.2. Typically, there is a singularity at $\zeta = A$, but very often the resulting function can be continued analytically to a wider region of the complex plane. We will now work out some examples of Borel transforms displaying two typical behaviors at the singularity: a simple pole and a logarithmic branch cut.

Example 3.1 Let us consider the series

$$\varphi(z) = \sum_{n=0}^{\infty} (-1)^n n! z^n, \qquad (3.2.14)$$

which corresponds to a growth of the form (3.2.5) with $A = -1$. In this case, the Borel transform is

$$\widehat{\varphi}(\zeta) = \sum_{n=0}^{\infty} (-1)^n \zeta^n, \tag{3.2.15}$$

which is a series with radius of convergence $\rho = 1$. However, it is an elementary fact that this series can be continued analytically to a meromorphic function with a single pole at $\zeta = -1$, namely

$$\widehat{\varphi}(\zeta) = \frac{1}{1+\zeta}. \tag{3.2.16}$$

In this case, the singularity of the Borel transform is a pole at $\zeta = A = -1$.   □

Example 3.2   Consider now the series

$$\varphi(z) = \sum_{k=0}^{\infty} \frac{\Gamma(k+b)}{\Gamma(b)} A^{-k} z^k, \tag{3.2.17}$$

where b is not an integer. The Borel transform is given by

$$\widehat{\varphi}(\zeta) = \sum_{k=0}^{\infty} \frac{\Gamma(k+b)}{k!\Gamma(b)} A^{-k} \zeta^k = (1 - \zeta/A)^{-b}, \tag{3.2.18}$$

which has a branch cut singularity at $\zeta = A$. Similarly, the series

$$\varphi(z) = \sum_{k=1}^{\infty} \Gamma(k) A^{-k} z^k \tag{3.2.19}$$

leads to the Borel transform

$$\widehat{\varphi}(\zeta) = -\log(1 - \zeta/A), \tag{3.2.20}$$

which has a logarithmic singularity at $\zeta = A$. This can be regarded as the $b = 0$ case of (3.2.17).   □

Let us now suppose that the Borel transform $\widehat{\varphi}(\zeta)$ has an analytic continuation to a neighborhood of the positive real axis, in such a way that the Laplace transform

$$s(\varphi)(z) = \int_0^{\infty} e^{-\zeta} \widehat{\varphi}(z\zeta) \, d\zeta = z^{-1} \int_0^{\infty} e^{-\zeta/z} \widehat{\varphi}(\zeta) \, d\zeta, \tag{3.2.21}$$

exists in some region of the complex z-plane. In this case, we say that the series $\varphi(z)$ is *Borel summable* and $s(\varphi)(z)$ is called the *Borel resummation* of $\varphi(z)$. Notice that, by construction, $s(\varphi)(z)$ has an asymptotic expansion around $z = 0$ which coincides with the original series $\widehat{\varphi}(\zeta)$, since

$$s(\varphi)(z) = z^{-1} \sum_{n \geq 0} \frac{a_n}{n!} \int_0^\infty e^{-\zeta/z} \zeta^n d\zeta = \sum_{n \geq 0} a_n z^n. \tag{3.2.22}$$

This procedure makes it possible in principle to reconstruct a well-defined function $s(\varphi)(z)$ from the asymptotic series $\varphi(z)$ (at least for some values of z). In many cases of interest in physics, the formal series $\varphi(z)$ is the asymptotic expansion of a well-defined function $f(z)$ (for example, $f(z)$ might be the ground state energy of a quantum system as a function of the coupling z, while $\varphi(z)$ is its asymptotic expansion). It might then happen that the Borel resummation $s(\varphi)(z)$ agrees with the original function $f(z)$, and in this favorable case, the Borel resummation reconstructs the original non-pertubative answer. There are analyticity conditions on the function $f(z)$ that guarantee that $s(\varphi)(z) = f(z)$ in some region. However, in practice, it is not always easy to use these conditions, since they require detailed information about $f(z)$.

Sometimes it is useful to perform the Laplace transform (3.2.21) along an arbitrary direction in the complex plane, specified by an angle θ. We then introduce the generalized Borel resummation along θ,

$$s_\theta(\varphi)(z) = \int_0^{e^{i\theta}\infty} e^{-\zeta} \widehat{\varphi}(z\zeta) \, d\zeta. \tag{3.2.23}$$

Example 3.3 In Example 3.1, the Borel transform extends to an analytic function on $\mathbb{C}\backslash\{-1\}$, and the integral (3.2.21) is

$$s(\varphi)(z) = \int_0^\infty \frac{e^{-\zeta}}{1 + z\zeta} d\zeta, \tag{3.2.24}$$

which exists for all $z \geq 0$. $\qquad\square$

In practice, even if the series $\varphi(z)$ is Borel summable, one only knows a few coefficients in its expansion, and this makes it very difficult to continue the Borel transform analytically to a neighborhood of the positive axis. We need a practical method to find accurate approximations to the resulting function. A useful method is to use *Padé approximants*. Given a series

$$\varphi(z) = \sum_{k=0}^\infty a_k z^k, \tag{3.2.25}$$

its Padé approximant $[l/m]_\varphi$, where l, m are positive integers, is the rational function

$$[l/m]_\varphi(z) = \frac{p_0 + p_1 z + \cdots + p_l z^l}{q_0 + q_1 z + \cdots + q_m z^m}, \quad (3.2.26)$$

where q_0 is fixed to 1, and one requires that

$$\varphi(z) - [l/m]_\varphi(z) = \mathcal{O}(z^{l+m+1}). \quad (3.2.27)$$

This fixes the coefficients involved in (3.2.26).

Given a series $\varphi(z)$, we can use Padé approximants to reconstruct the analytic continuation of its Borel transform. There are various methods to do this, but one simple approach is to use the following Padé approximant,

$$\mathcal{P}_n^\varphi(\zeta) = \left[[n/2]/[(n+1)/2]\right]_{\widehat\varphi}(\zeta) \quad (3.2.28)$$

which requires knowledge of the first $n + 1$ coefficients of the original series. The integral

$$s(\varphi)_n(z) = z^{-1} \int_0^\infty e^{-\zeta/z} \mathcal{P}_n^\varphi(\zeta) d\zeta \quad (3.2.29)$$

gives an approximation to the Borel resummation of the series (3.2.21), which can be systematically improved by increasing n. The quantity (3.2.29), which combines the Padé approximant with the Borel resummation, is sometimes called *Borel–Padé resummation*.

Example 3.4 *The quartic integral.* A simple example of this procedure is again the quartic integral (1.3.11). The Borel transform of the series (3.2.11) is given by

$$\widehat\varphi(\zeta) = \frac{2K(k)}{\pi(1 + 4\zeta)^{1/4}}, \qquad k^2 = \frac{1}{2} - \frac{1}{2\sqrt{1 + 4\zeta}}, \quad (3.2.30)$$

where $K(k)$ is the elliptic integral of the first kind. This function has a branch point at $\zeta = A = -1/4$. The presence of a singularity at this point in the ζ-plane can also be deduced by comparing the asymptotic growth (1.3.25) with the expression (3.2.5), and by remembering that the value of A gives the location of the singularity. We can compute (3.2.29) for increasing values of n and verify that the results give better and better approximations to the quartic integral (1.3.11), see Table 3.1. In this case, the Borel resummation reproduces the original non-perturbative object (1.3.11). On the other hand, for these values of the coupling, optimal truncation is not very good: the best approximation comes from keeping just the first two terms in the series, i.e. $N = 1$, and one finds

$$\varphi_1(0.2) = 0.85, \qquad \varphi_1(0.4) = 0.7. \quad (3.2.31)$$

□

Table 3.1 *The Borel–Padé resummations (3.2.29) of the asymptotic series (3.2.11) for the quartic integral $I(g)$ (1.3.11)*

The values are computed for two different values of g, $g = 0.2$ and $g = 0.4$, and for increasing values of n. In the last line we give the numerical result for $I(g)$. All numbers are presented with ten significant digits. For each value of n, we underline the digits in the result of (3.2.29) which agree with the numerical result.

n	$s(\varphi)_n(0.2)$	$s(\varphi)_n(0.4)$
10	0.9079854376	0.8576207823
20	0.9079847776	0.8576086008
30	0.9079847774	0.8576085854
$I(g)$	0.9079847774	0.8576085853

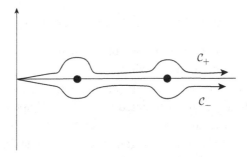

Figure 3.3 The paths $C_\pm$ avoiding the singularities of the Borel transform from above (respectively, below).

Suppose now that $\widehat{\varphi}(\zeta)$ has singularities on the *positive* real axis, and that it can be extended along a neigborhood of the positive real axis as a meromorphic or multi-valued function which decreases sufficiently fast at infinity. An example would be the function (3.2.18) when A is real and positive, corresponding to a series which is factorially divergent and *non-alternating*. Then, the integral (3.2.21) is in principle ill defined, but we can define closely related integrals by deforming the contour in (3.2.21) appropriately. A useful choice is to consider contours $C_\pm$ that avoid the singularities and branch cuts by following paths slightly above or below the positive real axis, as in Fig. 3.3. We then define the *lateral Borel resummations* by

$$s_\pm(\varphi)(z) = z^{-1} \int_{C_\pm} e^{-\zeta/z} \widehat{\varphi}(\zeta) d\zeta. \tag{3.2.32}$$

Note that, even if all the coefficients of the original series are real (as we are assuming here), the lateral Borel resummations are in general complex due to the contour deformation. Their difference is purely imaginary and it is encoded in the discontinuity function

$$\text{disc}(\varphi)(z) = s_+(\varphi)(z) - s_-(\varphi)(z). \tag{3.2.33}$$

This discontinuity gives information on the branch cut structure of the function which is reconstructed by the Borel resummations.

In some cases, the singularities of the Borel transform will take place at other points in the complex plane, along rays forming an angle θ with the positive real axis. In this case, we can consider paths $\mathcal{C}_{\theta\pm}$ slightly above and below the ray, and define the generalized lateral Borel resummations as

$$s_{\theta\pm}(\varphi)(z) = z^{-1} \int_{\mathcal{C}_{\theta\pm}} e^{-\zeta/z}\widehat{\varphi}(\zeta)d\zeta. \tag{3.2.34}$$

The discontinuity in this case is then defined as

$$\text{disc}_\theta(\varphi)(z) = s_{\theta+}(\varphi)(z) - s_{\theta-}(\varphi)(z). \tag{3.2.35}$$

One can also use the method of Padé approximants to calculate these lateral resummations (3.2.32) with high precision.

The discontinuity functions (3.2.35) make contact with the calculation of non-perturbative effects that we have presented in previous chapters. For example, the quartic integral (1.3.11) leads to a divergent series whose Borel transform has a singularity at $\zeta = -1/4$, i.e. on the negative real axis. The relevant lateral Borel resummations are then made along the direction $\theta = \pi$, and the discontinuity $\text{disc}_\pi(\varphi)(-g)$, with $g > 0$ is, up to an overall factor, the function that we considered in (1.3.17). Similarly, in the analysis of unstable potentials in QM, we established that the path integral around the instanton saddle gives the discontinuity of the partition function along the relevant direction in the complex plane of the coupling constant. In the case of the quartic oscillator, the direction was again the negative real axis. Therefore, we expect the discontinuity function to contain information about the non-perturbative aspects of the original series, i.e. about exponentially small contributions of the form $e^{-A/z}$.

It turns out that the structure of these non-perturbative effects is encoded in the singularity structure of the Borel transform $\widehat{\varphi}(\zeta)$. More precisely, we will now show that the behavior of the Borel transform near the singularity at $\zeta = A$ determines the term of order $e^{-A/z}$ in the discontinuity function. This is a rather remarkable result. Remember that the Borel transform was obtained by considering a formal series which appears in the *perturbative* sector of the theory. However, this series, when extended to a function in the complex ζ-plane, also knows about the *non-perturbative sector*! This means that there is a deep connection between the perturbative and the non-perturbative regimes.

In order to establish this connection, let us assume that the singular behavior of the Borel transform near the singularity at $\zeta = A > 0$ is a branch cut, of the form

$$\widehat{\varphi}(A + \xi) = (-\xi)^{-b} \sum_{n \geq 0} \hat{c}_n \xi^n + \cdots, \tag{3.2.36}$$

where the dots indicate non-singular terms. The Borel transform in (3.2.18) is a particular example of this, more general, form. We can now plug this expression into

$$s_+(\varphi)(z) - s_-(\varphi)(z) = z^{-1} \oint_\gamma e^{-\zeta/z} \widehat{\varphi}(\zeta) d\zeta, \tag{3.2.37}$$

where $\gamma = C_+ - C_-$ is a contour which can be deformed around the singularity/branch cut of $\varphi(z)$ at $\zeta = A$. The non-singular terms do not contribute to the contour integral, and after setting $\zeta = A + \xi$ we have to evaluate

$$\oint_\gamma (-\xi)^{-b} \xi^n e^{-\xi/z} d\xi = \left(e^{\pi i b} - e^{-\pi i b}\right) z^{n-b+1} \int_0^\infty u^{n-b} e^{-u} du$$

$$= 2i \sin(\pi b) z^{n-b+1} \Gamma(n+1-b), \tag{3.2.38}$$

where the discontinuity of $(-\xi)^{-b}$ leads to the difference of phases in the second term of the equation. We conclude that

$$\mathrm{disc}(\varphi)(z) = 2i \sin(\pi b) e^{-A/z} z^{-b} \sum_{n=0}^{\infty} c_n z^n, \tag{3.2.39}$$

where

$$c_n = \Gamma(n+1-b) \hat{c}_n. \tag{3.2.40}$$

A similar calculation can be done when the singularity is a simple pole or a logarithmic branch cut: if

$$\widehat{\varphi}(A + \xi) = -\frac{a}{\xi} - \log(\xi) \sum_{n \geq 0} \hat{c}_n \xi^n + \cdots, \tag{3.2.41}$$

we find

$$\mathrm{disc}(\varphi)(z) = 2\pi i e^{-A/z} \left(\frac{a}{z} + \sum_{n=0}^{\infty} c_n z^n \right), \tag{3.2.42}$$

where $c_n, \hat{c}_n$ are related by (3.2.40) for $b = 0$. Notice that, due to this relationship, the series appearing in (3.2.41) is the Borel transform of the series appearing in the discontinuity. This observation can be extended to the case with arbitrary b by defining the Borel transform of a series of the form

$$\varphi(z) = \sum_{n \geq 0} a_n z^{n-b} \tag{3.2.43}$$

by

$$\widehat{\varphi}(\zeta) = \sum_{n \geq 0} \frac{a_n}{\Gamma(n+1-b)} \zeta^{n-b}. \tag{3.2.44}$$

In the above calculation we have implicitly assumed that $\widehat{\varphi}(\zeta)$ has a single singularity, at $\zeta = A$. However, in general, there will be singularities at $\zeta = A_1, A_2, \ldots$. The discontinuity will then have contributions from all these singularities, which will lead to exponentially small effects of the form $e^{-A_k/z}$. If all the A_k are strictly positive, the largest non-perturbative effect will correspond to the singularity which is closest to the origin.

Example 3.5 In order to illustrate the above considerations, we will consider the following formal power series

$$\varphi(z) = \sum_{n=0}^{\infty} \frac{\Gamma\left(n+\frac{1}{2}\right)^2}{n!\,\Gamma\left(\frac{1}{2}\right)^2} z^n. \tag{3.2.45}$$

This is the asymptotic expansion of the function

$$f(z) = e^{-1/2z} \sqrt{\pi/z} I_0\left(\frac{1}{2z}\right), \tag{3.2.46}$$

where I_0 is the modified Bessel function. The Borel transform of (3.2.45) is a hypergeometric function,

$$\widehat{\varphi}(\zeta) = {}_2F_1\left(\frac{1}{2}, \frac{1}{2}, 1; \zeta\right). \tag{3.2.47}$$

This function has a logarithmic branch cut starting at $\zeta = A = 1$, and the singularity structure around this point is given by (3.2.41) with $a = 0$ (no pole) and

$$\hat{c}_n = \frac{(-1)^n}{\pi} \frac{\Gamma\left(n+\frac{1}{2}\right)^2}{n!^2\,\Gamma\left(\frac{1}{2}\right)^2}. \tag{3.2.48}$$

These coefficients can be read from the discontinuity equation for the hypergeometric functions,

$$ {}_2F_1\left(\frac{1}{2}, \frac{1}{2}, 1; \zeta + i\epsilon\right) - {}_2F_1\left(\frac{1}{2}, \frac{1}{2}, 1; \zeta - i\epsilon\right) = 2i \, {}_2F_1\left(\frac{1}{2}, \frac{1}{2}, 1; 1 - \zeta\right). \tag{3.2.49}$$

Indeed, if we set $\zeta = 1 + \xi$, and we use that $\log(\xi + i\epsilon) - \log(\xi - i\epsilon) = -2\pi i$, we find that

$$2\pi i \sum_{n \geq 0} \hat{c}_n \xi^n = 2i \, {}_2F_1\left(\frac{1}{2}, \frac{1}{2}, 1; -\xi\right). \tag{3.2.50}$$

Therefore, the coefficients $\hat{c}_n$ are precisely, up to a sign $(-1)^n$ and an overall factor $1/\pi$, those of the expansion around $\xi = 0$ of

$$_2F_1\left(\frac{1}{2}, \frac{1}{2}, 1; \xi\right),$$ (3.2.51)

and this leads to (3.2.48). The discontinuity function in this case is then given by

$$\mathrm{disc}(\varphi)(z) = 2\mathrm{i}e^{-1/z}\varphi_1(z),$$ (3.2.52)

where

$$\varphi_1(z) = \sum_{n=0}^{\infty}(-1)^n\frac{\Gamma\left(n + \frac{1}{2}\right)^2}{n!\Gamma\left(\frac{1}{2}\right)^2}z^n.$$ (3.2.53)

This formal power series is the asymptotic expansion of the modified Bessel function

$$e^{-1/2z}\sqrt{1/\pi z}K_0\left(\frac{1}{2z}\right).$$ (3.2.54)

□

3.3 Large order behavior and Borel transforms

We have seen that the Borel transform of the perturbative series has information about the non-perturbative sector, through the local behavior near the singularities. One particular consequence of this connection between the perturbative and the non-perturbative sectors is that the large order behavior of the coefficients of the perturbative series is controlled by the leading non-perturbative contribution. This was first noticed by Bender and Wu in the case of the quartic anharmonic oscillator, but it is a quite general fact. In our framework, this connection can be stated by saying that the large n behavior of the coefficients a_n of the series (3.2.1) is determined by the local behavior of its Borel transform in the nearby singularity. This is just a consequence of an old theorem by Darboux applied to the Borel transform $\widehat{\varphi}(\zeta)$ and it can already be seen in our previous examples. In all of them, the value of A in (3.2.5) gives the location of the singularity in the Borel plane. However, a much more precise statement can be made, as we are about to see: the full $1/n$ asymptotics of a_n is encoded in the series describing the behavior of $\widehat{\varphi}(\zeta)$ around $z = A$.

Let us then assume that the Borel transform $\widehat{\varphi}(\zeta)$ has a single singularity at $\zeta = A > 0$, and that it has the local expansion (3.2.36) around it. Around $\zeta = 0$ the coefficients of the Borel transform are given by the Cauchy formula

$$\frac{a_k}{k!} = \frac{1}{2\pi\mathrm{i}}\oint_{C_0}\frac{\widehat{\varphi}(\zeta)}{\zeta^{k+1}}\mathrm{d}\zeta,$$ (3.3.1)

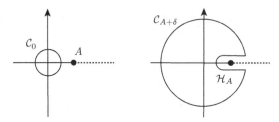

Figure 3.4 Contour deformation in (3.3.2).

where C_0 is a small circle around $\zeta = 0$. Let us choose a $\delta > 0$. We now enlarge the contour C_0 to a contour $C_{A+\delta} \cup \mathcal{H}_A$, where $C_{A+\delta}$ is a circle of radius $A + \delta$, minus an arc, and $\mathcal{H}_A$ is a Hankel contour centered around A, see Fig. 3.4. By deforming the contour we find

$$\frac{a_k}{k!} = \frac{1}{2\pi i} \oint_{C_{A+\delta}} \frac{\widehat{\varphi}(\zeta)}{\zeta^{k+1}} d\zeta + \frac{1}{2\pi i} \oint_{\mathcal{H}_A} \frac{\widehat{\varphi}(\zeta)}{\zeta^{k+1}} d\zeta. \tag{3.3.2}$$

The first integral can be estimated to be of order $\mathcal{O}((A + \delta)^{-k})$. Since the leading large k asymptotics goes like A^{-k}, and $A + \delta > A$, this is a subleading, exponentially small correction as k grows large, and it does not contribute to the leading $1/k$ asymptotics. On the other hand,

$$\frac{1}{2\pi i} \oint_{\mathcal{H}_A} \frac{\widehat{\varphi}(\zeta)}{\zeta^{k+1}} d\zeta = \frac{\sin(\pi b)}{\pi} \sum_{n \geq 0} \hat{c}_n \int_0^\delta \frac{\xi^{n-b}}{(A + \xi)^{k+1}} d\xi \tag{3.3.3}$$

where we have set $\zeta = A + \xi$ and we did a discontinuity calculation similar to the one in (3.2.38). An easy estimate shows that, at fixed n,

$$\int_0^\delta \frac{\xi^{n-b}}{(A + \xi)^{k+1}} d\xi = \int_0^\infty \frac{\xi^{n-b}}{(A + \xi)^{k+1}} d\xi + \mathcal{O}\left((A + \delta)^{-k}\right)$$

$$= A^{n-b-k} \frac{\Gamma(k + b - n)\Gamma(n - b + 1)}{\Gamma(k + 1)} + \mathcal{O}\left((A + \delta)^{-k}\right). \tag{3.3.4}$$

We conclude that the asymptotics of a_k is given by

$$a_k \sim \frac{\sin(\pi b)}{\pi} \sum_{n \geq 0} A^{n-b-k} c_n \Gamma(k + b - n). \tag{3.3.5}$$

The above derivation can be simplified if the Borel transform $\widehat{\varphi}(z)$ decays sufficiently fast at infinity, so that we can take $\delta \to \infty$. If there are further singularities along the real axis, one can include them systematically if one knows the singular behavior of the Borel transform in their neighborhood. They lead to subleading, exponentially small corrections to the asymptotics that we have obtained.

It is convenient to absorb the overall factor appearing in (3.2.39) in the definition of the coefficients c_n, and write

$$\text{disc}(\varphi)(z) = \text{i}e^{-A/z}z^{-b}\sum_{n=0}^{\infty}c_n z^n, \tag{3.3.6}$$

so that the large order asymptotics is given by

$$a_k \sim \frac{1}{2\pi}A^{-b-k}\Gamma(k+b)$$
$$\times\left[c_0 + \frac{c_1 A}{k+b-1} + \frac{c_2 A^2}{(k+b-2)(k+b-1)} + \cdots\right]. \tag{3.3.7}$$

It can easily be seen that, when $b = 0$, this formula gives the result which is obtained for a Borel transform with a logarithmic singularity. If the singularity occurs on the negative real axis, $\zeta = -A$ with $A > 0$, as it does in many interesting cases, the above derivation can be repeated *verbatim*, by using the discontinuity along the direction $\theta = \pi$. Let us write this discontinuity as

$$-\text{disc}_\pi(\varphi)(-z) = \text{i}e^{-A/z}z^{-b}\sum_{n=0}^{\infty}c_n z^n, \qquad z > 0. \tag{3.3.8}$$

The overall minus sign is due to the fact that, along the negative real axis, the discontinuity, as we defined it, computes the difference between the functions below the axis and above the axis, which is the opposite convention that we have used so far (and in particular differs in a minus sign from the discontinuity computed in for example (1.3.17)). Then, the large order behavior is given by

$$a_k \sim \frac{1}{2\pi}(-1)^{k+1}A^{-b-k}\Gamma(k+b)$$
$$\times\left[c_0 + \frac{c_1 A}{k+b-1} + \frac{c_2 A^2}{(k+b-2)(k+b-1)} + \cdots\right], \tag{3.3.9}$$

and the original series is alternating.

Example 3.6 Let us illustrate these results with the quartic integral (1.3.11). The leading order behavior of the discontinuity is given in (1.3.22). The singularity of the Borel transform occurs at $\zeta = -1/4$. The result (3.3.9) gives, in this case,

$$a_k \sim \frac{1}{\pi\sqrt{2}}(-1)^k 4^k \Gamma(k), \tag{3.3.10}$$

which is indeed the asymptotic behavior found in (1.3.24). □

3.4 The quartic anharmonic oscillator

We now consider a quantum mechanical example: the quartic anharmonic oscillator. As we pointed out in Chapter 1, the thermal partition function, which is defined
by a Euclidean path integral, can be regarded as a generalization of a function
defined by an integral. The energy of the ground state, $E_0(g)$, when computed in
perturbation theory, leads to an asymptotic series of the form:

$$\varphi(g) = \sum_{k=0}^{\infty} a_k g^k, \quad a_0 = \frac{1}{2}. \tag{3.4.1}$$

Since this system is unstable for $g < 0$, we expect this series to have zero radius of
convergence, by Dyson's argument. Of course, in this case we do not have closed
form expressions for the coefficients a_k, therefore we cannot compute the Borel
transform of this series explicitly. However, we have some analytic information
about the ground state energy $E_0(g)$ as a function of g, namely we know that there
should be a branch cut starting at $g = 0$, along the negative real axis, and we have
computed the discontinuity of $E_0(g)$, at one-loop, by using instanton methods. The
result was written down in (1.7.12). By comparing it to (3.3.8), we find,

$$b = \frac{1}{2}, \quad c_0 = 4\sqrt{\frac{2}{\pi}}, \quad A = 4/3. \tag{3.4.2}$$

Plugging these values into (3.3.9) we obtain the following formula for the large
order behavior,

$$a_k \sim (-1)^{k+1} \frac{\sqrt{6}}{\pi^{3/2}} \left(\frac{3}{4}\right)^k \Gamma\left(k + \frac{1}{2}\right), \tag{3.4.3}$$

which is the famous result of Bender and Wu. Notice that, if we compute the discontinuity at higher loops, which amounts to calculating the path integral around
the instanton solution to higher order in the coupling constant expansion, we will
find explicit values for the coefficients c_n, $n \geq 1$, appearing in (3.3.8), and these
higher-loop coefficients lead to $1/k$ corrections to the large order behavior (3.4.3).

In the case of the anharmonic oscillator, we can give another argument for the
large order behavior (3.4.3) which does not use Borel transforms (this is in fact the
original argument of Bender and Wu). First, we introduce the function

$$f(z) = \frac{1}{z}(E_0(z) - a_0) = \sum_{k=0}^{\infty} f_k z^k, \quad f_k = a_{k+1}. \tag{3.4.4}$$

As a function on the complex z-plane it has the following properties.

1. It is analytic with a cut along $(-\infty, 0)$. This is due to the discontinuity that we
   computed explicitly in terms of instantons.

2. At the origin it behaves like

$$\lim_{z \to 0} zf(z) = 0. \tag{3.4.5}$$

This is because the series (3.4.1) is asymptotic.
3. At infinity it goes like

$$|f(z)| \sim |z|^{-2/3}. \tag{3.4.6}$$

The first and the second properties can be proved rigorously. The last property follows from a simple scaling argument: at large g, we have that

$$H \approx \frac{p^2}{2} + g\frac{q^4}{4}. \tag{3.4.7}$$

If we rescale $q \to g^{-1/6}q$, we have

$$H \to g^{1/3}\left(\frac{p^2}{2} + \frac{q^4}{4}\right), \tag{3.4.8}$$

therefore the ground state energy will scale with g as

$$E_0(g) \approx Cg^{1/3}, \qquad g \to \infty, \tag{3.4.9}$$

where C is the energy of the ground state of the Hamiltonian $p^2/2 + q^4/4$ appearing in (3.4.8).

Now, let C_z be a contour around a point $z \in \mathbb{C}$ away from the branch cut, as in the left hand side of Fig. 3.5. Cauchy's theorem gives

$$f(z) = \frac{1}{2\pi i} \oint_{C_z} dx \frac{f(x)}{x - z}. \tag{3.4.10}$$

We can now deform the contour to encircle the branch cut in the negative real axis, as in the right hand side of Fig. 3.5. The contributions from the contours at infinity C_R and around the origin C_ϵ vanish, thanks to (3.4.6) and (3.4.5), respectively.

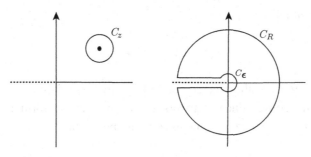

Figure 3.5 Contour deformation from the contour C_z around $z \in \mathbb{C}$. The dashed line represents the branch cut along the negative real axis.

The only remaining contribution comes from the lines which are parallel to the branch cut,

$$f(z) = -\frac{1}{2\pi i} \int_{-\infty}^{0} \frac{\text{disc}_\pi(f)(x)}{x - z} dx.$$ (3.4.11)

In terms of the original quantity, the ground state energy, we have

$$E_0(g) = a_0 - \frac{g}{2\pi i} \int_{-\infty}^{0} \frac{\text{disc}_\pi(E_0)(g')}{g'(g' - g)} dg'.$$ (3.4.12)

Since

$$\frac{g}{g'} \frac{1}{g' - g} = \sum_{k \geq 0} \frac{g^{k+1}}{(g')^{k+2}},$$ (3.4.13)

we find the integral representation

$$a_k = -\frac{(-1)^{k+1}}{2\pi i} \int_0^\infty \frac{\text{disc}_\pi(E)(-z)}{z^{k+1}} dz, \quad k \geq 1.$$ (3.4.14)

This is equivalent to the expression obtained above by using Borel transforms.

The main conclusion of this analysis is that, indeed, the perturbative series for the anharmonic oscillator has zero radius of convergence, as expected from Dyson's argument: for negative coupling the theory becomes unstable, so analyticity at $g = 0$ is impossible. Moreover, an analysis of this instability, in terms of instanton configurations, makes it possible to give a precise and quantitative characterization of the asymptotics of perturbation theory.

It is instructive to verify the large order behavior of the coefficients, (3.4.3), by computing a large number of terms in the perturbative series and studying how they grow when k is large. Equivalently, we can compute

$$Q_k = (-1)^{k+1} \frac{\pi^{3/2}}{\sqrt{6}} \left(\frac{3}{4}\right)^{-k} \frac{a_k}{\Gamma\left(k + \frac{1}{2}\right)},$$ (3.4.15)

which behaves, at large k, as

$$Q_k = 1 + \mathcal{O}\left(\frac{1}{k}\right).$$ (3.4.16)

Due to the tails in $1/k$, the convergence of Q_k to 1 is relatively slow. One way to accelerate the convergence is to do a Richardson transform of the original sequence. Given a sequence with the asymptotic behavior

$$s_k \sim \sum_{n=0}^{\infty} \frac{\sigma_n}{k^n},$$ (3.4.17)

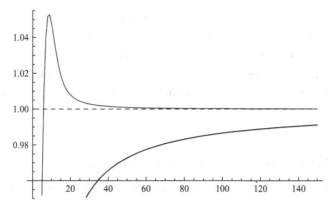

Figure 3.6 The bottom line shows the joined points of the sequence Q_k, which is defined in (3.4.15) from the coefficients a_k of the perturbative series of the ground state energy for the quartic oscillator. The top line is its first Richardson transform, with accelerated convergence to the expected value 1.

its Nth Richardson transform is defined by

$$s_k^{(N)} = \sum_{\ell=0}^{N} \frac{s_{k+\ell}(k+\ell)^N (-1)^{\ell+N}}{\ell!(N-\ell)!}. \qquad (3.4.18)$$

The effect of this transformation is to remove the first $N-1$ subleading tails in (3.4.17), and therefore it leads to a sequence which convergences to σ_0 much faster than the original one. We can now test (3.4.3) by computing the a_k with the recursion (1.2.28) explained in Chapter 1, use this sequence to calculate the sequence Q_k, and finally use Richardson transforms to accelerate the convergence. The result of this numerical experiment is shown in Fig. 3.6, which uses the first 150 values of a_k. The two lines are the (joined) sequences of points Q_k (bottom) and its first Richardson transform (top). The convergence to 1 is quite fast, and verifies the arguments connecting the large order behavior of the perturbative series with the instanton calculation.

It is possible to show that the factorial growth of a_k is due to the factorial growth of the number of Feynman diagrams involved in the calculation of a_k. Indeed, recall from Section 1.2 that the a_k can be computed as a sum over connected quartic graphs, and we have the formula (1.2.19). The total number of disconnected quartic graphs is simply given by the quartic integral (1.3.11), which as $k \to \infty$ behaves like (see (1.3.25))

$$4^{2k} k!, \qquad (3.4.19)$$

displaying a factorial growth. One might think that there would be a substantial reduction in this number when we consider connected diagrams, but a detailed analysis shows that this is not the case: at large k, the quotient of the number of connected and disconnected diagrams differs from 1 only in $\mathcal{O}(1/k)$ corrections. We conclude that there are $\sim k!$ diagrams that contribute to a_k. It turns out that the Feynman integrals associated to the graphs grow only exponentially with k, therefore at large k the growth of a_k is dominated by the growth in the corresponding number of Feynman diagrams, and this ultimately leads to (3.4.3).

3.5 Instantons and large order behavior in quantum theories

As we have seen in the éxample of the quartic oscillator, perturbative series in quantum theories diverge factorially, so their Borel transforms are analytic in a neighborhood of the origin. What are the possible sources of the singularities in the Borel transform? In the case of the quartic oscillator, the discontinuity in the energy is given by an instanton calculation, and the singularity in the Borel plane A is nothing but the action of the instanton. This is expected to be a generic feature: if a quantum theory admits an instanton configuration ϕ_* with finite action $S(\phi_*)$, the Borel transform of the perturbative expansion of an observable will have a singularity at $S(\phi_*)$. In particular, if a quantum theory admits non-trivial instanton configurations, the perturbative series has a zero radius of convergence.

Although this general result has not been proved, all available results indicate that it is a general property of quantum theories. An heuristic argument for this can be given, based on the analogy between Euclidean path integrals and functions defined by ordinary integrals. As we saw in the case of the quartic integral (1.3.11), the divergence of the series around the trivial saddle point at $z = 0$ is controlled by the non-trivial saddle points $z_{1,2}$ in (1.3.19). It is not difficult to show that this is a generic feature of functions defined by integrals, of the form

$$I(\hbar) = \int_{\mathcal{C}} g(z)\mathrm{e}^{-f(z)/\hbar}\mathrm{d}z, \tag{3.5.1}$$

where $\mathcal{C}$ is a contour in the complex plane which leads to a well-defined integral. The asymptotic expansion of such an integral around a given saddle point z_0 leads typically to an asymptotic series, and the Borel transform of this series has singularities at $f(z_n)$, where z_n, $n = 1, 2, \ldots$ are the other saddle points. If Euclidean path integrals behave in this respect like ordinary integrals, we should expect that the Borel transform of the perturbative series derived from them will have singularities at $S(\phi_*)$, as we claimed above. The arguments of Bender and Wu on the quartic oscillator have been generalized to QFT starting with the work of Lipatov. However, in general renormalizable QFT, not all singularities in the Borel transform are due to instantons. As in the case of the quartic oscillator, instantons encode the large order behavior of the perturbative series due to the growth in the number of diagrams that contribute at each order. There are other, very different sources of factorial divergence in perturbation theory, encoded in the so-called renormalons, which are in principle not described by the usual semiclassical saddle points, as we will see in the next chapter.

Using the general principle stated above, we can now discuss various possible behaviors that can arise in quantum theory (and in particular in QM), and how they are reflected in the large order behavior of perturbation theory.

Stable vacua

If we expand around a stable vacuum (like the absolute minimum of a potential in QM), there are no instanton solutions with positive action, therefore the Borel transform of the perturbative series has no singularities on the positive real axis. The perturbative series is then Borel summable, and in many cases the Borel resummation of the series reconstructs the non-perturbative result (this has been proved rigorously in the case of a harmonic oscillator perturbed by an even potential, like the quartic oscillator). In order to find the location of the singularities of the Borel transform, one has to consider instanton solutions for other values of the coupling constant. This is what happens in the case of the quartic, anharmonic oscillator with positive coupling $g > 0$. Instantons are obtained by looking at solutions with negative coupling constant $g = -\lambda$, $\lambda > 0$, and they determine the large order behavior of the original perturbative series.

It is instructive to use the method of Borel resummation and Padé approximants to calculate the ground state energy of the quartic oscillator $E_0(g)$. As in the case of the quartic integral, we calculate the integrals (3.2.29) for increasing values of n. The perturbative series is given by (1.2.18),

$$\varphi(g) = \frac{1}{2} + \sum_{n=1}^{\infty} a_n \left(\frac{g}{4}\right)^n. \qquad (3.5.2)$$

In this case, the ground state energy can be computed numerically from the Schrödinger equation. In Table 3.2 we compare the result obtained from (3.2.29), for different values of n, with the numerical result obtained from the Schrödinger equation, for two different values $g = 2$ and $g = 4$ of the coupling constant. As in the case of the quartic integral, the convergence to the non-perturbative result is very good.

Unstable vacua

If we consider the perturbation series around an *unstable* minimum there is always an instanton with real, positive action mediating the decay of the particle. This

Table 3.2 *The Borel–Padé resummations (3.2.29) of the asymptotic series (3.5.2)*
for the ground state energy of the quartic oscillator

Two values of the coupling constant are considered, $g = 2$ and $g = 4$, for
increasing values of n. In the last line we give the numerical result for $E_0(g)$. All
numbers are presented with ten significant digits. For each value of n, we
underline the digits in the result of (3.2.29) which agree with the numerical result.

n	$s(\varphi)_n(2)$	$s(\varphi)_n(4)$
10	0.6961229466	0.8031738769
25	0.6961758330	0.8037716736
50	0.6961758208	0.8037706511
$E_0(g)$	0.6961758208	0.8037706512

is what happens for example in the quartic oscillator with negative coupling constant, or in the cubic oscillator. The perturbative series will *not* be Borel summable,
since the Borel transform will have singularities on the positive real axis, and the
behavior of the Borel transform near these singularities controls the behavior of
perturbation theory at large order. In the case of unstable potentials in QM, lateral
Borel resummations have an interesting physical meaning. If we consider the perturbative series for the ground state energy, the lateral resummations will have a
small imaginary part:

$$E_{0,\pm}(g) = \operatorname{Re} E_0(g) \pm i \operatorname{Im} E_0(g), \qquad \operatorname{Im} E_0(g) \approx e^{-A/g}. \tag{3.5.3}$$

The decay rate is given by,

$$\Gamma = 2 \left| \operatorname{Im} E_0(g) \right|, \tag{3.5.4}$$

and represents the probability of decay of the particle in the unstable vacuum.
Therefore, the fact that the lateral Borel resummations of the perturbative series
are complex is precisely what is needed in order to capture the physics of the problem. Notice that, since the lateral resummations can be done along two different
paths, the sign of the imaginary part of the energy is ambiguous, as happened in
our instanton calculations.

Example 3.7 *The cubic oscillator.* Let us consider the cubic potential (1.7.13).
Standard perturbation theory gives a series for the ground state energy $E_0(g)$ of the
form

$$\varphi(g) = \sum_{k=0}^{\infty} a_k g^{2k}, \tag{3.5.5}$$

therefore the relevant coupling constant is rather g^2. The coefficients a_k, with $k \geq 1$, all have the same sign (negative), and the series is non-alternating. This corresponds to the fact that the action of the instanton, computed in (1.7.17), is real and positive, and the series $\varphi(g)$ is not Borel summable. The one-instanton contribution to the imaginary part of the ground state energy was computed in (1.7.20). Using the values,

$$b = \frac{1}{2}, \quad A = \frac{2}{15}, \quad c_0 = \frac{2}{\sqrt{\pi}}, \tag{3.5.6}$$

we find that the large order behavior of the series in (3.5.5) is

$$a_k \sim -\frac{(60)^{k+1/2}}{(2\pi)^{3/2}2^{3k}}\Gamma(k+1/2). \tag{3.5.7}$$

The overall minus sign appearing here is due to the correct resolution of the sign ambiguity in extracting the square root in (1.4.56). One can perform lateral Borel resummations of the series (3.5.5) and calculate in this way the decay rate of the energy level. □

Complex instantons

One of the lessons we learned from the quartic oscillator with positive coupling constant is that, in order to understand the source of the factorial divergence of its perturbative series, one has to look for instantons in the unstable case, i.e. in the case in which the coupling constant is negative. This is due to the fact that the pole of the Borel transform of the series appearing in the original problem is on the negative axis. Semiclassically, this corresponds to the action of an instanton in which the coupling constant is negative. In general, the poles of the Borel transform can be anywhere in the complex plane. Therefore, if they have an instanton interpretation, they should correspond to *complex* instanton solutions, with complex actions. These solutions can be obtained by complexifying the couplings of the problem, or by considering a complexified phase space where q, p are complex variables. The complex instantons might not be directly relevant to the physics of the problem (since they do not mediate actual tunneling behavior), but they control the large order behavior of the perturbative series, as one can see in examples.

Note that, when the singularities of the Borel transform are complex, the perturbative series is Borel summable. If the original series is real, singularities appear in complex conjugate pairs, and in order to study the large order behavior of perturbation theory, one has to take into account the contribution of the pair of complex conjugate singularities which is closest to the origin. Let us denote these singularities by A and $\bar{A}$. In the contour deformation of Fig. 3.4 we will pick contributions from both of them, and we have to add their contributions into (3.3.7). This has an interesting consequence on the behavior of the series. Let us write

$$A = |A|e^{-i\theta_A}, \qquad c_0 = |c_0|e^{i\theta_c}, \tag{3.5.8}$$

where A and c_0 are the parameters appearing in (3.3.6). The large order behavior is then obtained by adding the contribution of the two instantons, and reads

$$a_k \sim \Gamma(k+b)|A|^{-k-b} \cos\left((k+b)\theta_A + \theta_c\right). \tag{3.5.9}$$

We then have an *oscillatory* behavior for the series of perturbative coefficients. Conversely, an oscillatory behavior in a perturbative series is a clear indication that the relevant instanton is complex.

Example 3.8   As our first example, let us consider the potential

$$W(q) = \frac{1}{8g} \sinh^2\left(2\sqrt{g}q\right), \tag{3.5.10}$$

which is stable. One can use stationary perturbation theory to calculate the perturbative series for the ground state energy, and one finds,

$$\frac{1}{2} + \frac{g}{2} - \frac{g^2}{2} + \cdots . \tag{3.5.11}$$

Note that this potential can be obtained from the periodic potential (1.10.29) by rotating $\sqrt{g} \to i\sqrt{g}$. The perturbative series (3.5.11) and (1.10.31) are related by this rotation, and we pass from a non-alternating series to an alternating series. Since the potential (3.5.10) is stable, there are no instantons for real g. However, if we regard q as a complex coordinate, we see that along the imaginary axis our potential (3.5.10) becomes, up to a sign, the periodic potential (1.10.29). We can then look for instanton configurations satisfying (1.4.3), with $E = 0$, in which q is imaginary. It is easy to verify that the complexified version of the instanton solution of the periodic potential

$$q(t) = \frac{i}{\sqrt{g}} \tan^{-1}\left(e^t\right) \tag{3.5.12}$$

is also a solution, interpolating between the extrema of the potential along the imaginary axis. This example is very similar to the quartic oscillator, in which changing the sign of the quartic coupling takes us from a stable potential with no instantons, to an unstable potential with instantons.   □

Example 3.9   Consider a particle situated at the origin of the potential

$$\frac{1}{g} W\left(\sqrt{g}q\right), \qquad W(q) = \frac{1}{2}q^2 - \gamma q^3 + \frac{1}{2}q^4. \tag{3.5.13}$$

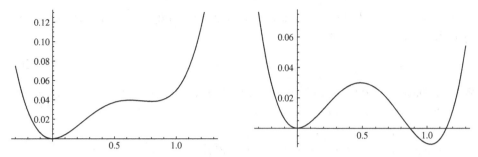

Figure 3.7 On the left: the potential (3.5.13) for $\gamma = 0.95$, where the origin is a stable minimum. On the right: the potential (3.5.13) for $\gamma = 1.01$; the origin is now unstable.

The ground state energy has the series expansion,

$$\sum_{n\geq 0} a_n g^n. \tag{3.5.14}$$

There are two different situations here (see Fig. 3.7).

1. For $|\gamma| > 1$, the origin is not an absolute minimum, which is in fact at

$$q_0 = \frac{3\gamma + \sqrt{-8+9\gamma^2}}{4}. \tag{3.5.15}$$

2. For $|\gamma| < 1$, the origin is the absolute minimum.

In the first case, $|\gamma| > 1$, the vacuum is quantum mechanically unstable, and there is an instanton given by a trajectory from $q = 0$ to the turning point

$$q_+ = \gamma - \sqrt{\gamma^2 - 1}. \tag{3.5.16}$$

The action of this instanton is given by A/g, where

$$A = 2\int_0^{q_+} (2W(q))^{1/2}dq = -\frac{2}{3} + \gamma^2 - \frac{1}{2}\gamma(\gamma^2 - 1)\log\frac{\gamma+1}{\gamma-1}. \tag{3.5.17}$$

Since

$$\int_0^{q_+} \left(\frac{1}{\sqrt{2W(q)}} - \frac{1}{q}\right)dq = -\log\left[\frac{1}{2}(\gamma^2 - 1)^{1/2}q_+\right], \tag{3.5.18}$$

the one-loop prefactor appearing in (3.3.6) can be read from (1.7.8) and it is given by

$$c_0 = \frac{2}{\pi^{1/2}}(\gamma^2 - 1)^{-1/2}. \tag{3.5.19}$$

In the second case, $|\gamma| < 1$, we have to *analytically continue* the results of the first case. The instanton considered above is now complex, and it leads to a complex trajectory. In fact, there are two complex conjugate instantons, described by trajectories going from $q = 0$ to

$$q = \gamma \pm i\sqrt{1 - \gamma^2}. \tag{3.5.20}$$

We then have to add the contributions of both instantons to obtain the contribution to the large order behavior

$$a_k \sim -\frac{1}{\pi^{3/2}}\Gamma(k + 1/2)\left[A^{-k-1/2}i(1 - \gamma^2)^{-1/2} - \overline{A}^{-k-1/2}i(1 - \gamma^2)^{-1/2}\right]$$

$$= \frac{2}{\pi^{3/2}}\Gamma(k + 1/2)(1 - \gamma^2)^{-1/2}\mathrm{Im}\left(A^{-k-1/2}\right). \tag{3.5.21}$$

For $\gamma = 0$ (the quartic potential) we find,

$$\mathrm{Im}\left(A^{-k-1/2}\right) = (-1)^{1+k}\left(\frac{3}{2}\right)^{k+1/2}, \tag{3.5.22}$$

and the final result for the asymptotics is

$$a_k \sim \frac{(-1)^{k+1}\sqrt{6}}{\pi^{3/2}}\left(\frac{3}{2}\right)^k, \tag{3.5.23}$$

which agrees with the previous result after taking into account the different normalization of the quartic coupling g, which introduces an extra factor 2^k. $\qquad\square$

The non-Borel summable case

We have seen that, in the case of unstable potentials in QM, the fact that the series is not Borel summable indicates that particles will eventually tunnel, and the imaginary part of the energies appearing in the lateral Borel resummations is essentially decay rates. However, in some situations, the perturbative series is not Borel summable, yet the potential is stable and the energy eigenvalues are real. This is what happens for example in the case of the double-well potential. Let us denote the perturbative series computing the energy of the ground state by,

$$E^{(0)}(g) = \sum_{k \geq 0} a_k g^k. \tag{3.5.24}$$

The very first coefficients of this series are written down in (1.8.3). The a_k appearing here grow factorially, and the series is not Borel summable (all coefficients except the leading term have negative sign). In this case, lateral Borel resummations of $E^{(0)}(g)$ cannot lead to the true answer, since they have an imaginary part:

$$E_{\pm}^{(0)}(g) = \mathrm{Re}E_{\pm}^{(0)}(g) \pm i\,\mathrm{Im}E_{\pm}^{(0)}(g). \tag{3.5.25}$$

It turns out that it is still possible to extract the exact ground state energy $E_0(g)$ from Borel resummation, but we have to consider the *full series* of instanton corrections:

$$E^{(0)}(g) + E^{(1)}(g) + E^{(2)}(g) + \cdots \tag{3.5.26}$$

where $E^{(n)}(g)$ denotes the series obtained by perturbative expansion around the n-instanton. The first terms in these series have been computed in (1.9.68), and their general structure was written down in (1.9.73). If we fix the instanton number n and the power k of the logarithmic term, the resulting series in g, with coefficients $e_{N,nkl}$ for $l = 0, 1, \ldots$, is factorially divergent and it is not Borel summable. We should then consider lateral Borel resummations of all the series involved in the multi-instanton expansion,

$$E_\pm(g) = E_\pm^{(0)}(g) + E_\pm^{(1)}(g) + E_\pm^{(2)}(g) + \cdots . \tag{3.5.27}$$

Alternatively, we can define the lateral Borel resummations by a procedure of analytic continuation. We first consider the series $E^{(n)}(g)$ for *negative g*. In this case, the series are Borel summable, and this is in turn related to the fact that the instanton corrections are only well defined for negative g, as we saw in Chapter 1. The resummed series $E_\pm^{(n)}(g)$ is then given by an analytic continuation of this Borel sum from g negative to $g = |g| \pm i0$, see Fig. 3.8. Note that, with this definition, the two-instanton configuration computed in (1.9.68) picks an imaginary part, since

$$\log\left(-\frac{2}{g}\right) \rightarrow \log\left(\frac{2}{|g|}\right) \pm \pi i, \tag{3.5.28}$$

therefore

$$\operatorname{Im} E_\pm^{(2)}(g) \approx \pm \left(\frac{2}{g}\right) \frac{e^{-1/(3g)}}{2\pi} \operatorname{Im}\left[\log\left(\frac{2}{g}\right)\right] = \pm \frac{1}{g} e^{-1/(3g)}. \tag{3.5.29}$$

We now impose the physical requirement that $E_\pm(g)$ must be *independent* of the resummation prescription and *real* (since it is the energy of a bound state). This means that imaginary parts must *cancel* in the total sum (3.5.27). At leading order

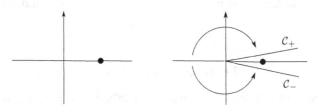

Figure 3.8 Lateral resummations can be obtained as two different analytic continuations of a Borel sum along the negative real axis (where there are no singularities) to the positive real axis.

in the instanton expansion, this implies in particular that the imaginary part of $E_{\pm}^{(0)}$ must be equal but opposite in sign to the imaginary part of the two-instanton contribution $\mathrm{Im}\, E_{\pm}^{(2)}$:

$$\mathrm{Im}\, E_{\pm}^{(0)}(g) = -\mathrm{Im}\, E_{\pm}^{(2)}(g) \Rightarrow \mathrm{Im}\, E^{(0)}(g) \approx -\frac{1}{g}e^{-1/(3g)}. \qquad (3.5.30)$$

This resolves the puzzle that we noted just after equation (1.9.69): the imaginary part of $E^{(2)}(g)$ is not only a problem, but is rather necessary if we want to make sense of the energy. The Borel resummation of the one-instanton contribution also has an imaginary part, but it is proportional to $e^{-1/(2g)}$ and cancels against the third-instanton contribution, so we do not have to consider it at this order. The cancellation (3.5.30) determines the large order behavior of the perturbative series, since (3.5.30) gives the discontinuity (3.3.6). One then finds, from (3.3.7), the large order behavior

$$a_k \sim -\frac{1}{\pi}3^{k+1}\Gamma(k+1), \qquad k \gg 1, \qquad (3.5.31)$$

which can be tested against the explicit results for the perturbative series, in this way providing a confirmation of the cancellation mechanism (3.5.30).

The cancellation between perturbative and non-perturbative contributions appearing in the double-well potential also occurs for other potentials in QM, like the periodic potential analyzed in Section 1.10. Moreover, it has been argued to be relevant in more general situations in quantum theory. These situations involve non-Borel summable series which however should lead to well-defined, non-perturbative real quantities, and they include realistic examples in QFT. We will briefly address some of these issues in our discussion of renormalons in Yang–Mills theory, in the next chapter.

One general consequence of the above analysis is that, when the perturbation series is not Borel summable, non-perturbative effects are not intrinsically defined: they depend on the choice of lateral resummation one uses for the perturbative series. Different choices in the lateral resummation of the perturbative series lead to different choices in the lateral resummation of the non-perturbative effects, in such a way that the final answer is unchanged and agrees with the underlying physical quantity.

3.6 Bibliographical notes

The connection between the divergence of perturbation theory and unstable vacua was pointed by Dyson in [82]. Bender and Wu, in a remarkable series of papers [29, 30], established the divergent character of the perturbative series for the quartic oscillator and established a quantitative, precise connection between the large order behavior of perturbation theory and tunneling effects in that example. Their original work is reviewed in [26], while [169] gives a mathematical perspective. The book [122] is a reprint volume with many relevant papers and an extensive list of references on large order behavior. Pedagogical introductions to this problem can be found in [77, 137].

For excellent surveys of classical asymptotics, see the books [28, 143]. The general theory of Borel transforms and Padé approximants, as applied to divergent series in physics, is reviewed for example in [43, 199]. My exposition is deeply influenced by the theory of resurgence of Écalle, which is presented for example in [56, 164]. Pedagogical introductions to the ideas of resurgence for a physics audience can be found in [14, 76, 137]. The connection between the analytic structure of the Borel transform and the large order behavior of perturbation theory is discussed in, for example, [70] and [19]. The example in the text based on Bessel functions is taken from [80]. The monograph [199] discusses the large order behavior of perturbation theory and its connection to instantons, both in QM and in QFT.

The argument that the factorial large order behavior of a_k is due to the growth in the number of Feynman diagrams is reviewed in [26]. In [31], a detailed statistical analysis of Feynman diagrams leads to a direct derivation of (3.4.3). The growth of connected and disconnected diagrams in various theories is studied in [27].

The behavior of perturbative series for the quartic and cubic oscillators, as well as their Borel resummations, are studied in detail in [10, 43, 66, 100, 115]. References to the relevant mathematical literature can be found in [43], including rigorous proofs that perturbation theory is Borel summable in the case of even perturbations of the quantum mechanical harmonic oscillator.

The role of complex instantons is emphasized in [18, 19]. The example of the hyperbolic potential is discussed in [22, 171]. The potential of Example 3.9 is discussed in [39]. The cancellation of non-perturbative ambiguities in the double-well potential is analyzed in detail in [200, 201]. A simpler example of this phenomenon, in the Painlevé II equation, is studied in [136].

Studies of large order behavior due to instantons in QFT started with the work of Lipatov [125], see for example [38] and the collection [122].

4

Non-perturbative aspects of Yang–Mills theories

4.1 Introduction

Non-Abelian gauge theories are the fundamental building block of the Standard Model of elementary particles. In the case of pure Yang–Mills (YM) theory (describing the interaction of gluons) and QCD (describing the interactions of gluons and fermions in the fundamental representation of the gauge group), a purely perturbative approach is only useful at high energies, due to the fundamental property of asymptotic freedom. In fact, at low energies, YM theory displays a number of phenomena which cannot be seen in perturbation theory. The first phenomenon is the existence of a mass gap. This means that, when the theory is quantized in a very large volume, the ground state energy is strictly positive. The second phenomenon is confinement: if we consider non-dynamical quarks in the theory, the energy between a quark and an antiquark grows linearly with their separation. Therefore, the spectrum of physical states of the theory consists of color singlets. Finally, in QCD we have the phenomenon of chiral symmetry breaking, i.e. the spontaneous breaking of chiral symmetry, which we will review in some detail below. All of these phenomena resist analytic treatment in conventional YM/QCD theories, and in particular in conventional perturbation theory, although some of them can be addressed in simpler toy models. It is then of the highest interest to develop non-perturbative techniques in the study of non-Abelian gauge theories.

In this chapter we will study some non-perturbative aspects in YM theories. As we saw in the study of instantons in QM, in order to understand non-perturbative effects it is useful to focus on quantities which are invisible in perturbation theory, like tunneling effects. Therefore, we will start our study of non-perturbative effects in YM theories by focusing on the dependence on the theta angle, which is a purely non-perturbative effect. We will then focus on instanton solutions in YM theories. Unfortunately, in conventional YM theories, and in contrast to QM and scalar field theory, there is no reliable semiclassical approximation to the Euclidean partition function of YM theory based on instantons, due to IR effects. In order to make sense of instanton calculus in this theory we need an IR cutoff. We will then develop instanton calculus at one-loop for YM theories defined on compact manifolds, where the semiclassical approximation makes sense. Other IR cutoffs (like the presence of Higgs fields) also lead to calculable instanton effects, but we will not consider them here.

4.2 Basics of Yang–Mills theories

In this book we assume the reader has some basic background on YM theories and their quantization. In this section, we will set up our conventions and review some basic facts.

To construct a YM theory we have to pick up a gauge group G, which will be taken to be a simple Lie group, with Lie algebra $\mathbf{g}$. The dimension of G will be denoted by $d(G)$. The generators of the Lie algebra, T_a, $a = 1, \ldots, d(G)$, are chosen to be Hermitian matrices and satisfy the commutation relations

$$[T_a, T_b] = \mathrm{i} f_{abc} T_c, \qquad a, b = 1, \ldots, d(G), \tag{4.2.1}$$

where f_{abc} are the structure constants of the algebra. For $SU(2)$, for example, we can take

$$T_a = \frac{1}{2}\sigma_a, \qquad a = 1, 2, 3, \tag{4.2.2}$$

where σ_a are the Pauli matrices, and the structure constants are

$$f_{abc} = \epsilon_{abc}. \tag{4.2.3}$$

The generators of the Lie algebra can be chosen to be orthogonal with respect to the inner product defined by the trace of their product:

$$\mathrm{Tr}(T_a T_b) = \alpha \delta_{ab}. \tag{4.2.4}$$

The overall normalization is encoded in the coefficient α. In this book we will mostly use the popular normalization $\alpha = \frac{1}{2}$, so that

$$\mathrm{Tr}(T_a T_b) = \frac{1}{2}\delta_{ab}. \tag{4.2.5}$$

The basic field in a YM theory is the YM connection, or gluon field,

$$A_\mu = A_\mu^a T_a, \tag{4.2.6}$$

which is a Lie algebra valued vector field or, equivalently, it is a field in the adjoint representation of the Lie algebra. Fields in a YM theory (for example matter fields) are labelled by representations r of the Lie algebra. For each representation

r of dimension $d(r)$ we have a matrix representation of the generators $\left(T_a^r\right)_j^i$, $i, j = 1, \ldots, d(r)$, and we recover the basis T_a introduced above when r is the fundamental representation. The covariant derivative acting on a field ϕ in the representation r is defined as

$$D_\mu \phi = \partial_\mu \phi - iA_\mu^a T_a^r \phi. \tag{4.2.7}$$

We will denote the adjoint representation by $r = G$. In this case, the matrices T_a^G have components

$$\left(T_a^G\right)_{bc} = if_{bac}. \tag{4.2.8}$$

If a field ϕ is in the adjoint representation, we can write it as in (4.2.6),

$$\phi = \phi_a T_a, \tag{4.2.9}$$

and the covariant derivative has components

$$\left(D_\mu \phi\right)_a = \partial_\mu \phi_a + f_{abc} A_\mu^b \phi_c. \tag{4.2.10}$$

Two important Lie algebraic quantities which are ubiquitous in one-loop calculations are $C(r)$ and $C_2(r)$, defined by the equations

$$
\begin{aligned}
C(r)\delta_{ab} &= \mathrm{Tr}\left(T_a^r T_b^r\right), \\
C_2(r)\delta_j^i &= \left(T_a^r\right)_k^i \left(T_a^r\right)_j^k.
\end{aligned} \tag{4.2.11}
$$

$C_2(r)$ is often called the quadratic Casimir of the representation r. These two quantities are not independent, since after tracing them one finds the relationship

$$C(r)d(G) = C_2(r)d(r), \tag{4.2.12}$$

and in particular they are equal for the adjoint representation $r = G$, where they can be evaluated in terms of the structure constants of the Lie algebra as

$$C_2(G)\delta_{ab} = f_{acd} f_{bcd}. \tag{4.2.13}$$

In particular,

$$f_{abc} f_{abc} = d(G)C_2(G). \tag{4.2.14}$$

In the case of $G = SU(N)$, one has, for the fundamental representation $r = \text{fund}$,

$$C(\text{fund}) = \frac{1}{2}, \tag{4.2.15}$$

while for the adjoint representation we have

$$C_2\left(SU(N)\right) = N. \tag{4.2.16}$$

The YM field strength, or curvature of the connection, takes values in the adjoint representation of the algebra, and it is defined as

$$F_{\mu\nu} = i[D_\mu, D_\nu] = \partial_\mu A_\nu - \partial_\nu A_\mu - i[A_\mu, A_\nu]. \qquad (4.2.17)$$

Its components in the T_a basis are

$$F^a_{\mu\nu} = \partial_\mu A^a_\nu - \partial_\nu A^a_\mu + f_{abc} A^b_\mu A^c_\nu. \qquad (4.2.18)$$

Sometimes it is useful to collect the components of the connection A_μ and the field strength $F_{\mu\nu}$ into a one-form and a two-form, respectively, with values in the Lie algebra,

$$A = A_\mu dx^\mu, \qquad F = \frac{1}{2} F_{\mu\nu} dx^\mu \wedge dx^\nu. \qquad (4.2.19)$$

Our convention for the exterior differential is the following: if

$$\psi = \psi_{\mu_1\mu_2\cdots\mu_p} dx^{\mu_1} \wedge dx^{\mu_2} \wedge \cdots \wedge dx^{\mu_p}, \qquad (4.2.20)$$

is a p-form, then

$$d\psi = \partial_\mu \psi_{\mu_1\mu_2\cdots\mu_p} dx^\mu \wedge dx^{\mu_1} \wedge \cdots \wedge dx^{\mu_p}. \qquad (4.2.21)$$

In this language, we have

$$F = dA - iA \wedge A, \qquad (4.2.22)$$

where d is the exterior differential and $\wedge$ denotes the standard wedge product of differential forms.

A gauge transformation acts on a gauge connection as

$$A_\mu(x) \to A^U_\mu(x) = U(x)A_\mu(x)U^\dagger(x) + iU(x)\partial_\mu U^\dagger(x), \qquad (4.2.23)$$

where $U(x)$ is a function of spacetime which takes values in the Lie group G. If we write

$$U = e^{i\phi}, \qquad \phi = \phi^a T_a, \qquad (4.2.24)$$

we have that, infinitesimally,

$$\delta A_\mu = D_\mu \phi. \qquad (4.2.25)$$

The YM field strength transforms as

$$F_{\mu\nu}(x) \to F^U_{\mu\nu}(x) = U(x)F_{\mu\nu}U^\dagger(x), \qquad (4.2.26)$$

and infinitesimally we have

$$\delta F_{\mu\nu} = i[\phi, F_{\mu\nu}]. \qquad (4.2.27)$$

The Lagrangian for pure YM theory is given by

$$\mathcal{L}_{\mathrm{YM}} = -\frac{1}{2g_0^2}\mathrm{Tr}\left(F_{\mu\nu}F^{\mu\nu}\right) = -\frac{1}{4g_0^2}F_{\mu\nu}^a F^{\mu\nu a}, \tag{4.2.28}$$

where g_0 is the bare coupling constant. It is invariant under gauge transformations, due to (4.2.23) and the cyclic property of the trace. The equation of motion derived from this Lagrangian is

$$D^\mu F_{\mu\nu} = 0. \tag{4.2.29}$$

Sometimes it is more convenient to use rescaled fields, in such a way that the coupling constant appears only in the interaction vertices. In this convention, one redefines

$$\hat{A}_\mu = \frac{1}{g_0}A_\mu, \tag{4.2.30}$$

and the field strength is defined by

$$\hat{F}_{\mu\nu}^a = \partial_\mu \hat{A}_\nu^a - \partial_\nu \hat{A}_\mu^a + g_0 f^{abc}\hat{A}_\mu^b \hat{A}_\nu^c, \tag{4.2.31}$$

so that

$$\mathcal{L}_{\mathrm{YM}} = -\frac{1}{4}\hat{F}_{\mu\nu}^a \hat{F}^{\mu\nu a}. \tag{4.2.32}$$

So far we have presented our results in Minkowski space. In the study of instantons it will be crucial to consider the Euclidean version of the theory. As usual, we perform a Wick rotation by redefining the temporal coordinate,

$$x_0 = -ix_4. \tag{4.2.33}$$

A vector field in Euclidean space will be defined as

$$A_i^{\mathrm{E}} = -A^i, \quad i = 1, 2, 3, \qquad A_4^{\mathrm{E}} = -iA_0, \tag{4.2.34}$$

and the field strength is then

$$F_{ij}^{a\mathrm{E}} = F_{ij}^a, \qquad F_{0j}^{a\mathrm{E}} = -iF_{4j}^a. \tag{4.2.35}$$

With these conventions, the Euclidean Lagrangian is

$$\mathcal{L}_{\mathrm{YM}}^{\mathrm{E}} = \frac{1}{4g_0^2}\left(F_{\mu\nu}^{a\mathrm{E}}\right)^2, \tag{4.2.36}$$

and the weight of a field configuration in the Euclidean path integral is $e^{-S_{\mathrm{YM}}^{\mathrm{E}}}$, where

$$S_{\mathrm{YM}}^{\mathrm{E}} = \int \mathrm{d}^4 x\, \mathcal{L}_{\mathrm{YM}}^{\mathrm{E}} \tag{4.2.37}$$

is the Euclidean action.

At the quantum level, Yang–Mills theory is renormalizable, and it famously exhibits a *running coupling constant* and *asymptotic freedom*. The relation between the bare coupling constant g_0^2 and the renormalized coupling constant g^2, in dimensional regularization and in the $\overline{\text{MS}}$ scheme, is given by

$$g_0^2 = \mu^{2\epsilon} \left\{ g^2 + \sum_{k=1}^{\infty} a_k(g^2) \epsilon^{-k} \right\}, \qquad (4.2.38)$$

where the dimension is $d = 4 - 2\epsilon$ and μ is the renormalization mass. Up to order g^4, we have

$$a_1(g^2) = -\frac{g^4}{(4\pi)^2} \frac{11 C_2(G)}{3}, \qquad a_k = 0, \quad k \geq 2. \qquad (4.2.39)$$

This can be obtained by a one-loop computation.

The behavior of the coupling constant as we change the renormalization scale μ is governed by the beta function, which is defined by

$$\beta(g) = \mu \frac{\partial g}{\partial \mu} = -\sum_{n=0}^{\infty} \beta_n g^{2n+3} = -\beta_0 g^3 - \beta_1 g^5 + \cdots. \qquad (4.2.40)$$

The coefficients β_0, β_1 are independent of the regularization scheme. The one-loop coefficient β_0 at $\epsilon \to 0$ can be obtained from (4.2.38) and (4.2.39) and reads

$$\beta_0 = \frac{1}{(4\pi)^2} \frac{11 C_2(G)}{3}. \qquad (4.2.41)$$

The coefficient β_1 is

$$\beta_1 = \frac{34}{3} C_2^2(G). \qquad (4.2.42)$$

Since the first coefficient of the beta function is negative, the theory is asymptotically free. It follows from the running of the coupling constant that the quantity

$$\Lambda = \mu \left(\beta_0 g^2 \right)^{-\beta_1/(2\beta_0^2)} e^{-1/(2\beta_0 g^2)} \exp\left(-\int_0^g \left\{ \frac{1}{\beta(x)} + \frac{1}{\beta_0 x^3} - \frac{\beta_1}{\beta_0^2 x} \right\} dx \right) \qquad (4.2.43)$$

is in fact independent of μ and therefore defines a scale which is invariant under the renormalization group (RG). We will often use the one-loop approximation

$$\Lambda \approx \mu e^{-1/(2\beta_0 g^2)}. \qquad (4.2.44)$$

Λ is the so-called *dynamically generated scale* of YM theory, and it depends on the regularization scheme used to compute the beta function. The fact that a theory with a dimensionless coupling constant g generates a dimensionful scale Λ is called *dimensional transmutation*.

4.3 Topological charge and θ vacua

In YM theory, besides the standard YM action, there is another term that can be added to the action. This term is called the *topological charge* for reasons that will become clear in a moment, and it is given by

$$Q = \int q(x) \mathrm{d}^4 x, \tag{4.3.1}$$

where

$$q(x) = \frac{1}{32\pi^2} F^a_{\mu\nu} \widetilde{F}^{\mu\nu a} = \frac{1}{64\pi^2} \epsilon_{\mu\nu\rho\sigma} F^{\mu\nu a} F^{\rho\sigma a}, \tag{4.3.2}$$

and the dual field strength $\widetilde{F}^{\mu\nu a}$ is defined by

$$\widetilde{F}^a_{\mu\nu} = \frac{1}{2} \epsilon_{\mu\nu\rho\sigma} F^{\rho\sigma a}. \tag{4.3.3}$$

In (4.3.2) and (4.3.3), $\epsilon_{\mu\nu\rho\sigma}$ is the totally antisymmetric Levi-Civita symbol, with the convention $\epsilon_{0123} = 1$. It can be written, in terms of differential forms, as

$$q(x) \mathrm{d}^4 x = \frac{1}{8\pi^2} \mathrm{Tr}\,(F \wedge F). \tag{4.3.4}$$

The topological charge is a Lorentz invariant, gauge invariant, marginal operator of the theory, so it is natural to add it to the action and to take as our Lagrangian

$$\mathcal{L}_{\mathrm{YM},\theta} = \mathcal{L}_{\mathrm{YM}} - \theta q(x), \tag{4.3.5}$$

where θ is a new dimensionless parameter called the *theta angle*. Note that the operator $q(x)$ changes sign under a parity transformation

$$A_0(\mathbf{x}, t) \rightarrow A_0(-\mathbf{x}, t), \qquad A_i(\mathbf{x}, t) \rightarrow -A_i(-\mathbf{x}, t), \quad i = 1, 2, 3. \tag{4.3.6}$$

It is straightforward to generalize the topological charge to Euclidean signature. The definition of the dual field strength is the same as in (4.3.2), but now the Levi-Civita symbol satisfies $\epsilon_{1234} = 1$. Since $F^a_{0j} = iF^{a\mathrm{E}}_{4j}$, we get a $-i$ in the topological density, i.e.

$$q(x) = -iq(x)_\mathrm{E}, \tag{4.3.7}$$

where

$$q(x)_\mathrm{E} = \frac{1}{64\pi^2} \epsilon_{\mu\nu\rho\sigma} F^{\mathrm{E}\mu\nu a} F^{\mathrm{E}\rho\sigma a} \tag{4.3.8}$$

is the topological density in Euclidean signature. Since the Wick rotation also introduces an extra i, the Euclidean theory is defined by the Lagrangian,

$$\mathcal{L}^\mathrm{E}_{\mathrm{YM},\theta} = \mathcal{L}^\mathrm{E}_{\mathrm{YM}} - i\theta q(x)_\mathrm{E}. \tag{4.3.9}$$

The different observables of YM theory should in principle be sensitive to the theta angle. One such observable is the ground state energy density $E(\theta)$, or vacuum energy density. To compute this energy density, we proceed as in QM: we consider the Euclidean version of the theory, and we put it in a large, finite spacetime volume V. The partition function in the presence of a theta angle θ is defined as

$$Z(\theta) = \int [\mathcal{D}A] e^{-\int d^4 x \, \mathcal{L}^E_{YM,\theta}}. \tag{4.3.10}$$

One then has,

$$E(\theta) = -\lim_{V \to \infty} \frac{1}{V} \log Z(\theta). \tag{4.3.11}$$

This is just the analogue of (1.2.5) in field theory. Let us now expand the right hand side of (4.3.10) around $\theta = 0$. We have,

$$\frac{Z(\theta)}{Z(0)} = \sum_{n \geq 0} \frac{(i\theta)^n}{n!} \langle Q^n \rangle, \tag{4.3.12}$$

where

$$\langle X \rangle = \frac{1}{Z(0)} \int [\mathcal{D}A] e^{-\int d^4 x \, \mathcal{L}^E_{YM}} X \tag{4.3.13}$$

is the normalized average in the Euclidean path integral. In the large volume limit, the odd powers of θ in this expansion vanish. The reason is that, in order to compute their coefficients, we have to compute the average of operators involving odd powers of $q_E(x)$ in YM theory with $\theta = 0$. These operators are odd under parity reversal, and since in the path integral we integrate over gauge connections and their parity transformations, their average vanishes. In particular, $E(\theta)$ is a real quantity. Another important property of the vacuum energy density $E(\theta)$ is that it has an absolute minimum at $\theta = 0$, i.e.

$$E(0) \leq E(\theta), \qquad \theta \neq 0. \tag{4.3.14}$$

This is because, when $\theta \neq 0$, we are inserting in the Euclidean path integral (4.3.10) an oscillating function $e^{i\theta Q}$, and this leads to a smaller value for $Z(\theta)$ compared to the one obtained with no insertion, at $\theta = 0$. When expanded around this minimum, the energy density $E(\theta)$ leads to the following power series,

$$E(\theta) - E(0) = \frac{1}{2} \chi_t \theta^2 s(\theta), \tag{4.3.15}$$

where we have defined,

$$s(\theta) = 1 + \sum_{n=1}^{\infty} b_{2n} \theta^{2n}. \tag{4.3.16}$$

The coefficient χ_t is an important quantity and measures the leading dependence of $E(\theta)$ on the θ angle around $\theta = 0$, since

$$\chi_t = \left(\frac{d^2 E(\theta)}{d\theta^2} \right)_{\theta=0}. \tag{4.3.17}$$

This is a positive quantity, since $\theta = 0$ is a minimum, and it is called the *topological susceptibility*. It can be regarded as the infinite volume limit of the topological susceptibility at finite volume χ_t^V,

$$\chi_t = \lim_{V \to \infty} \chi_t^V, \tag{4.3.18}$$

where

$$\chi_t^V = \frac{\langle Q^2 \rangle}{V} = \frac{1}{V} \int_V d^4x \int_V d^4y \, \langle q_E(x) q_E(y) \rangle. \tag{4.3.19}$$

We can use translation invariance in the infinite volume limit to write the topological susceptibility as the integrated two-point function of the topological density,

$$\chi_t = \int \langle q_E(x) q_E(0) \rangle d^4x. \tag{4.3.20}$$

The Minkowksi signature version of the topological susceptibility can be obtained from (4.3.20) and (4.3.7),

$$\chi_t = -i \int \langle T q(x) q(0) \rangle d^4x, \tag{4.3.21}$$

where T denotes as usual time ordering. These expressions for the topological susceptibility, in terms of integrals of the two-point function of the topological density, should be taken with care. The reason is that $\langle q_E(x) q_E(0) \rangle$ diverges when $x \to 0$ as $|x|^{-8}$ (up to logarithmic terms due to the running coupling constant), and the integral in (4.3.20) is singular unless one specifies a prescription to integrate the singularity. In simplified models (like the large N $\mathbb{CP}^{N-1}$ model, which we will study in Section 6.3), it is possible to make sense of the integral above, but in general this is a difficult problem. In practice, one uses definitions of the topological susceptibility (both in the lattice and in the continuum) which are equivalent to (4.3.17) but can be computed unambiguously. We will not need these definitions here, and we refer the reader to the references at the end of this chapter for more details on this issue.

The function $E(\theta)$ has two important properties. As we will show in a moment, smooth field configurations with a finite action have quantized values of Q. Thus we expect $E(\theta)$ to be *periodic*, with period 2π:

$$E(\theta + 2\pi) = E(\theta). \tag{4.3.22}$$

This is the second property of $E(\theta)$. In the limit of infinite volume, smooth configurations of finite action are just a zero-measure set in the domain of integration of the path integral. Therefore, we might think that the value of $E(\theta)$ would be dominated by field configurations in which Q is not an integer. However, using a fully non-perturbative definition on the lattice, one can see that Q takes integer values for any discretized lattice configuration, and we have periodicity in θ in the continuum limit.

In the remainder of this section, we will always consider theories defined in Euclidean signature, and we will remove the subscripts and superscripts E for simplicity of notation. The rotation to Minkowski signature is obvious from the rules we have given before.

Although we have said that observables in YM theory should be sensitive to the θ parameter, this dependence is very subtle. The reason is that (4.3.2) is a total divergence,

$$q(x) = \partial_\mu K^\mu, \tag{4.3.23}$$

where

$$K_\mu = \frac{1}{16\pi^2} \epsilon^{\mu\nu\rho\sigma} A_\nu^a \left(\partial_\rho A_\sigma^a + \frac{1}{3} f_{abc} A_\nu^a A_\rho^b A_\sigma^c \right). \tag{4.3.24}$$

To see this, we write

$$
\begin{aligned}
q(x) &= \frac{1}{32\pi^2} \epsilon_{\mu\nu\rho\sigma} \mathrm{Tr}\left(F^{\mu\nu} F^{\rho\sigma} \right) \\
&= \frac{1}{8\pi^2} \epsilon_{\mu\nu\rho\sigma} \mathrm{Tr}\left[(\partial^\mu A^\nu - iA^\mu A^\nu)(\partial^\rho A^\sigma - iA^\rho A^\sigma) \right],
\end{aligned} \tag{4.3.25}
$$

where we used the fact that the totally antisymmetric symbol $\epsilon_{\mu\nu\rho\sigma}$ antisymmetrizes the indices μ, ν and ρ, σ. Since the trace has cyclic symmetry, we also have that

$$\epsilon_{\mu\nu\rho\sigma} \mathrm{Tr}\left(A^\mu A^\nu A^\rho A^\sigma \right) = 0, \tag{4.3.26}$$

therefore

$$q(x) = \frac{1}{8\pi^2} \epsilon_{\mu\nu\rho\sigma} \mathrm{Tr}\left(\partial^\mu A^\nu \partial^\rho A^\sigma - 2i\partial^\mu A^\nu A^\rho A^\sigma \right). \tag{4.3.27}$$

Next, we write

$$\epsilon_{\mu\nu\rho\sigma} \mathrm{Tr}\left(\partial^\mu A^\nu \partial^\rho A^\sigma \right) = \frac{1}{3} \partial_\rho \epsilon^{\mu\nu\rho\sigma} \mathrm{Tr}\left(A_\mu A_\nu A_\sigma \right). \tag{4.3.28}$$

We conclude that

$$q(x) = \frac{1}{8\pi^2} \partial_\mu \left\{ \epsilon^{\mu\nu\rho\sigma} \mathrm{Tr}\left(A_\nu \partial_\rho A_\sigma - \frac{2i}{3} A_\nu A_\rho A_\sigma \right) \right\}, \tag{4.3.29}$$

which, when written in components, leads to (4.3.23). The tensor (4.3.24) suggests introducing the three-form

$$\omega_{CS}(A) = \frac{1}{16\pi^2} A^a_\nu \left(\partial_\rho A^a_\sigma + \frac{1}{3} f_{abc} A^a_\nu A^b_\rho A^c_\sigma \right) dx^\nu \wedge dx^\rho \wedge dx^\sigma$$

$$= \frac{1}{8\pi^2} \text{Tr} \left(A \wedge dA - \frac{2i}{3} A \wedge A \wedge A \right), \qquad (4.3.30)$$

which is called the *Chern–Simons form*. The equality (4.3.29) can then be written, in terms of differential forms, as

$$\frac{1}{8\pi^2} \text{Tr}(F \wedge F) = d\omega_{CS}(A). \qquad (4.3.31)$$

One consequence of (4.3.23) is that the Fourier transform

$$\tilde{q}(p) = \int e^{ipx} q(x) d^4 x \qquad (4.3.32)$$

vanishes at zero momentum, since it is of the form $p^\mu \tilde{K}_\mu(p)$. On the other hand, the topological susceptibility can be written as

$$\chi_t = \lim_{k \to 0} U(k), \qquad (4.3.33)$$

where

$$U(k) = \int e^{ikx} \langle q(x)q(0) \rangle d^4 x = \int \frac{d^4 p}{(2\pi)^4} \langle \tilde{q}(k)\tilde{q}(p) \rangle. \qquad (4.3.34)$$

Since $\tilde{q}(0) = 0$, (4.3.33) vanishes order by order in perturbation theory. However, and as we will see in explicit computations in simplified models, it does not have to vanish in the full theory. Therefore, the topological susceptibility is a quantity which, like the imaginary part of the energy in an unstable quantum mechanical potential, vanishes in perturbation theory but it is non-zero non-perturbatively.

We now explain why any Euclidean Yang–Mills configuration of finite action on $\mathbb{R}^4$ leads to a quantized value of the topological charge. The condition of finite action gives constraints on the large distance behavior of the fields. In order to see how they must behave as $r \to \infty$, we notice that, schematically, the Euclidean action can be written as

$$S^E_{YM} \sim \int dr \, r^3 F^2. \qquad (4.3.35)$$

If we want this to be finite, the field strength $F_{\mu\nu}$ has to go faster than $1/r^2$ as $r \to \infty$. This means that $A(r)$ should decay at infinity faster than $1/r$. However, since a gauge connection which is a pure gauge transformation leads to a vanishing field strength, we can have the more general behavior

$$A_\mu \approx iU \partial_\mu U^\dagger + \mathcal{O}\left(r^{-2}\right), \qquad r \to \infty. \qquad (4.3.36)$$

The boundary at infinity $r \to \infty$ is a three-sphere $\mathbb{S}^3 \subset \mathbb{R}^4$, therefore any gauge connection with the behavior (4.3.36) defines a map from $\mathbb{S}^3$ to the gauge group, i.e.

$$U : \mathbb{S}^3 \to G. \tag{4.3.37}$$

This choice is clearly not gauge invariant. Let us suppose that we perform a gauge transformation $A_\mu \to A_\mu^V$, where V has a well-defined limit on the boundary at infinity, V_∞, which defines a map from $\mathbb{S}^3$ to G. Then, the behavior of A_μ^V at infinity is given by

$$A_\mu^V \approx \mathrm{i}(V_\infty U)\partial_\mu (V_\infty U)^\dagger + \mathcal{O}\left(r^{-2}\right), \qquad r \to \infty, \tag{4.3.38}$$

and defines a map

$$V_\infty U : \mathbb{S}^3 \to G. \tag{4.3.39}$$

However, this map is a smooth deformation of U. The reason for this is that the map V_∞ is induced by the smooth map V from $\mathbb{R}^4$ to G. By varying r, we find a smooth deformation from the constant map $V(0)$ to the map V_∞. The constant $V(0) \in G$ can itself be regarded as a smooth deformation of the identity, if we assume that G is connected, as we are doing. We conclude that $V_\infty U$ can be obtained from U by a smooth deformation. Therefore, when we look for configurations of finite action, we should look for maps of the form (4.3.37) up to a smooth deformation, due to gauge equivalence.

We would now like to classify these maps, up to smooth deformations. Fortunately, this is a well-studied problem in mathematics. We will first consider a slightly more general class of maps, of the form,

$$U : \mathbb{S}^d \to M, \tag{4.3.40}$$

where $\mathbb{S}^d$ is the d-dimensional sphere and M is a connected manifold. Two maps U_1, U_2 of the form (4.3.40) are said to be *homotopically equivalent* if they can be continuously deformed one into another, i.e. if there exists a continuous map

$$\Phi : \mathbb{S}^d \times [0, 1] \to M \tag{4.3.41}$$

such that

$$\Phi(\theta^i, 0) = U_1(\theta^i), \qquad \Phi(\theta^i, 1) = U_2(\theta^i). \tag{4.3.42}$$

Here, θ^i, $i = 1, \ldots, d$ are coordinates for $\mathbb{S}^d$. It is easy to see that this defines an equivalence relation in the space of maps, therefore we can divide this space into equivalence classes. The set of equivalence classes of maps is called the dth *homotopy group* $\pi_d(M)$, and the equivalence class of U will be denoted by $[U]$.

This set indeed carries a group structure. To define it, it is convenient to represent the sphere $\mathbb{S}^d$ by the topologically equivalent manifold

$$[0, 1]^d / \sim, \tag{4.3.43}$$

where the equivalence relation $\sim$ identifies all the points in the boundary of the d-dimensional hypercube $[0, 1]^d$. Now, let U_1, U_2 be two different maps from $\mathbb{S}^d$ (realized as (4.3.43)) to M, and let us consider the map

$$(U_1 \star U_2)\,(t_1, \ldots, t_d) = \begin{cases} U_1(2t_1, t_2, \ldots, t_d), & t_1 \in [0, 1/2], \\ U_2(2t_1 - 1, t_2, \ldots, t_d), & t_1 \in [1/2, 1]. \end{cases} \tag{4.3.44}$$

Then, one can define the product of equivalence classes as

$$[U_1] \star [U_2] = [U_1 \star U_2]. \tag{4.3.45}$$

It can be verified that it is well defined, i.e. it does not depend on the representatives. Given an element U, we also define

$$U^{-1}(t_1, t_2, \ldots, t_d) = U(1 - t_1, t_2, \ldots, t_d). \tag{4.3.46}$$

Finally, the unit element $[1]$ is defined as the equivalence class that contains the constant map $U(t_1, \ldots, t_d) = m_0 \in M$. One can show that the product (4.3.44) of homotopy classes defines a group, where $[1]$ is the unit element and the inverse of $[U]$ is given by $[U]^{-1} = [U^{-1}]$.

Let us now come back to our original problem of classifying the configurations of finite action of YM theory. As we have seen, this involves classifying maps from $\mathbb{S}^3$ to the gauge group G, up to smooth deformations. Mathematically, this is equivalent to computing the homotopy group $\pi_3(G)$. One can show that

$$\pi_3\,(G) = \mathbb{Z}, \tag{4.3.47}$$

for any compact, connected, simple Lie group, with the group structure given by the addition. Therefore, all configurations of finite action are classified by an integer number. In the case of $SU(2)$, this can be understood intuitively by using the fact that $SU(2)$ is, as a manifold, $\mathbb{S}^3$ (see for example (A.1)). We then have to classify the maps from $\mathbb{S}^3$ to $\mathbb{S}^3$, up to smooth deformations. These maps are classified by an integer ν which encodes the number of times that the $\mathbb{S}^3$ in the domain "wraps" the target $\mathbb{S}^3$. It is similar to the winding number which we found in the study of instantons in the periodic potential, which classifies maps (1.10.14) from $\mathbb{S}^1$ to $\mathbb{S}^1$, and we will also call it the winding number.

One interesting property of the winding number classifying maps from $\mathbb{S}^3$ to $SU(2)$ is that it can be written as the integral of a local functional of the map. To do this in a more general setting, let M_n be an odd-dimensional, compact manifold, i.e. $n = 2k + 1$, and let

$$U : M_n \rightarrow G. \tag{4.3.48}$$

Let $\theta^1, \ldots, \theta^n$ denote local coordinates on M_n. Then, the Cartan–Maurer invariant is defined as

$$I_{\text{CM}}[U] = \int_{M_n} d\theta^1 \cdots d\theta^u \, \epsilon^{i_1 \cdots i_u} \text{Tr} \left\{ U^{-1} \frac{\partial U}{\partial \theta^{i_1}} \cdots U^{-1} \frac{\partial U}{\partial \theta^{i_u}} \right\}. \tag{4.3.49}$$

This can be written in a more compact and convenient way by using the language of differential forms. We have,

$$dU = \frac{\partial U}{\partial \theta^i} d\theta^i, \tag{4.3.50}$$

which is a one-form on M_d taking values in G. Then, we have

$$I_{\text{CM}}[U] = \int_{M_n} \text{Tr} \left(U^{-1} dU \wedge U^{-1} dU \wedge \cdots \wedge U^{-1} dU \right), \tag{4.3.51}$$

where $U^{-1}dU$ appears n times. This is an integral over M_n of a differential form of degree n. The most important property of $I_{\text{CM}}[U]$ is that it is invariant under infinitesimal deformations of U, i.e.

$$I_{\text{CM}}[U + \delta U] = I_{\text{CM}}[U]. \tag{4.3.52}$$

To see this, notice that

$$\delta \left(U^{-1} dU \right) = -U^{-1} \delta U U^{-1} dU + U^{-1} d(\delta U) = U^{-1} d \left(\delta U U^{-1} \right) U. \tag{4.3.53}$$

Therefore,

$$\delta I_{\text{CM}}[U] = n \int_{M_n} \text{Tr} \left(U^{-1} dU \wedge U^{-1} dU \wedge \cdots \wedge U^{-1} d \left(\delta U U^{-1} \right) U \right)$$

$$= n \int_{M_n} \text{Tr} \left(dU \wedge U^{-1} \wedge \cdots \wedge dU \wedge U^{-1} \wedge d \left(\delta U U^{-1} \right) \right), \tag{4.3.54}$$

after using cyclicity of the trace. It is easy to see, by using the nilpotency of the exterior differential $d^2 = 0$, that the integrand in the second line is an exact differential

$$d \left(dU \wedge U^{-1} \wedge \cdots \wedge dU \wedge U^{-1} \wedge \delta U U^{-1} \right). \tag{4.3.55}$$

Note that when acting with d on the product of $d - 1 \, dU \wedge U$, we get an even number of terms with the same form but with alternating signs, which then add up to zero, so that only the action of d on $\delta U U^{-1}$ survives, giving back the integrand in the last line of (4.3.54). The integral of a total differential on a compact manifold vanishes by Stokes' theorem (i.e. by integration by parts). Therefore (4.3.54) is zero.

Example 4.1 As a simple example of the Maurer–Cartan invariant, let us consider the case $n = 1$, $M_1 = \mathbb{S}^1$, $G = U(1)$, and

$$U(\theta) = e^{i\ell\theta}. \tag{4.3.56}$$

We then have

$$I_{\text{CM}}[U] = \int_{\mathbb{S}^1} U^{-1}dU = \int_0^{2\pi} i\ell d\theta = 2\pi\ell i, \tag{4.3.57}$$

which is, up to a factor $2\pi i$, the winding number of the map (1.10.15) which we considered in our analysis of instantons in periodic potentials. □

Since $I_{\text{CM}}[U]$ does not change under small deformations of U, two maps U_1, U_2 which can be deformed continuously into each other will have the same value of the Cartan–Maurer invariant. If $M_n = \mathbb{S}^n$ is an odd-dimensional sphere, the invariant will be the same for all maps in the same homotopy class, and it can be used to characterize these classes in a simple way. In addition, the Cartan–Maurer invariant respects the product structure in the homotopy group, in the sense that it maps the product of two classes into the sum of their invariants. This is easily seen as follows: let U_1, U_2 be two maps. The Cartan–Maurer invariant of their product $I_{\text{CM}}[U_1 \star U_2]$ can be computed as follows. First, we do the integral over $0 \leq t_1 \leq 1/2$, and then we add the integral over $1/2 \leq t_1 \leq 1$. We find in this way,

$$I_{\text{CM}}[U_1 \star U_2] = I_{\text{CM}}[U_1] + I_{\text{CM}}[U_2]. \tag{4.3.58}$$

Similarly, if we take the constant map we obviously have $I_{\text{CM}}[1] = 0$. If ν is an integer, we can consider the homotopy class

$$[U^\nu] = \begin{cases} [U] \star \cdots \star [U], & \nu > 0, \\ [U]^{-1} \star \cdots \star [U]^{-1}, & \nu < 0, \end{cases} \tag{4.3.59}$$

where the products involve $|\nu|$ elements. The general result (4.3.58) implies in particular that

$$I_{\text{CM}}[U^\nu] = \nu I_{\text{CM}}[U]. \tag{4.3.60}$$

In the case of the maps from $\mathbb{S}^3$ to $SU(2)$, homotopy classes are classified by a single integer number, the winding number, since $\pi_3(SU(2)) = \mathbb{Z}$. We will now show that this number is equal to the Cartan–Maurer invariant, after an appropriate normalization. We can parametrize $SU(2)$ explicitly with Euler angles, as in Appendix A. This parametrization can be regarded as a one-to-one map U from $\mathbb{S}^3$ to $SU(2)$, which is written down explicitly in (A.4). In this case, the one-form $U^{-1}dU$ appearing in the Cartan–Maurer invariant is called the Maurer–Cartan form of the group $SU(2)$, and it is given in (A.6). It is in turn related to the vierbeins defined by the standard metric G on $\mathbb{S}^3$, as shown in (A.14). We conclude that

$$U^{-1}\partial_\mu U = ie^a_\mu \sigma^a, \qquad (4.3.61)$$

therefore

$$\mathrm{Tr}\left(U^{-1}\mathrm{d}U \wedge U^{-1}\mathrm{d}U \wedge U^{-1}\mathrm{d}U\right) = 24\pi^2 \Omega, \qquad (4.3.62)$$

where

$$\Omega = \frac{1}{2\pi^2}\sqrt{G}\,\mathrm{d}t_1 \wedge \mathrm{d}t_2 \wedge \mathrm{d}t_3 \qquad (4.3.63)$$

is the normalized volume form on $SU(2) = \mathbb{S}^3$, which satisfies

$$\int_{SU(2)} \Omega = 1. \qquad (4.3.64)$$

Therefore, when we parametrize $SU(2)$ by Euler angles, the integrand of the Cartan–Maurer invariant is simply proportional to the volume form of $\mathbb{S}^3$, and we have

$$\frac{1}{24\pi^2} I_{\mathrm{CM}}[U] = 1. \qquad (4.3.65)$$

We conclude that the map U is a representative of the homotopy class of maps with winding number $\nu = 1$. We can produce representatives of maps with arbitrary winding number by taking powers of this map, as in (4.3.59). From (4.3.60) we deduce that

$$\frac{1}{24\pi^2} I_{\mathrm{CM}}[U^\nu] = \nu. \qquad (4.3.66)$$

We will now finally show that, in the case of $G = SU(2)$, the topological charge Q is proportional to the Cartan–Maurer invariant, and equal to the winding number. Let us consider a gauge connection one-form which behaves on the sphere $\mathbb{S}^3$ at infinity as (4.3.36), i.e.

$$A \approx -\mathrm{i}\mathrm{d}U \wedge U^{-1}, \qquad (4.3.67)$$

where U defines a map of winding number ν. First of all, we notice that, at infinity, the field strength vanishes. Therefore we can replace the antisymmetrized derivative $2\partial_{[\mu}A^a_{\nu]}$ by

$$-f_{abc}A^b_\mu A^c_\nu. \qquad (4.3.68)$$

This means that, at infinity, the one-form K_μ in (4.3.24) can be replaced by

$$-\frac{1}{96\pi^2}\epsilon^{\mu\nu\rho\sigma} f_{abc} A^a_\nu A^b_\rho A^c_\sigma. \qquad (4.3.69)$$

Since the integrand of Q, $q(x)$, is an exact differential, we can evaluate Q by using the Stokes theorem and taking as our boundary the three-sphere at infinity. We find,

$$Q = -\frac{1}{96\pi^2} f_{abc} \int_{\mathbb{S}^3} A_i^a A_j^b A_k^c d\theta^i \wedge d\theta^j \wedge d\theta^k, \tag{4.3.70}$$

where θ^i are coordinates on $\mathbb{S}^3$. On the other hand, due to (4.3.67) the Cartan–Maurer invariant can be evaluated on the $\mathbb{S}^3$ at infinity as

$$I_{CM}[U] = i \int_{\mathbb{S}^3} \text{Tr} (A \wedge A \wedge A). \tag{4.3.71}$$

By comparing (4.3.70) with (4.3.71), and taking into account the normalization (4.2.5), we find that the topological charge is given by

$$Q = \frac{1}{24\pi^2} I_{CM}[U], \tag{4.3.72}$$

i.e. it is given by the winding number of the field configuration. We conclude that YM configurations with finite action fall apart into different *topological sectors*, characterized by the integer value of the topological charge, which is equal to the winding number of the map associated to the behavior of the connection at infinity.

Although most of our arguments have been developed in detail for the gauge group $SU(2)$ only, they can be generalized to any compact, connected, simple Lie group. In particular, the topological charge is an integer number in one-to-one correspondence with the integer classifying homotopy classes in (4.3.47). This is a consequence of a theorem of Bott which asserts that any map from $\mathbb{S}^3$ to G can be smoothly deformed into a map to an $SU(2)$ subgroup of G.

Since the topological charge is quantized, the partition function $Z(\theta)$ defined in (4.3.10) is periodic, as anticipated above. This means that it can be decomposed in Fourier modes:

$$Z(\theta) = \sum_{\nu=-\infty}^{\infty} e^{i\nu\theta} Z_\nu. \tag{4.3.73}$$

This decomposition is similar to the one we performed in the analysis of a periodic potential in QM, in (1.10.7). As in that case, the coefficients Z_ν are path integrals in the Euclidean version of the theory, in this case in the Euclidean YM theory with Lagrangian $\mathcal{L}_{YM}^E$. The Z_ν are obtained by performing the path integral over gauge fields with a *fixed* topological charge $Q = \nu$:

$$Z_\nu = \int [\mathcal{D}A]_{Q=\nu} \, e^{-\int d^4x \, \mathcal{L}_{YM}^E}. \tag{4.3.74}$$

Note that we can regard the normalized coefficient of the Fourier expansion in (4.3.73),

$$P_\nu = \frac{Z_\nu}{Z(0)}, \tag{4.3.75}$$

as the probability of finding a gauge field with charge ν, i.e. as the probability distribution of the topological charge. By using (4.3.18), we also find that the topological susceptibility at finite volume is given by

$$\chi_t^V = \frac{1}{V} \sum_\nu \nu^2 P_\nu = \frac{\langle \nu^2 \rangle}{V}, \tag{4.3.76}$$

which can then be interpreted as the second moment of the probability distribution (up to the volume factor.) This also gives a probabilistic interpretation to the coefficients $b_{2n}^V(\theta)$ appearing in (4.3.16): since they are given by the connected vevs of the topological charge, they vanish if the probability distribution is a Gaussian, of the form

$$P_\nu = \frac{1}{\sqrt{2\pi \langle \nu^2 \rangle}} \exp\left(-\frac{\nu^2}{\langle \nu^2 \rangle} \right), \tag{4.3.77}$$

and therefore they parametrize deviations from a simple Gaussian behavior.

As in the case of the periodic potential in QM, the structure (4.3.73) of the Euclidean partition function is telling us something about the vacuum structure of the theory. First of all, it says that YM theory has an infinite set of approximate degenerate vacua, $|\nu\rangle$, which are labelled by an integer ν. The true vacuum of the theory is then obtained as in (1.10.22), by forming a superposition

$$|\theta\rangle = \sum_{\nu=-\infty}^{\infty} e^{i\nu\theta} |\nu\rangle, \tag{4.3.78}$$

which is determined by the value of the theta angle. This vacuum is called the *theta vacuum*. In the problem of the periodic potential, Euclidean configurations with winding number ℓ describe tunneling between two approximate vacua separated by a distance ℓT. Similarly, Euclidean configurations of YM theory with winding number ν should describe the tunneling between two vacua $|\nu_1\rangle$, $|\nu_2\rangle$ with $\nu_2 - \nu_1 = \nu$.

The approximate degenerate vacua of YM theory can also be found by performing a Hamiltonian analysis of the theory in Minkowski space. In this analysis, one considers the theory on $\mathbb{R} \times \mathbb{S}^3$, where $\mathbb{R}$ represents the time direction and $\mathbb{S}^3$ the compactified space directions of $\mathbb{R}^3$. By calculating the Hamiltonian explicitly, as the integral of a density on $\mathbb{S}^3$, one finds that, in the gauge $A_0 = 0$, the vacua are given by gauge connections which are pure gauge. Therefore, they are associated to maps of the form (4.3.37), and they are classified by $\pi_3(G) = \mathbb{Z}$. This leads to the vacua appearing in (4.3.78) and labelled by ν.

4.4 Instantons in Yang–Mills theory

As we have seen in the case of QM, instantons contain non-perturbative information, and they can be incorporated in a semiclassical approximation to the Euclidean path integral. We will now look for instantons with topological charge ν in Euclidean Yang–Mills theory, since these might be used in principle as starting points for a semiclassical evaluation of the path integrals (4.3.74). In particular they might be useful to understand the dependence on the theta angle. In order to find instantons, we will first show that, in each of the topological sectors, there is a configuration which minimizes the action, and therefore solves the EOM. This will make it possible to find instantons as solutions to first order differential equations.

Let us consider gauge configurations with a fixed topological charge $Q = \nu$. We start with the obvious identity

$$\int d^4x \, \mathrm{Tr} \left\{ (F \pm \tilde{F})_{\alpha\beta} (F \pm \tilde{F})^{\alpha\beta} \right\} \geq 0. \tag{4.4.1}$$

Since $\tilde{F}_{\alpha\beta} \tilde{F}^{\alpha\beta} = F_{\alpha\beta} F^{\alpha\beta}$, we find

$$\frac{1}{2g_0^2} \int d^4x \, \mathrm{Tr} \left(F_{\alpha\beta} F^{\alpha\beta} \right) \pm \frac{1}{2g_0^2} \int d^4x \, \mathrm{Tr} \left(F_{\alpha\beta} \tilde{F}^{\alpha\beta} \right) \geq 0, \tag{4.4.2}$$

or equivalently

$$S_{\mathrm{YM}}^{\mathrm{E}} \pm \frac{8\pi^2 \nu}{g_0^2} \geq 0. \tag{4.4.3}$$

We conclude that

$$S_{\mathrm{YM}}^{\mathrm{E}} \geq \frac{8\pi^2 |\nu|}{g_0^2}. \tag{4.4.4}$$

To see when the inequality is saturated, we first note that $S_{\mathrm{YM}}^{\mathrm{E}}$ is positive. Therefore, if $\nu > 0$ is positive, we must have the negative sign in (4.4.3), and

$$F_{\alpha\beta} = \tilde{F}_{\alpha\beta}, \qquad S_{\mathrm{YM}}^{\mathrm{E}} = \frac{8\pi^2 \nu}{g_0^2}, \tag{4.4.5}$$

i.e. the gauge field is *self-dual* (SD). This is called an *instanton* of YM theory. Similarly, if $\nu < 0$ is negative, we have

$$F_{\alpha\beta} = -\tilde{F}_{\alpha\beta}, \qquad S_{\mathrm{YM}}^{\mathrm{E}} = -\frac{8\pi^2 \nu}{g_0^2}, \tag{4.4.6}$$

i.e. the gauge field is *anti-self-dual* (ASD) and this is called an *anti-instanton*. If any of these conditions holds, the corresponding gauge field minimizes the action for a fixed topological class with topological charge ν, and in particular it has to solve the EOM. However, in contrast to the standard EOM (4.2.29) of YM theory, which is of second order in derivatives, the (anti)self-duality conditions are *first order* equations. It is useful to notice that, if we regard each component of the field strength $F^m_{\alpha\beta}$, $m = 1, \ldots, d(G)$, as a 4×4 antisymmetric matrix whose entries are labelled by the indices α, β, the self-duality condition means that this matrix must be of the form

$$\begin{pmatrix} 0 & a & b & c \\ -a & 0 & c & -b \\ -b & -c & 0 & a \\ -c & b & -a & 0 \end{pmatrix}, \tag{4.4.7}$$

and in particular it has only three independent entries. Similarly, the anti-self-duality condition means that the matrix is of the form

$$\begin{pmatrix} 0 & a & b & c \\ -a & 0 & -c & b \\ -b & c & 0 & -a \\ -c & -b & a & 0 \end{pmatrix}. \tag{4.4.8}$$

We will now solve the equations (4.4.5) explicitly when the gauge group is $SU(2)$ and the topological charge is $\nu = 1$ (the one-instanton solution). To do this, let us first be more explicit concerning the asymptotic form of the gauge connection, which we will write as

$$A_\mu = iU \partial_\mu U^\dagger. \tag{4.4.9}$$

The solution for U with $\nu = 1$, and restricted to the boundary $\mathbb{S}^3$ at infinity, is given explicitly in (A.4) in terms of Euler angles. We can extend it to $\mathbb{R}^4$ minus the origin by regarding these angles as angular coordinates on $\mathbb{R}^4$, and assuming that U is independent of the distance to the origin r. The resulting expression is given simply by

$$U = \frac{x_4 + i\mathbf{x} \cdot \boldsymbol{\sigma}}{r}, \tag{4.4.10}$$

where

$$\mathbf{x} = (x_1, x_2, x_3), \qquad \boldsymbol{\sigma} = (\sigma_1, \sigma_2, \sigma_3), \tag{4.4.11}$$

and σ_i, $i = 1, 2, 3$ are the Pauli matrices. Since

$$\partial_4 U = -\frac{x_4}{r^2} U + \frac{1}{r}, \qquad \partial_k U = -\frac{x_k}{r^2} U + \frac{i\sigma_k}{r}, \qquad k = 1, 2, 3, \tag{4.4.12}$$

we find

$$A_4 = -\frac{\mathbf{x} \cdot \boldsymbol{\sigma}}{r^2},$$

$$A_k = \frac{1}{r^2}(x_4 \sigma_k + \epsilon_{k\ell m} x_\ell \sigma_m), \quad k = 1, 2, 3, \tag{4.4.13}$$

where we used that

$$\sigma_j \mathbf{x} \cdot \boldsymbol{\sigma} = x_k + i\epsilon_{k\ell m} x_\ell \sigma_m, \tag{4.4.14}$$

and $\epsilon_{k\ell m}$ is the totally antisymmetric tensor with $\epsilon_{123} = 1$. Let us write the gauge connection in components,

$$A_\mu = \frac{1}{2}\sigma_a A_\mu^a. \tag{4.4.15}$$

We define the 't Hooft matrices or 't Hooft symbols $\eta_{\mu\nu}^a$, $a = 1, 2, 3$, as three antisymmetric matrices with entries,

$$\eta_{ij}^a = \epsilon_{aij}, \quad \eta_{i4}^a = \delta_{ai}, \tag{4.4.16}$$

where $i, j = 1, 2, 3$. Explicitly, we have

$$\eta_{\mu\nu}^1 = \begin{pmatrix} 0 & 0 & 0 & 1 \\ 0 & 0 & 1 & 0 \\ 0 & -1 & 0 & 0 \\ -1 & 0 & 0 & 0 \end{pmatrix}, \quad \eta_{\mu\nu}^2 = \begin{pmatrix} 0 & 0 & -1 & 0 \\ 0 & 0 & 0 & 1 \\ 1 & 0 & 0 & 0 \\ 0 & -1 & 0 & 0 \end{pmatrix},$$

$$\eta_{\mu\nu}^3 = \begin{pmatrix} 0 & 1 & 0 & 0 \\ -1 & 0 & 0 & 0 \\ 0 & 0 & 0 & 1 \\ 0 & 0 & -1 & 0 \end{pmatrix}. \tag{4.4.17}$$

Notice that these matrices are self-dual. They also satisfy the following useful properties,

$$\epsilon_{abc}\eta_{\mu\rho}^b \eta_{\nu\sigma}^c = \delta_{\mu\nu}\eta_{\rho\sigma}^a - \delta_{\mu\sigma}\eta_{\rho\nu}^a - \delta_{\rho\nu}\eta_{\mu\sigma}^a + \delta_{\rho\sigma}\eta_{\mu\nu}^a,$$

$$\epsilon_{\mu\nu\rho\sigma}\eta^{a\sigma\lambda} = -\delta_\mu^\lambda \eta_{\nu\rho}^a + \delta_\nu^\lambda \eta_{\mu\rho}^a - \delta_\rho^\lambda \eta_{\mu\nu}^a. \tag{4.4.18}$$

These identities are easily established by exploiting symmetry properties and using the explicit expressions provided above. In terms of these matrices, we can write

$$A_\mu^a = 2\eta_{\mu\nu}^a \frac{x^\nu}{r^2}. \tag{4.4.19}$$

This asymptotic form suggests the following ansatz for the connection

$$A_\mu^a = 2\eta_{\mu\nu}^a \frac{x^\nu}{r^2} f(r^2), \tag{4.4.20}$$

where

$$f(r^2) \to 1, \qquad r \to \infty. \tag{4.4.21}$$

Also, regularity at the origin requires that

$$f(r^2) \approx r^2, \qquad r \to 0. \tag{4.4.22}$$

We will now use this ansatz to solve the self-duality conditions (4.4.5). A direct computation gives

$$\partial_\mu A_\nu^a = -2 \left\{ \frac{f}{r^2} \eta_{\mu\nu}^a + \frac{2\eta_{\nu\rho}^a x_\mu x^\rho}{r^4} (f - r^2 f') \right\}, \tag{4.4.23}$$

where the $'$ in f means derivative with respect to r^2. Also, by using the first equation in (4.4.18), we find

$$\epsilon_{abc} A_\mu^b A_\nu^c = \frac{4f^2}{r^2} \left\{ \eta_{\mu\nu}^a + \frac{\eta_{\nu\rho}^a x_\mu x^\rho - \eta_{\mu\rho}^a x_\nu x^\rho}{r^2} \right\}. \tag{4.4.24}$$

It follows that

$$F_{\mu\nu}^a = -4 \left\{ \eta_{\mu\nu}^a \frac{f(1-f)}{r^2} + \frac{\eta_{\nu\rho}^a x_\mu x^\rho - \eta_{\mu\rho}^a x_\nu x^\rho}{r^4} (f(1-f) - r^2 f') \right\}. \tag{4.4.25}$$

The dual field strength can be computed by using the first property listed in (4.4.18), and reads

$$\widetilde{F}_{\mu\nu}^a = -4 \left\{ \eta_{\mu\nu}^a f' - \frac{\eta_{\nu\rho}^a x_\mu x^\rho - \eta_{\mu\rho}^a x_\nu x^\rho}{r^4} (f(1-f) - r^2 f') \right\}. \tag{4.4.26}$$

We conclude that the self-duality condition for the field strength is satisfied if $f(r^2)$ satisfies the first order differential equation

$$f(1-f) - r^2 \frac{df}{dr^2} = 0. \tag{4.4.27}$$

The non-trivial solution to this equation with the above boundary conditions is

$$f(r^2) = \frac{r^2}{r^2 + \rho^2}. \tag{4.4.28}$$

This gives the one-instanton solution of $SU(2)$ YM theory, which can be written as

$$A_\mu = \frac{r^2}{r^2 + \rho^2} i U \partial_\mu U^\dagger, \tag{4.4.29}$$

where U is given in (4.4.10).

The one-instanton solution that we have just constructed interpolates between the trivial vacuum $f = 0$ at the origin and the homotopically non-trivial gauge transformation with $\nu = 1$ as $r \to \infty$, since at large r it is indeed of the form

(4.3.36). In (4.4.28) ρ is an integration constant which can be regarded as the *size* of the instanton. There is an interesting contrast between the instantons of YM theory and the instantons of the scalar theory studied in Chapter 3. The size of a "bubble" in the scalar theory was fixed by the parameters of the potential, while the size of an instanton in YM theory is a free parameter. It is yet another example of a collective coordinate, and as it is always the case, its existence is due to a symmetry of the theory. In this case, the symmetry is the scale invariance of the classical YM action.

In writing the ansatz (4.4.20) and (4.4.28) we have already fixed some integration constants: the above configuration is centered at the origin, but by using translation invariance one can write down a more general solution,

$$A_\mu^a = 2\eta_{\mu\nu}^a \frac{(x - x_0)^\nu}{(x - x_0)^2 + \rho^2}, \qquad (4.4.30)$$

where x_0^μ can be regarded as the position of the center of the instanton. This gives four extra collective coordinates. The field strength of the instanton is given by

$$F_{\mu\nu}^a = -4\eta_{\mu\nu}^a \frac{\rho^2}{\left((x - x_0)^2 + \rho^2\right)^2}, \qquad (4.4.31)$$

which is manifestly self-dual due to the self-duality properties of the 't Hooft symbols $\eta_{\mu\nu}^a$.

In order to construct an anti-instanton, we have to solve instead the anti-self-duality condition (4.4.6). The solution with $\nu = -1$ is obtained by introducing the anti-self-dual analogue of the 't Hooft matrices, $\bar{\eta}_{\mu\nu}^a$. They are antisymmetric matrices defined by

$$\bar{\eta}_{ij}^a = \epsilon_{aij}, \qquad \bar{\eta}_{i4}^a = -\delta_{ai}, \qquad (4.4.32)$$

where $i, j = 1, 2, 3$. Their explicit expressions are

$$\bar{\eta}_{\mu\nu}^1 = \begin{pmatrix} 0 & 0 & 0 & -1 \\ 0 & 0 & 1 & 0 \\ 0 & -1 & 0 & 0 \\ 1 & 0 & 0 & 0 \end{pmatrix}, \quad \bar{\eta}_{\mu\nu}^2 = \begin{pmatrix} 0 & 0 & -1 & 0 \\ 0 & 0 & 0 & -1 \\ 1 & 0 & 0 & 0 \\ 0 & 1 & 0 & 0 \end{pmatrix},$$

$$\bar{\eta}_{\mu\nu}^3 = \begin{pmatrix} 0 & 1 & 0 & 0 \\ -1 & 0 & 0 & 0 \\ 0 & 0 & 0 & -1 \\ 0 & 0 & 1 & 0 \end{pmatrix}, \qquad (4.4.33)$$

and they are anti-self-dual. They satisfy the properties (4.4.18) after changing $\epsilon_{\mu\nu\rho\sigma} \to -\epsilon_{\mu\nu\rho\sigma}$. The anti-instanton is then obtained by replacing $\eta_{\mu\nu}^a$ by $\bar{\eta}_{\mu\nu}^a$ in (4.4.30), (4.4.31), i.e. it is given by

$$B_\mu = \frac{1}{2}\sigma_a B_\mu^a, \qquad B_\mu^a = 2\bar{\eta}_{\mu\nu}^a \frac{x^\nu}{r^2 + \rho^2}, \tag{4.4.34}$$

where we have set its center to be at the origin. It can also be written as

$$B_\mu = \frac{r^2}{r^2 + \rho^2} i U^\dagger \partial_\mu U. \tag{4.4.35}$$

One can now try to construct explicit instanton solutions with higher topological charge, or multi-instantons. It is relatively easy to generalize the construction above and write down such a solution, but the most general ν-instanton solution on Euclidean $\mathbb{R}^4$ requires the so-called Atiyah–Drinfeld–Hitchin–Manin (ADHM) formalism, and we refer the interested reader to the references at the end of this chapter for further details on this.

In order to construct instantons in gauge groups of higher rank, one can embed the solution found above inside the larger group. For example, if we denote by $A_\mu^{SU(2)}$ the 2×2 Hermitian matrix representing the $SU(2)$ instanton with topological charge $\nu = 1$, we can obtain an $SU(N)$ instanton by considering the field

$$A_\mu^{SU(N)} = \begin{pmatrix} 0_{N-2} & 0_2 \\ 0_2 & A_\mu^{SU(2)} \end{pmatrix}. \tag{4.4.36}$$

Here, 0_ℓ is the $\ell \times \ell$ matrix which has vanishing entries. This is an $SU(N)$ field which solves the self-duality equation and has topological charge $\nu = 1$ as well. Other, more complicated embeddings give instantons with higher topological charge. The general instanton solution in $SU(N)$ with topological charge $|\nu| > 1$ can also be constructed by using the ADHM formalism.

4.5 Instanton calculus

In QM and in the theory of scalar fields, instanton calculus can be developed by simply expanding the Euclidean path integral around a solution to the Euclidean EOM. However, in theories which are classically scale invariant, like YM theory, one often needs an infrared (IR) cutoff to make sense of instanton calculus. This is because the size of an instanton ρ is a collective coordinate, and the integration over instanton sizes in the path integral leads to an IR divergence, as we will see in this section. There are many ways to introduce such a cutoff. The most natural one is to consider the theory in a spacetime with a finite, small volume V. In this case, since the size of the instanton cannot be greater than the characteristic scale of spacetime $V^{1/4}$, the integral over the instanton size has a natural cutoff. Another possibility is to consider the theory at finite temperature. Yet another situation in which there is a natural cutoff is to have a Higgs field with a large vev which sets the scale. In all these cases, instanton calculus is well defined, and one can obtain semiclassical expressions for the different quantities of interest in a QFT as a sum over instanton sectors.

In this section we will consider Euclidean YM theory at finite volume. A convenient and precise way to put the theory in a finite volume is to define it on a compact, Riemannian manifold M, of volume V, since in this case we do not have to worry about boundary effects. The case of Euclidean $\mathbb{R}^4$ can be regarded as a limiting case of this situation: if M is the four-sphere $\mathbb{S}^4$ of radius R, endowed with the standard metric, the Euclidean space $\mathbb{R}^4$ is recovered in the limit of infinite radius $R \to \infty$.

To consider Euclidean YM theory on M, we have to generalize some of the ingredients which go into the construction of the theory. The basic field is the gauge connection A_μ. When working on a general manifold, a connection can be defined in a general way by introducing a principal bundle, but we will use a more down-to-earth approach and consider a covering of M by patches or coordinate charts $\{\mathcal{U}_\alpha\}_{\alpha \in \mathcal{A}}$. As usual in the theory of differentiable manifolds, we will assume that each of these charts is diffeomorphic to an open set of $\tilde{\mathcal{U}}_\alpha$ of $\mathbb{R}^4$, i.e. that there exists a diffeomorphism

$$\phi_\alpha : \mathcal{U}_\alpha \subset M \to \tilde{\mathcal{U}}_\alpha \subset \mathbb{R}^4. \tag{4.5.1}$$

These diffeomorphisms endow $\mathcal{U}_\alpha$ with local coordinates $x_\mu^{(\alpha)}$. To give a connection on M, we have to give a family of one-forms $A^{(\alpha)}$ on each of the $\mathcal{U}_\alpha$. For this to make sense globally, we have to make sure that, in the intersection of two patches, say $\mathcal{U}_\alpha, \mathcal{U}_\beta$, the corresponding one-forms agree. This does not mean that they are equal, since they could differ by a gauge transformation. Therefore, we require that the corresponding one-forms $A^{(\alpha)}$, $A^{(\beta)}$ are related as

$$A^{(\alpha)} = U_{\alpha\beta} A^{(\beta)} U_{\alpha\beta}^\dagger + i U_{\alpha\beta} \mathrm{d} U_{\alpha\beta}^\dagger, \tag{4.5.2}$$

where

$$U_{\alpha\beta} : \mathcal{U}_\alpha \cap \mathcal{U}_\beta \to G \tag{4.5.3}$$

are local gauge transformations. Equivalently, by using the diffeomorphisms ϕ_α, we can define the connection on M by defining one-forms $\tilde{A}^{(\alpha)}$ with values in the Lie algebra $\mathbf{g}$ on each of the open sets $\tilde{\mathcal{U}}_\alpha$ of $\mathbb{R}^4$. These one-forms are related to the $A^{(\alpha)}$ by pullback, i.e.

$$\phi_\alpha^\star \left(\tilde{A}^{(\alpha)} \right) = A^{(\alpha)}. \tag{4.5.4}$$

The relationship (4.5.2) implies that, after changing coordinates from the coordinate system on $\tilde{\mathcal{U}}_\alpha$ to the one on $\tilde{\mathcal{U}}_\beta$, the connection $\tilde{A}^{(\alpha)}$ can be written as

$$\left(\phi_\alpha \circ \phi_\beta^{-1} \right)^\star \left(\tilde{A}^{(\alpha)} \right) = \tilde{U}_{\alpha\beta} \tilde{A}^{(\beta)} \tilde{U}_{\alpha\beta}^\dagger + i \tilde{U}_{\alpha\beta} \mathrm{d} \tilde{U}_{\alpha\beta}^\dagger, \tag{4.5.5}$$

where

$$\widetilde{U}_{\alpha\beta} = \left(\phi_\beta^{-1}\right)^\star \left(U_{\alpha\beta}\right). \tag{4.5.6}$$

Of course, the equation (4.5.5) just means that $\widetilde{A}^{(\alpha)}$ and $\widetilde{A}^{(\beta)}$ are related by the change of variables $\phi_\alpha \circ \phi_\beta^{-1}$ relating the two patches. Similarly, the field strength F can be defined in terms of local patches: on each patch $\mathcal{U}_\alpha$ we define a two-form $F^{(\alpha)}$ with values in $\mathbf{g}$, and in the intersection of patches the gluing rule is

$$F^{(\alpha)} = U_{\alpha\beta} F^{(\beta)} U_{\alpha\beta}^\dagger. \tag{4.5.7}$$

Another thing that we have to generalize is the covariant derivative. If α_μ is a covariant tensor (equivalently, a one-form) taking values in the Lie algebra of G, the covariant derivative is now defined as,

$$\nabla_\mu \alpha_\nu = \partial_\mu \alpha_\nu - \mathrm{i}[A_\mu, \alpha_\nu] - \Gamma_{\mu\nu}^\rho \alpha_\rho, \tag{4.5.8}$$

where $\Gamma_{\mu\nu}^\rho$ is the Christoffel symbol of the metric $g_{\mu\nu}$. In other words, the covariant derivative contains both the gauge connection A_μ and the Levi-Civita connection. In particular, the commutator of covariant derivatives will now contain both the field strength $F_{\mu\nu}$ and the Riemann tensor (or its contractions). For example,

$$[\nabla_\mu, \nabla_\nu] \alpha^\nu = -\mathrm{i}[F_{\mu\nu}, \alpha^\nu] - R_{\mu\nu} \alpha^\nu. \tag{4.5.9}$$

We are therefore considering YM theory on a curved space with a background gravitational field. The Euclidean YM action on M is defined as

$$S_{\mathrm{YM}}^{\mathrm{E}} = \frac{1}{2g_0^2} \int_M \mathrm{d}^4 x \sqrt{g} \mathrm{Tr}\left(F_{\mu\nu} F^{\mu\nu}\right), \tag{4.5.10}$$

where $F_{\mu\nu}$ is given by (4.2.17), and g denotes as usual the determinant of the inverse metric $g^{\mu\nu}$, which is also used to raise the indices. We have denoted the YM coupling constant as g_0, since it is a bare coupling constant. It should not be confused with the metric g. It is possible to write the Euclidean YM action in a more invariant way by using the Hodge dual operator $\star$. This operator is defined on any Riemannian manifold. In local coordinates it acts like

$$\star \left(\mathrm{d}x^{\mu_1} \wedge \cdots \wedge \mathrm{d}x^{\mu_p}\right) = \frac{\sqrt{g}}{(d-p)!} \epsilon^{\mu_1 \cdots \mu_p}{}_{\mu_{p+1} \cdots \mu_d} \mathrm{d}x^{\mu_{p+1} \cdots \mu_d}, \tag{4.5.11}$$

where d is the dimension of the manifold M, and we raise the indices of the Levi-Civita symbol with the inverse metric. This operator maps p-forms into $(d-p)$-forms, and it is invertible, since $\star^2 = (-1)^{p(d-p)}$ acting on p-forms. One finds,

$$F \wedge \star F = F_{\mu\nu} F^{\mu\nu} \sqrt{g} \mathrm{d}^4 x, \tag{4.5.12}$$

therefore we can write

$$S^{\mathrm{E}}_{\mathrm{YM}} = \frac{1}{2g_0^2} \int_M \mathrm{Tr}\,(F \wedge \star F). \tag{4.5.13}$$

The EOM derived from this action is

$$\nabla^\mu F_{\mu\nu} = 0. \tag{4.5.14}$$

The topological charge can be written on a general manifold M as

$$Q = \frac{1}{8\pi^2} \int_M \mathrm{Tr}\,(F \wedge F). \tag{4.5.15}$$

The (anti)self-duality conditions which define instantons can also be generalized to an arbitrary Riemannian four-manifold by using the Hodge operator:

$$F = \pm \star F. \tag{4.5.16}$$

The argument presented in the previous section, showing that (anti)self-dual configurations minimize the YM action for each topological class, can now be immediately generalized. We conclude that (Euclidean) YM gauge theory can be defined on any Riemannian four-manifold, and that there is a notion of (anti)instanton on such manifolds.

Example 4.2 *Gauge fields on the four-sphere.* An interesting example of this construction are gauge fields on the four-sphere $\mathbb{S}^4$. It is well known that the d-dimensional sphere $\mathbb{S}^d$ of radius R, defined by

$$\sum_{i=1}^{d+1} r_i^2 = R^2, \tag{4.5.17}$$

can be covered by two patches:

$$\mathcal{U}^{(N)} = \mathbb{S}^d \backslash \{(0,\dots,0,-R)\}, \qquad \mathcal{U}^{(S)} = \mathbb{S}^d \backslash \{(0,\dots,0,R)\}. \tag{4.5.18}$$

The first patch, which we will call the northern patch, excludes the south pole, while the second patch, which we will call the southern patch, excludes the north pole. On the northern patch we define the coordinates

$$x_\mu = \frac{R r_\mu}{R + r_{d+1}}, \qquad \mu = 1,\dots,d, \tag{4.5.19}$$

which give a diffeomorphism ϕ_N between $\mathcal{U}^{(N)}$ and $\widetilde{\mathcal{U}}^{(N)} = \mathbb{R}^d$. In the southern patch we have

$$y_\mu = \frac{R r_\mu}{R - r_{d+1}}, \qquad \mu = 1,\dots,d, \tag{4.5.20}$$

which give a diffeomorphism ϕ_S between $\mathcal{U}^{(S)}$ and $\tilde{\mathcal{U}}^{(S)} = \mathbb{R}^d$. These two maps are stereographic projections from the south pole and the north pole, respectively. The relationship between these coordinates on $\tilde{\mathcal{U}}^{(N)} \cap \tilde{\mathcal{U}}^{(S)}$ is

$$x_\mu = R^2 \frac{y_\mu}{y^2}, \tag{4.5.21}$$

where we denote

$$x^2 = \sum_{\mu=1}^{d} x_\mu^2, \qquad y^2 = \sum_{\mu=1}^{d} y_\mu^2. \tag{4.5.22}$$

Let us now consider the one-instanton solution of $SU(2)$ YM theory on $\mathbb{R}^4 = \tilde{\mathcal{U}}^{(N)}$,

$$\tilde{A}^{(N)} = \frac{x^2}{x^2 + \rho^2} iU\,dU^\dagger, \tag{4.5.23}$$

where U is given in (4.4.10). Through the stereographic projection (4.5.19), (4.5.23) defines a one-form $A^{(N)}$ on the northern patch. Similarly, we define the following one-form on $\mathbb{R}^4 = \tilde{\mathcal{U}}^{(S)}$,

$$\tilde{A}^{(S)} = \frac{y^2}{y^2 + (R^2/\rho)^2} iU^\dagger\,dU, \tag{4.5.24}$$

which gives by pullback a one-form $A^{(S)}$ on the southern patch of the four-sphere. After changing coordinates from x to y, the form (4.5.23) reads,

$$\left(\phi_N \circ \phi_S^{-1}\right)^* \left(\tilde{A}^{(N)}\right) = \frac{R^4}{R^4 + \rho^2 y^2} iU\,dU^\dagger. \tag{4.5.25}$$

This can be written as

$$\left(\phi_N \circ \phi_S^{-1}\right)^* \left(\tilde{A}^{(N)}\right) = U\tilde{A}^{(S)}U^\dagger + iU\,dU^\dagger, \tag{4.5.26}$$

which has the structure indicated in (4.5.5). Therefore, the two forms (4.5.23), (4.5.24) define a global connection on the four-sphere, which is the one-instanton solution. $\qquad\square$

The quantum YM theory can also be extended to this more general setting. The basic reason is that the renormalizability properties depend only on the short-distance behavior of the fields, and not on the long-distance properties: on a Riemannian four-manifold, the short-distance structure is by construction identical to that on $\mathbb{R}^4$. We will indeed verify explicitly that the quantum theory is well-defined at one-loop.

Since all fields we will deal with are, geometrically, differential forms with values in the Lie algebra, it is useful to define a metric

$$\langle a, b \rangle = \frac{1}{2\pi g_0^2} \int_M 2\text{Tr}\,(a \wedge \star b). \tag{4.5.27}$$

For example, if a and b are both one-forms with values in $\mathbf{g}$, one has,

$$\langle a, b \rangle = \frac{1}{2\pi g_0^2} \int_M 2\text{Tr}\,(a_\mu b_\nu)\, g^{\mu\nu} \sqrt{g} d^4 x. \tag{4.5.28}$$

The overall factor $(2\pi g_0^2)^{-1}$ guarantees that the modes are appropriately normalized, as we will see in a moment.

We are now ready to study the semiclassical approximation to the path integral around a non-trivial background. We will consider a generic background $\overline{A}$ which solves the classical EOM (4.2.29), and we will split the quantum connection A into the background $\overline{A}$, plus a fluctuation α:

$$A = \overline{A} + \alpha. \tag{4.5.29}$$

To calculate the action evaluated on such a gauge configuration, we first compute

$$F_{\mu\nu}(\overline{A} + \alpha) = F_{\mu\nu}(\overline{A}) + \nabla_\mu \alpha_\nu - \nabla_\nu \alpha_\mu - i[\alpha_\mu, \alpha_\nu]. \tag{4.5.30}$$

Here, ∇_μ is the covariant derivative in the background field $\overline{A}$. We then find that

$$S_{\text{YM}}^{\text{E}}(\overline{A} + \alpha) = S_{\text{YM}}^{\text{E}}(\overline{A}) + S(\alpha), \tag{4.5.31}$$

where

$$S(\alpha) = \frac{1}{2g_0^2} \int_M d^4 x \sqrt{g} \left\{ \text{Tr}\left(\nabla_\mu \alpha_\nu - \nabla_\nu \alpha_\mu\right)^2 + f^{abc} F_{\mu\nu}^a(\overline{A}) \alpha^{\mu b} \alpha^{\mu c} \right\} + \mathcal{O}(\alpha^3), \tag{4.5.32}$$

and we have only kept quadratic terms in α since we will only be interested in the one-loop evaluation of the path integral. Higher terms in α start contributing at two loops and higher. Notice that there is no linear term in α, since $\overline{A}$ solves the EOM. After integrating by parts, we find that the quadratic part of $S(\alpha)$ reads

$$\frac{1}{2g_0^2} \int_M d^4 x \sqrt{g} \left\{ -\alpha_\mu^a \left(\nabla_\nu \nabla^\nu \alpha^\mu\right)^a + \alpha_\mu^a \left(\nabla^\mu \nabla^\nu \alpha_\nu\right)^a - \alpha^{\mu a}\left([\nabla_\mu, \nabla_\mu]\alpha^\mu\right)^a \right.$$
$$\left. + f^{abc} F_{\mu\nu}^a(\overline{A}) \alpha^{\mu b} \alpha^{\nu c} \right\}. \tag{4.5.33}$$

Using (4.5.9) we find

$$S(\alpha) = \pi \langle \alpha, \Delta_1 \alpha \rangle + \mathcal{O}(\alpha^3), \tag{4.5.34}$$

where the operator Δ_1 is defined by

$$(\Delta_1 \alpha)_\mu = -\nabla_\nu \nabla^\nu \alpha_\mu + \nabla_\mu \nabla^\nu \alpha_\nu + R_{\mu\nu} \alpha^\nu + 2i[F_{\mu\nu}(\overline{A}), \alpha^\nu] \qquad (4.5.35)$$

and we have used the metric (4.5.27). This operator has the following properties:

$$(\Delta_1 \nabla \phi)_\mu = 0, \qquad (4.5.36)$$

i.e. it annihilates vector fields which are given by infinitesimal gauge transformations, and

$$\nabla^\mu (\Delta_1 \alpha)_\mu = 0. \qquad (4.5.37)$$

These properties can be verified explicitly by commuting the covariant derivatives and taking into account that $F_{\mu\nu}(\overline{A})$ satisfies the EOM (4.5.14). Equation (4.5.36) follows from the gauge covariance of this equation. Notice that, if we decompose the field α_ν in eigenmodes of Δ_1 which are orthonormal with respect to the metric (4.5.27),

$$\alpha_\nu = \sum_n c_n \omega_\mu^{(n)}, \qquad \Delta_1 \omega_\mu^{(n)} = \lambda_n \omega_\mu^{(n)}, \qquad \langle \omega_\mu^{(n)}, \omega_\mu^{(m)} \rangle = \delta^{nm}, \qquad (4.5.38)$$

we find

$$\pi \langle \alpha, \Delta_1 \alpha \rangle = \pi \sum_n \lambda_n c_n^2, \qquad (4.5.39)$$

and

$$\int \prod_n dc_n \exp\left(-\pi \langle \alpha, \Delta_1 \alpha \rangle\right) = \prod_n \lambda_n^{-1/2} = (\det(\Delta_1))^{-1/2}. \qquad (4.5.40)$$

Therefore, when the metric is defined as in (4.5.27), the appropriately normalized measure can be taken to be simply the one appearing in (4.5.40). In this discussion of the normalization of the modes, we have ignored subtleties related to zero modes, but this will be addressed at due time.

Since we are doing a calculation in a gauge theory, we have to choose a gauge. Our choice will be the standard, covariant gauge,

$$g_{\overline{A}}(\alpha) = \nabla^\mu \alpha_\mu = 0, \qquad (4.5.41)$$

where $g_{\overline{A}}$ is the gauge fixing function. We recall that in the standard Fadeev–Popov (FP) gauge fixing one first defines

$$\Delta_{\overline{A}}^{-1}(\alpha) = \int \mathcal{D}U \, \delta\left(g_{\overline{A}}(\alpha^U)\right), \qquad (4.5.42)$$

where the integration over U is over all gauge transformations, and then inserts into the path integral

$$1 = \left[\int \mathcal{D}U \, \delta \left(g_{\overline{A}} (\alpha^U) \right) \right] \Delta_{\overline{A}} (\alpha). \tag{4.5.43}$$

However, the standard gauge fixing procedure has to be taken with care when the background connection $\overline{A}$ has a non-trivial isotropy group $\mathcal{H}$. The isotropy group of a connection $\overline{A}$ is the subgroup of gauge transformations which leave $\overline{A}$ invariant, i.e. $U \in \mathcal{H}$ if

$$\overline{A}^U = \overline{A}. \tag{4.5.44}$$

Infinitesimally, a gauge transformation $U = e^{i\phi}$ belongs to $\mathcal{H}$ if

$$\nabla_\mu \phi = 0, \tag{4.5.45}$$

where the covariant derivative is evaluated in the background of $\overline{A}$. We will assume in what follows that the isotropy group $\mathcal{H}$ consists of *constant* gauge transformations that leave $\overline{A}$ invariant,

$$U \overline{A} U^\dagger = \overline{A}. \tag{4.5.46}$$

Therefore, they are in one-to-one correspondence with a subgroup of G which we will denote by H. When there is a non-trivial isotropy group, the gauge fixing condition does not fix the gauge completely, since

$$g_{\overline{A}}(\alpha^U) = U g_{\overline{A}}(\alpha) U^\dagger, \qquad U \in \mathcal{H}, \tag{4.5.47}$$

i.e. the basic assumption that the gauge fixing condition cuts the gauge orbit only once is not true: there is a residual symmetry given by the isotropy group. Another way to see this is that the standard FP determinant vanishes due to zero modes. In fact, the standard calculation of (4.5.42) (which is valid if the isotropy group of $\overline{A}$ is trivial) gives

$$\Delta_{\overline{A}}^{-1} (\alpha) = \left| \det \frac{\delta g_{\overline{A}}(\alpha^U)}{\delta U} \right|^{-1} = \left| \det \nabla^\mu \nabla_\mu^A \right|^{-1}. \tag{4.5.48}$$

Here, in the covariant derivative with the superscript A, the gauge connection involved is the total one (background plus fluctuations). When the background is the trivial one, $\overline{A} = 0$, the above result is the standard one for the covariant gauge fixing (4.5.41). At one-loop order we have to set $\alpha = 0$ in this determinant and we have to consider the determinant of the operator $\nabla^\mu \nabla_\mu$. However, if the isotropy group $\mathcal{H}$ is non-trivial, the operator ∇_μ has non-trivial zero modes, due to (4.5.45), and the FP procedure is ill defined. The correct way to proceed in the calculation of (4.5.42) is to split the integration over the gauge group. One way to see this is to look at the algebra of gauge transformations and write $U = e^{i\phi}$, where ϕ takes values in the Lie algebra. The operator ∇_μ is a differential operator acting on the space of maps ϕ, and it has a non-trivial kernel. We can use the metric (4.5.27) to split this space into the kernel and its orthogonal complement,

$$(\text{Ker } \nabla_\mu) \oplus (\text{Ker } \nabla_\mu)^\perp. \tag{4.5.49}$$

Corresponding to this splitting, we have a splitting of the integral over the gauge group into two pieces. The first piece leads to an integration over the isotropy group, whose algebra is the first summand in (4.5.49). This just leads to a factor of $\text{vol}(\mathcal{H})$. The second piece gives an integration over the remaining part of the gauge transformations, whose algebra is the second summand of (4.5.49). This leads to the standard FP determinant (4.5.48) but with the zero modes removed. We then find,

$$\Delta_{\overline{A}}^{-1}(\alpha) = \text{vol}(\mathcal{H}) \left| \det \nabla^\mu \nabla_\mu^A \right|_{(\text{Ker } \nabla_\mu)^\perp}^{-1}. \tag{4.5.50}$$

As usual, the determinant appearing here can be written as a path integral over ghost fields, with action

$$S_{\text{ghosts}}(C, \overline{C}, \alpha) = \langle \overline{C}, \nabla^\mu \nabla_\mu^A C \rangle, \tag{4.5.51}$$

where $C, \overline{C}$ are Grassmannian fields taking values in

$$(\text{Ker } \nabla_\mu)^\perp. \tag{4.5.52}$$

The action for the ghosts can be divided into a kinetic term plus an interaction term between the ghost fields and the fluctuation α:

$$S_{\text{ghosts}}(C, \overline{C}, \alpha) = \langle \overline{C}, \nabla^\mu \nabla_\mu C \rangle - i \langle \overline{C}, \nabla^\mu [\alpha_\mu, C] \rangle. \tag{4.5.53}$$

The modified FP gauge fixing then leads to the path integral around the background $\overline{A}$

$$Z_{\overline{A}}(M) = e^{-S_{\text{YM}}^{\text{E}}(\overline{A})} \int \mathcal{D}\alpha \, e^{-S(\alpha)} \Delta_{\overline{A}}(\alpha) \delta \left(\nabla^\mu \alpha_\mu \right)$$

$$= \frac{e^{-S_{\text{YM}}^{\text{E}}(\overline{A})}}{\text{vol}(\mathcal{H})} \int \mathcal{D}\alpha \, \delta \left(\nabla^\mu \alpha_\mu \right) \int_{(\text{Ker } \nabla_\mu)^\perp} \mathcal{D}C \mathcal{D}\overline{C} \, e^{-S(\alpha) - S_{\text{ghosts}}(C, \overline{C}, \alpha)}. \tag{4.5.54}$$

We now have to analzye the delta constraint on α. First of all, every one-form with values in $\mathbf{g}$ can be decomposed as the sum of an element of $\text{Ker } \nabla^\mu$ (acting on one-forms), and an element of its orthogonal complement with respect to the metric (4.5.27). But, by integration by parts, one can see that this orthogonal complement is the image of the operator ∇_μ, acting on scalar fields with values in $\mathbf{g}$. Therefore, the fluctuation field α_μ can be decomposed as

$$\alpha_\mu = \nabla_\mu \phi + \alpha'_\mu, \tag{4.5.55}$$

where

$$\phi \in \left(\operatorname{Ker} \nabla_\mu\right)^\perp, \quad \alpha'_\mu \in \operatorname{Ker} \nabla^\mu. \tag{4.5.56}$$

The presence of the operator ∇_μ in the change of variables (4.5.55) leads to a non-trivial Jacobian. Indeed, the norm of α_μ, computed with the metric (4.5.27), is

$$\|\alpha_\mu\|^2 = \langle \phi, \Delta_0 \phi \rangle + \|\alpha'_\mu\|^2, \tag{4.5.57}$$

where

$$\Delta_0 = -\nabla^\mu \nabla_\mu \tag{4.5.58}$$

is an operator acting on functions (i.e. zero-forms) taking values in $\mathbf{g}$. The measure in the functional integral becomes

$$\mathcal{D}\alpha = \left(\det' \Delta_0\right)^{1/2} \mathcal{D}\phi \, \mathcal{D}\alpha', \tag{4.5.59}$$

where the $'$ indicates that we are removing zero modes. Notice that the operator in (4.5.59) is positive-definite, so the square root of its determinant is well defined. We also have that

$$\delta \left(\nabla^\mu \alpha_\mu\right) = \delta \left(\Delta_0 \phi\right) = \left(\det' \Delta_0\right)^{-1} \delta(\phi), \tag{4.5.60}$$

which is a straightforward generalization of the standard formula

$$\delta(ax) = \frac{1}{|a|} \delta(x). \tag{4.5.61}$$

We conclude that the delta function, together with the Jacobian in (4.5.59), lead to the following factor in the path integral:

$$\left(\det' \Delta_0\right)^{-1/2}. \tag{4.5.62}$$

In addition, the delta function sets $\phi = 0$. It only remains to carry out the integration over the transverse gauge fluctuations α'_μ in $\operatorname{Ker} \nabla^\mu$, which we relabel $\alpha'_\mu \to \alpha_\mu$. The final result for the gauge fixed path integral is then

$$Z_{\bar{A}}(M) = \frac{e^{-S_{\mathrm{YM}}^{\mathrm{E}}(\bar{A})}}{\operatorname{vol}(\mathcal{H})} \int_{\operatorname{Ker} \nabla^\mu} \mathcal{D}\alpha_\mu \left(\det' \Delta_0\right)^{-1/2}$$
$$\times \int_{(\operatorname{Ker} \nabla_\mu)^\perp} \mathcal{D}C \mathcal{D}\bar{C} \, e^{-S(\alpha) - S_{\mathrm{ghosts}}(C, \bar{C}, \alpha)}. \tag{4.5.63}$$

This is the starting point to perform gauge fixed perturbation theory.

In order to present an explicit result for the one-loop approximation to the above path integral, let us first calculate vol($\mathcal{H}$). The isotropy group $\mathcal{H}$ is the space of constant functions, taking values in a subgroup $H \subset G$ of the gauge group. Each generator of its Lie algebra has a norm given by $2\text{Tr}(T_a^2)$, times the spacetime factor appearing in (4.5.27):

$$\left(\frac{1}{2\pi g_0^2} \int_M \star 1\right)^{1/2} = \left(\frac{\text{vol}(M)}{2\pi g_0^2}\right)^{1/2}. \tag{4.5.64}$$

Therefore,

$$\text{vol}(\mathcal{H}) = \left(\frac{\text{vol}(M)}{2\pi g_0^2}\right)^{\dim(H)/2} \text{vol}(H). \tag{4.5.65}$$

At one-loop, we get the determinant of Δ_0 coming from the ghost fields, which together with (4.5.62) gives

$$\left(\det' \Delta_0\right)^{1/2}. \tag{4.5.66}$$

Finally, we obtain

$$Z_{\overline{A}}^{\text{one-loop}}(M) = \frac{e^{-S_{\text{YM}}^{\text{E}}(\overline{A})}}{\text{vol}(H)} \left(\frac{2\pi g_0^2}{\text{vol}(M)}\right)^{\dim(H)/2}$$

$$\times \int_{\text{Ker}\,\nabla^\mu} \mathcal{D}\alpha_\mu \left(\det' \Delta_0\right)^{1/2} e^{-\pi\langle\alpha, \Delta_1\alpha\rangle}. \tag{4.5.67}$$

It remains to perform the integration over the transverse gauge fluctuations α_μ, with $\nabla^\mu \alpha_\mu = 0$. Note first that, due to (4.5.37), the operator Δ_1 can be restricted to the space Ker ∇^μ. We now have to be careful, since Δ_1 will have zero modes in this space. This is due to the fact that, in general, the background $\overline{A}$ belongs to a family of solutions, and this leads automatically to zero modes, as we found already in the analysis of instantons in QM. For example, in the case of the gauge theory instanton with topological charge 1, there is a family of solutions parametrized by five collective coordinates: four for the center of the instanton, and one for the size of the instanton. Let us denote by $\gamma_r, r = 1, \ldots, m$ the set of collective coordinates characterizing the instanton solution $\overline{A}_\mu^{\gamma_r}$. If we vary this solution with respect to the γ_r, we obtain a one-form with values in **g** which can be decomposed as in (4.5.55),

$$\delta \overline{A}_\mu^{\gamma_r} = \sum_{r=1}^m \left(Z_\mu^r + \nabla_\mu \Lambda^r\right) \delta\gamma_r, \tag{4.5.68}$$

where

$$\nabla^\mu Z_\mu^r = 0. \tag{4.5.69}$$

Since $\overline{A}_\mu^{\gamma_r}$ solves the EOM for any γ_r, it follows that Z_μ^r are zero modes of the operator Δ_1:

$$\Delta_1 Z^r = 0. \tag{4.5.70}$$

This can be proved as in the derivation which led to (1.4.40) in the case of instantons in QM, by taking the derivative with respect to γ_r in the EOM. Let us now denote by ω^r the orthonormal zero modes, and by c_r the corresponding coordinates in the expansion of a general fluctuation. As in the derivation of (1.4.46) in QM, we have

$$\sum_{r=1}^m Z_\mu^r \delta\gamma_r = \sum_{r=1}^m \omega_\mu^r \delta c_r. \tag{4.5.71}$$

By calculating the norm squared of these one-forms with the metric (4.5.27), we find

$$\sum_{r=1}^m (\delta c_r)^2 = \frac{1}{2\pi g_0^2} \sum_{r,s=1}^m \delta\gamma_r J_{rs} \delta\gamma_s, \tag{4.5.72}$$

where

$$J_{rs} = 2 \int_M \mathrm{Tr}\left(Z_r \wedge \star Z_s\right). \tag{4.5.73}$$

It follows that the relation between the canonical measure in terms of the c_r and the measure expressed in terms of the collective coordinates $\delta\gamma_r$ is

$$\prod_{r=1}^m dc_r = \left(\frac{1}{2\pi g_0^2}\right)^{m/2} (\det J)^{1/2} \prod_{r=1}^m d\gamma_r. \tag{4.5.74}$$

The remaining modes can be integrated and they lead to the inverse square root of the determinant of the operator Δ_1, restricted to transverse gauge fluctuations, and with the zero modes removed. We will denote it by

$$\left(\det{}' \Delta_1\right)^{-1/2}. \tag{4.5.75}$$

This determinant can be expressed in a way which is useful for concrete calculations. Let us consider the operator

$$(\Delta_1^t \alpha)_\mu = (\Delta_1 \alpha)_\mu - \nabla_\mu \nabla^\nu \alpha_\nu. \tag{4.5.76}$$

Notice that, when restricted to tranverse one-forms satisfying $\nabla^\nu \alpha_\nu = 0$, it agrees with Δ_1, and it has the following property,

$$\Delta_1^t \left(\nabla_\mu \phi\right) = \nabla_\mu \left(\Delta_0 \phi\right). \tag{4.5.77}$$

This means that, if ϕ is an eigenstate of Δ_0, $\nabla_\mu \phi$ will be an eigenstate of Δ_1^t with the same eigenvalue. Let us consider the spectrum of the operator Δ_1^t acting on all adjoint-valued one-forms α_μ. Due to the orthogonal decomposition (4.5.55), the spectrum of Δ_1^t is given by the spectrum of Δ_1 acting on tranverse fields, together with the spectrum of Δ_1^t acting on forms of $\nabla_\mu \phi$. Due to (4.5.77), this is equal to the spectrum of the Laplacian Δ_0 on adjoint-valued zero-forms. We conclude that

$$\left(\det' \Delta_1\right)^{-1/2} = \left(\det' \Delta_1^t\right)^{-1/2} \left(\det' \Delta_0\right)^{1/2}. \tag{4.5.78}$$

We finally obtain the following expression for the one-loop approximation to the Euclidean YM path integral around the background $\overline{A}$:

$$Z_{\overline{A}}^{\text{one-loop}}(M) = \frac{e^{-S_{\text{YM}}^E(\overline{A})}}{\text{vol}(H)} \left(\text{vol}(M)\right)^{-\dim(H)/2} \left(\frac{1}{2\pi g_0^2}\right)^{m/2 - \dim(H)/2}$$

$$\times \int \prod_{r=1}^{m} d\gamma_r \ (\det J)^{1/2} \left(\det' \Delta_0\right) \left(\det' \Delta_1^t\right)^{-1/2}. \tag{4.5.79}$$

Notice that the determinants of the operators will depend in general on the collective coordinates, so we have to keep them inside the integral over γ_r. It is convenient to normalize this expression by the path integral around the trivial gauge background $\overline{A} = 0$. In this case, the isotropy group is the full gauge group G and one obtains

$$\frac{Z_{\overline{A}}^{\text{one-loop}}(M)}{Z_0^{\text{one-loop}}(M)} = e^{-S_{\text{YM}}^E(\overline{A})} \text{vol}(G/H) \left(\text{vol}(M)\right)^{d_{G/H}/2} \left(\frac{1}{2\pi g_0^2}\right)^{m/2 + d_{G/H}/2}$$

$$\times \int \prod_{r=1}^{m} d\gamma_r \ (\det J)^{1/2} \left(\frac{\det' \Delta_0}{\det' \Delta_0^{(0)}}\right) \left(\frac{\det' \Delta_1^t}{\det' \Delta_1^{t,(0)}}\right)^{-1/2}. \tag{4.5.80}$$

Here, $\Delta_0^{(0)}, \Delta_1^{t(0)}$ are the operators Δ_0, Δ_1^t in the background of the trivial connection $\overline{A} = 0$, and

$$d_{G/H} = \dim(G) - \dim(H). \tag{4.5.81}$$

Notice that this quotient is the one-loop approximation to the semiclassical calculation on the manifold M of the normalized probability

$$\overline{P}_\nu = \frac{P_\nu}{P_0}, \tag{4.5.82}$$

where P_ν has been introduced in (4.3.75).

The above expression seems to reduce the calculation of the one-loop approximation to the path integral to a well-defined quantity. However, as in the expression (2.2.27) for decay rates in scalar field theory, the determinants of the operators appearing in (4.5.80) are divergent. We expect that, by renormalizing the theory and using the renormalized coupling constant, the divergences will cancel and a finite expression will be obtained. We will now make this precise and obtain a manifestly well-defined and finite expression for (4.5.80), which can in fact be evaluated in some simple cases.

To do this, we regulate the determinants of the operators by using the dimensional regularization procedure explained in Appendix B.4: instead of working on the original manifold M, we extend the theory to a manifold $X = M \times Y_p$, where Y_p is a $2p$-dimensional manifold which can be taken to be $Y_p = (\mathbb{S}^2)^p$, the pth power of the two-sphere. The instanton solution extends to this space, by requiring the extra components of the gauge connection to vanish, and the four-dimensional coordinates are taken to be independent of the extra dimension and equal to the instanton configuration on M. After the calculation is done, one does, as in standard dimensional regularization, an analytic continuation to negative dimensions and sets $p = -\epsilon$. The logarithms of the determinants of the operators can be computed in this regularization scheme and studied when $\epsilon \to 0$. Their divergence is now manifest and it is given by a pole at $\epsilon = 0$. The finite part is given by the zeta function regularization of the determinant. The computation is explained in detail in Appendix B, where we also give the relevant background on heat kernels and zeta functions. The result is given in (B.112). Notice now that, for an instanton configuration of topological charge ν in $X = M \times (\mathbb{S}^2)^p$, the action is

$$S_{\text{YM}}(\overline{A}) = (4\pi r^2)^p \frac{8\pi^2 \nu}{g_0^2} \tag{4.5.83}$$

where the first factor is just the volume of the extra $(\mathbb{S}^2)^p$. We now use (B.112) to analyze the divergent part of (4.5.80) in our one-loop analysis. It is given by the exponential of

$$\frac{r^{-2\epsilon}}{\epsilon} \frac{11 \nu C_2(G)}{6} - (4\pi r^2)^{-\epsilon} \frac{8\pi^2 \nu}{g_0^2}. \tag{4.5.84}$$

It is easy to see that the divergence is absorbed if we use the renormalized coupling constant as computed in *flat space*, i.e. the expression (4.2.38). One finds, as $\epsilon \to 0$,

$$-\frac{8\pi^2 \nu}{g^2} + \frac{11 \nu C_2(G)}{6} \log\left(4\pi \mu^2\right) \tag{4.5.85}$$

and the dependence on r, the radius of the auxiliary sphere, has dropped out. Notice that, since the structure of divergences only depends on the UV behavior of the theory, it is the same in the background of an instanton and on a general manifold M, therefore the renormalization of the coupling constant in flat space eliminates all the divergences, at least at one-loop. We then find, at the end of the day,

$$\frac{Z_A^{\text{one-loop}}(M)}{Z_0^{\text{one-loop}}(M)} = \text{vol}(G/H)\,(\text{vol}(M))^{d_{G/H}/2}\left(\frac{1}{2\pi g^2}\right)^{m/2+d_{G/H}/2}$$

$$\times \exp\left\{-\frac{8\pi^2 v}{g^2} + \frac{11 v C_2(G)}{6}\left(\log\left(4\pi\mu^2\right) - \gamma_{\text{E}}\right) + \frac{C_2(G)v}{6}\right\}$$

$$\times \int \prod_{r=1}^{m} d\gamma_r\,(\det J)^{1/2}$$

$$\times \exp\left[\frac{1}{2}\zeta'_{\Delta_1'}(0) - \zeta'_{\Delta_0}(0) - \frac{1}{2}\zeta'_{\Delta_1^{t,(0)}}(0) + \zeta'_{\Delta_0^{(0)}}(0)\right]. \qquad (4.5.86)$$

This is our final expression for the normalized partition function in the background of an instanton on a four-manifold M. When expressed in terms of the renormalized coupling at the scale μ, it is manifestly finite. The dependence on μ is easy to understand, since the final result must be invariant under the renormalization group. Therefore, once we know that the factor $e^{-8\pi^2 v/g^2}$ appears in the answer, in view of (4.2.44) it must combine with

$$\mu^{(4\pi)^2\beta_0 v} \qquad (4.5.87)$$

to have an RG invariant quantity. It is reassuring that we reproduce this dependence in an explicit calculation. Note in particular that the calculation in Appendix B.4 leading to (B.112) determines the value of β_0 through heat kernel techniques.

One can now try to evaluate (4.5.86) for particular cases. A relatively simple and workable case occurs when $M = \mathbb{S}^4$, a four-sphere of radius R, the gauge group is $G = SU(n)$, and we consider an instanton of topological charge $v = 1$. In this case the instanton configuration can be easily obtained from the results derived above, since the four-sphere can be mapped to $\mathbb{R}^4$ by using the stereographic projection. One can then take the known instanton solution on $\mathbb{R}^4$ for $v = 1$ and map it to $\mathbb{S}^4$. The number of zero modes or collective coordinates m is five, since there are four coordinates for the center of the instanton and one coordinate for the size of the instanton, ρ. The isotropy group of the instanton configuration is quite large in this case: if we embed the one-instanton solution of $SU(2)$ into $SU(N)$ as in (4.4.36), it is clear that constant gauge transformations of the form

$$U = \begin{pmatrix} M_{N-2} & 0_2 \\ 0_2 & a\mathbf{1}_2 \end{pmatrix}, \qquad (4.5.88)$$

leave the gauge connection (4.4.36) invariant. Here, the condition that $U \in SU(N)$, and in particular that $\det(U) = 1$, implies that $a \in \mathbb{C}$ with $|a| = 1$, and that M can be any $U(N-2)$ matrix with determinant $\det(M) = a^{-2}$. The isotropy group is then given by

$$H = U(N-2) \qquad (4.5.89)$$

and we find

$$d_{G/H} = 4N - 5. \tag{4.5.90}$$

It follows that $m + d_{G/H} = 4N$. It turns out that this formula can be generalized to an instanton on the four-sphere with arbitrary v, and one finds

$$m + d_{G/H} = 4C_2(G)v. \tag{4.5.91}$$

The most difficult part of the calculation of (4.5.86) involves the zeta function regularization of the determinants, but this has been done by Lüscher, generalizing previous computations by 't Hooft, Bernard, and others. However, the dependence on the radius of the sphere R is easy to understand from dimensional considerations: since the final answer for (4.5.86) is a dimensionless quantity, the dependence on R must combine with the dependence on μ to guarantee this. It follows that

$$\overline{P}_1^{\text{one-loop}} = K(N)(\mu R)^{11N/3} g^{-4N} e^{-8\pi^2/g^2}, \tag{4.5.92}$$

where the value of $K(N)$ is known explicitly. One can now use an RG analysis to write this result in terms of RG invariant quantities. The probability $\overline{P}_1$ is clearly an observable, therefore it must be invariant under the RG and satisfies the equation,

$$\left(\mu \frac{\partial}{\partial \mu} + \beta(g) \frac{\partial}{\partial g} \right) \overline{P}_1 = 0. \tag{4.5.93}$$

This means that, in the function $\overline{P}_1(\mu R, g)$, one can trade a change in μR by a change in g. If we define the coupling constant $\overline{g}(\mu R, g)$ by the equation

$$\log(\mu R) = - \int_g^{\overline{g}} \frac{dx}{\beta(x)}, \tag{4.5.94}$$

we see that $\overline{g}$ satisfies

$$\left(\mu \frac{\partial}{\partial \mu} + \beta(g) \frac{\partial}{\partial g} \right) \overline{g}(\mu R, g) = 0, \tag{4.5.95}$$

therefore any function where the dependence on μR and g occurs solely through $\overline{g}$ satisfies the RG equation. We conclude that

$$\overline{P}_1(\mu R, g) = \overline{P}_1 \left(1, \overline{g}(\mu R, g) \right), \tag{4.5.96}$$

since the right hand side satisfies the RG equation and they both agree at the point $\mu R = 1$. We then conclude, from (4.5.96) and (4.5.92), that

$$\overline{P}_1 = K(N)\overline{g}^{-4N} e^{-8\pi^2/\overline{g}^2} \left(1 + \mathcal{O}(\overline{g}^2) \right). \tag{4.5.97}$$

An additional interest of introducing $\bar{g}(\mu R, g)$ is that it actually depends only on R and the dynamically generated scale of QCD, Λ, which was defined in (4.2.43). Indeed, by using (4.5.94) we can write

$$\int_0^g \left\{ \frac{1}{\beta(x)} + \frac{1}{\beta_0 x^3} - \frac{\beta_1}{\beta_0^2 x} \right\} dx = \log(\mu R) + \int_0^{\bar{g}} \left\{ \frac{1}{\beta(x)} + \frac{1}{\beta_0 x^3} - \frac{\beta_1}{\beta_0^2 x} \right\} dx$$

$$+ \int_{\bar{g}}^g \left\{ \frac{1}{\beta_0 x^3} - \frac{\beta_1}{\beta_0^2 x} \right\} dx, \qquad (4.5.98)$$

and we obtain the following expression in terms of the coupling $\bar{g}$,

$$- \log(\Lambda R) = \frac{1}{2\beta_0 \bar{g}^2} + \frac{\beta_1}{2\beta_0^2} \log\left(\beta_0 \bar{g}^2\right)$$

$$+ \int_0^{\bar{g}} \left\{ \frac{1}{\beta(x)} + \frac{1}{\beta_0 x^3} - \frac{\beta_1}{\beta_0^2 x} dx \right\}. \qquad (4.5.99)$$

This equation makes very explicit the asymptotic freedom of YM theory: as $R \to 0$, i.e. in the limit of small volume, the renormalized coupling constant $\bar{g}$ goes to zero as

$$\bar{g}^2 = -\frac{1}{2\beta_0 \log(\Lambda R)} - \frac{\beta_1}{4\beta_0^3} \frac{\log(-2\log(\Lambda R))}{(\log(\Lambda R))^2} + \cdots. \qquad (4.5.100)$$

We can finally plug this expression into (4.5.97) to obtain an expression for $\bar{P}_1$ in terms of the radius of the four-sphere and the dynamically generated scale, in the limit of $R \to 0$, where the semiclassical approximation becomes exact,

$$\bar{P}_1 \approx \beta_0^{2N} K(N) (\Lambda R)^{11N/3} (-2\log(\Lambda R))^{5N/11}. \qquad (4.5.101)$$

In writing this expression, we used the explicit form for the two-loop beta function coefficient β_1 in (4.2.42).

One consequence of (4.5.92) is that the one-loop instanton contribution *diverges* in the limit where $R \to \infty$, i.e. it diverges on the Euclidean space $\mathbb{R}^4$. Therefore, instanton calculus on flat space in the loop approximation is simply ill defined, and this is one of the reasons why we have developed instanton calculus on a four-manifold of finite volume. Another way of seeing the divergence is by considering the limit $R \to \infty$ of the *integrand* of (4.5.80) (sometimes called the density of the instanton). This limit is well defined, and one finds from dimensional analysis and translation invariance that

$$\frac{d\bar{P}_1}{d\rho d^4 x_0} = \frac{C(N)}{\rho^5} (\mu\rho)^{11N/3} g^{-4N} e^{-8\pi^2/g^2(\mu)} \left(1 + \mathcal{O}(g^2)\right), \qquad (4.5.102)$$

where $C(N)$ is a constant (first determined for $SU(2)$ by 't Hooft). The factor of $1/\rho^5$ arises for dimensional reasons, since the left hand side has the dimensions of $1/[\text{length}]^5$. Notice that the factor

$$(\mu\rho)^{11N/3} e^{-8\pi^2/g^2(\mu)} \tag{4.5.103}$$

can be predicted by dimensional analysis and RG invariance, without further ado. The integration over the center of the instanton gives the volume of spacetime V, but the integration over the size of the instanton ρ diverges at large ρ. This is the famous "IR embarrassment" of instanton calculus in YM theory, which makes it a difficult analytic tool to work with in the absence of an IR cutoff. Of course, in the regime $\rho \to \infty$ the density (4.5.102) is not really the right answer. As the instanton size becomes large, the running coupling constant defined at the scale set by ρ, $\bar{g}(\mu\rho, g)$, enters the strong coupling regime, and the weak coupling approximation that we used in our instanton calculation simply breaks down. One might think that there is some dynamical mechanism that cuts off the integration over the size of an instanton in $\mathbb{R}^4$, in such a way that the one-instanton partition function becomes finite and still gives the dominant, non-perturbative contribution. If this were the case, the theta dependence of the vacuum energy density would come from the terms $\nu = \pm 1$ in (4.3.73), which by parity invariance should be complex conjugate to each other, and one should find

$$E(\theta) - E(0) \approx 2 \left(1 - \cos\theta\right) e^{-8\pi^2/g^2}. \tag{4.5.104}$$

However, lattice simulations do not seem to favor this dependence on the theta angle at infinite volume.

4.6 Renormalons

As we saw in Chapter 3, instantons dominate the large order behavior of perturbation theory in QM. There are other simple quantum models where this is still the case, in the sense that the large order of perturbation theory is determined by the factorial growth of the number of diagrams. However, in *renormalizable* QFTs, the large order behavior of perturbation theory seems to be dominated by another type of divergence called *renormalon* divergences, or renormalons for sort. Renormalon divergences also lead to a factorial growth in perturbation theory. However, this is not due to the proliferation of diagrams, but to the integrals over momenta in special classes of Feynman diagrams, which grow factorially with the loop order.

The existence of renormalons can be argued by diagrammatic analysis and by other indirect arguments based for example on the operator product expansion (OPE). Here we will present a general argument, based solely on the RG equations, which indicates the possible presence of non-perturbative effects which are not of the instanton type. To show that these effects are actually present, one needs more sophisticated techniques which go beyond the scope of this book. More detailed references can be found in the bibliographical notes at the end of this chapter.

Let us consider a generic, RG invariant observable in YM theory. Its asymptotic expansion around $g = 0$ can be computed in perturbation theory, but in addition we should consider possible non-perturbative corrections, so we will write it as

$$\varphi(g) = \varphi_{\mathrm{p}}(g) + \varphi_{\mathrm{np}}(g), \tag{4.6.1}$$

where

$$\varphi_{\mathrm{p}}(g) = \sum_{n=0}^{\infty} a_n g^{2(n+1)} \tag{4.6.2}$$

is the perturbative contribution, and $\varphi_{\mathrm{np}}(g)$ is the non-perturbative contribution. This observable will also depend on an energy scale Q^2, as well as on the renormalization scale μ. Since both $\varphi(g)$ and $\varphi_{\mathrm{p}}(g)$ are separately RG invariant, the same should hold for $\varphi_{\mathrm{np}}(g)$: it must satisfy the equation

$$\left(\mu \frac{\partial}{\partial \mu} + \beta(g) \frac{\partial}{\partial g} + \gamma(g) \right) \varphi_{\mathrm{np}}(g) = 0 \tag{4.6.3}$$

where

$$\gamma(g) = \gamma_1 g^2 + \cdots \tag{4.6.4}$$

is the anomalous dimension associated to the observable. As a consequence of this equation, and assuming an exponentially small dependence on the coupling constant, $\varphi_{\mathrm{np}}(g)$ must be of the form

$$\varphi_{\mathrm{np}}(g) = C \left(\frac{\mu^2}{Q^2} \right)^{d/2} g^{\delta} \exp\left(-\frac{d}{2\beta_0 g^2} \right) (1 + \mathcal{O}(g)), \tag{4.6.5}$$

where d is a dimension associated to the observable under consideration. Indeed, it can easily be checked that this functional form satisfies the RG equation (4.6.3), since

$$\mu \frac{\partial}{\partial \mu} \varphi_{\mathrm{np}}(g) = d \varphi_{\mathrm{np}}(g), \tag{4.6.6}$$

and

$$\beta(g) \frac{\partial}{\partial g} \varphi_{\mathrm{np}}(g) = -\left(\beta_0 g^3 + \beta_1 g^5 + \cdots \right) \left(\frac{d}{\beta_0 g^3} + \frac{2\delta}{g} + \cdots \right) \varphi_{\mathrm{np}}(g)$$

$$= -\left\{ d + \left(2\delta\beta_0 + \frac{d\beta_1}{\beta_0} \right) g^2 + \cdots \right\} \varphi_{\mathrm{np}}(g). \tag{4.6.7}$$

Therefore, RG invariance requires

$$\delta = \frac{\gamma_1}{2\beta_0} - \frac{d}{2} \frac{\beta_1}{\beta_0^2}. \tag{4.6.8}$$

We can now perform an analysis of the series (4.6.2) and its non-perturbative correction in (4.6.1) by using the general ideas of Chapter 3. We consider a Borel transform of the form

$$\widehat{\varphi}_{\mathrm{p}}(\zeta) = \sum_{n=0}^{\infty} \frac{a_n}{n!} \zeta^n, \tag{4.6.9}$$

i.e. we regard g^2 as our coupling constant. From our general analysis, we know that a non-perturbative effect of the form (4.6.5) is associated to a singularity in the Borel plane at

$$A = \frac{d}{2\beta_0}. \tag{4.6.10}$$

This singularity is called the *IR renormalon singularity*. Since, in our conventions, $\beta_0 > 0$, this singularity is on the positive real axis and leads to an obstruction to Borel summability. This means that we have to perform lateral Borel resummations with a discontinuity of the form (3.2.39), times an extra factor of g^2 which comes from the fact that the perturbative series (4.6.2) starts at g^2. The ambiguity in the Borel resummation of the perturbative series is imaginary and of order

$$\left(g^2\right)^{1-b} \mathrm{e}^{-A/g^2}. \tag{4.6.11}$$

As in the calculation of the ground state energy for the double-well potential, the full observable (4.6.1) must be real, and this requires that the imaginary part in (4.6.11) *cancels* against an imaginary part coming from the non-perturbative contribution (4.6.5). This imaginary part must appear in the coefficient C in (4.6.5), and the ambiguity in the choice of the sign of this imaginary part must be correlated with the ambiguity in the lateral Borel resummations, in such a way that the cancellation takes place. As a consequence of this cancellation, and by comparing the exponent of the leading terms in g^2, we deduce that the exponent b in (4.6.11) must be given by

$$1 - b = \delta. \tag{4.6.12}$$

We then see that, by using RG arguments, we can relate perturbative and non-perturbative effects in a non-trivial way. In particular, we find that the location of the IR renormalon involves the one-loop coefficient of the beta function. The value of d depends on the observable under scrutiny.

What about the role of instantons in YM theory, and their effect on the large order behavior of perturbation theory? It has been argued that, just as the large order behavior in the double-well potential is due to a two-instanton configuration made out of an instanton–anti-instanton pair, the leading singularity in the Borel plane due to instantons corresponds to a configuration with total topological charge zero but with action equal to twice the action of a YM instanton ($n = 2$ in (4.4.4)):

$$S = \frac{16\pi^2}{g^2}.$$ (4.6.13)

This would lead to a singularity in the Borel plane at

$$\zeta = 16\pi^2.$$ (4.6.14)

However, for low values of d, renormalons are more important than instantons, since they lead to singularities which are closer to the origin. Indeed, for pure $SU(N)$ YM, the IR renormalon singularity is located at

$$\zeta = 16\pi^2 \frac{3d}{22N}.$$ (4.6.15)

The effect of this singularity in the large order behavior of perturbation theory has been recently observed in observables with $d = 1$ and $d = 4$.

4.7 Bibliographical notes

Useful introductions to gauge theories can be found in many textbooks. Our favorite ones are [12, 154, 158], but there are many others.

The θ dependence in YM theories is reviewed in [182]. Useful references on this topic can also be found in [83, 128, 141, 177]. Precise definitions of the topological susceptibility, in the continuum and in the lattice, are discussed for example in [97, 129, 130]. A detailed study of the two-point function of the topological density in the large N $\mathbb{CP}^{N-1}$ model, which makes manifest some of the subtleties of this quantity and its integral, can be found in [181]. The quantized nature of the topological charge for gauge configurations with finite action is explained in many references, for example [50, 157]. A particularly detailed presentation, which we have followed here, can be found in [186]. The theta vacuum of YM theory was discovered in [45, 112]. A different route to the theta vacuum, using the Hamiltonian quantization of the theory, can be found in [12, 110, 111].

Instantons in YM theory were first found in [25]. There are many useful reviews of instantons in YM theories. Introductory reviews can be found in Chapter 9 of [50], in the book [167], and in the more recent lectures [176, 178]. The review [75] is more advanced and covers the ADHM formalism first introduced in [16], as well as applications to supersymmetric gauge theories. The review [160] covers foundational material and applications to QCD. The study of Euclidean YM theory on general four-manifolds has become an important direction in modern geometry; a useful introduction can be found in [62], while the book [73] is a comprehensive reference. The calculation of quantum effects due to instantons was started in a classic paper by 't Hooft [175], which triggered a large literature. Quantum effects of instantons on manifolds of finite volume are considered in [151, 163] and reviewed in [1]. For the regularization of the determinants involved in the instanton calculation, we follow [127]. The partition function at one-loop in the background of a one-instanton on the four-sphere, for gauge group $SU(n)$, is computed in all details in [128], and it generalizes previous results in [24, 34, 46]. A lattice calculation of the theta dependence of the vacuum energy in $SU(3)$ YM theory, disagreeing with naive instanton expectations, can be found in [98].

A comprehensive review of renormalons can be found in [32]. The argument for the structure of IR renormalon singularities based on the RG equation is due to Parisi [152] and is reviewed in [105]. The instanton–anti-instanton singularity in YM theory is dicussed in [36]. Direct evidence for the role of IR renormalons in the large order behavior of perturbation theory in YM theories has been found in [23].

5

Instantons and fermions

5.1 Introduction

So far, in our study of instanton physics, we have focused on theories with bosonic fields only. In this chapter we will analyze instanton effects in theories with fermionic fields. We will first consider an extension of the quantum-mechanical models studied in Chapter 1, and in particular we will look at supersymmetric QM. This is our only incursion into supersymmetry in this book, but it has some important lessons for instanton physics: first of all, it gives yet another example of a phenomenon which is purely non-perturbative, namely the breaking of supersymmetry in some quantum mechanical models. Second, our study of these simplified models will give us a first taste of one important aspect of instanton calculus in the presence of fermions: the existence of zero modes for the fermion fields.

After this detailed study of instantons and fermions in a simplified setting, we will move on to QCD. We will first study chiral symmetries due to the presence of quark flavors, and we will present a quick introduction to the effective Lagrangian of QCD at low energies. This is a subject which is both elegant and relevant to the phenomenology of QCD. Finally, we will discuss the $U(1)$ problem in QCD and how it can be solved in principle by taking into account the anomaly in the axial current. In Chapter 7 we will make this more precise, at large N, through the Witten–Veneziano formula.

5.2 Instantons in supersymmetric Quantum Mechanics

We will now study a very interesting variant of the quantum mechanical models that were considered in Chapter 1: supersymmetric QM. We will restrict ourselves to one-dimensional models.

General aspects

In order to introduce supersymmetric QM, we have to add fermionic coordinates to the standard bosonic coordinate q. In terms of operators, this means that on top of the usual bosonic operators $\hat{q}$, $\hat{p}$, we have to introduce Grassmann operators $\hat{\psi}_{1,2}$, obeying the anticommutation relations,

$$\{\hat{\psi}_\alpha, \hat{\psi}_\beta\} = \delta_{\alpha\beta}. \tag{5.2.1}$$

These operators are formally Hermitian,

$$\hat{\psi}_\alpha^\dagger = \hat{\psi}_\alpha, \qquad \alpha = 1, 2. \tag{5.2.2}$$

It is also useful to consider the creation and annihilation operators

$$\hat{\psi}_\pm = \frac{1}{\sqrt{2}}\left(\hat{\psi}_1 \pm i\hat{\psi}_2\right), \tag{5.2.3}$$

which satisfy

$$\{\hat{\psi}_+, \hat{\psi}_-\} = 1, \qquad \hat{\psi}_\pm^2 = 0, \tag{5.2.4}$$

and

$$\hat{\psi}_\pm^\dagger = \hat{\psi}_\mp. \tag{5.2.5}$$

This algebra can be represented by the matrices

$$\hat{\psi}_+ = \begin{pmatrix} 0 & 1 \\ 0 & 0 \end{pmatrix}, \qquad \hat{\psi}_- = \begin{pmatrix} 0 & 0 \\ 1 & 0 \end{pmatrix}. \tag{5.2.6}$$

Wavefunctions are then represented by vector-valued objects,

$$\Psi(x) = \begin{pmatrix} \phi_1(x) \\ \phi_2(x) \end{pmatrix}. \tag{5.2.7}$$

We have the following representation for the operators:

$$\hat{\psi}_1 = \frac{1}{\sqrt{2}}\begin{pmatrix} 0 & 1 \\ 1 & 0 \end{pmatrix}, \qquad \hat{\psi}_2 = \frac{1}{\sqrt{2}}\begin{pmatrix} 0 & -i \\ i & 0 \end{pmatrix}, \tag{5.2.8}$$

and their commutator is given by

$$[\hat{\psi}_1, \hat{\psi}_2] = \frac{i}{2}\sigma_3. \tag{5.2.9}$$

The operator

$$F = \frac{1 - \sigma_3}{2} \tag{5.2.10}$$

can be regarded as the fermion number: eigenstates of σ_3 with eigenvalue ± 1 have fermion number 0 or 1, respectively, and acting with a fermionic operator on one of these eigenstates changes the fermion number.

We will define the fermionic version of QM by the Hamiltonian,

$$\hat{H} = \frac{1}{2}\hat{p}^2 + W(\hat{q}) - \frac{i}{2}Y(\hat{q})[\hat{\psi}_1, \hat{\psi}_2]. \tag{5.2.11}$$

When acting on wavefunctions of the form (5.2.7), the Hamiltonian is given by the operator

$$\hat{H} = -\frac{1}{2}\frac{\partial^2}{\partial q^2} + W(q) + \frac{1}{2}Y(q)\sigma_3. \tag{5.2.12}$$

In matrix notation, we have

$$H = \begin{pmatrix} -\partial_q^2 + W(q) + Y(q)/2 & 0 \\ 0 & -\partial_q^2 + W(q) - Y(q)/2 \end{pmatrix}. \tag{5.2.13}$$

Since σ_3 commutes with the Hamiltonian, we can diagonalize them simultaneously. Therefore, we can study the spectrum by considering wavefunctions of the form

$$\begin{pmatrix} \phi_1(q) \\ 0 \end{pmatrix}, \quad \begin{pmatrix} 0 \\ \phi_2(q) \end{pmatrix}. \tag{5.2.14}$$

Let us assume that the functions $W(q)$, $Y(q)$ appearing in (5.2.12) satisfy

$$W(q) = \frac{1}{2}\omega^2(q), \qquad Y(q) = \omega'(q), \tag{5.2.15}$$

where $\omega(q)$ is called the *superpotential*. In this case, the above quantum mechanical system has an additional fermionic symmetry called *supersymmetry* (SUSY). There are two equivalent ways to see this. In the Hamiltonian picture, we note that there are two conserved fermionic charges, or *supercharges*, defined by

$$\begin{aligned} \hat{Q}_+ &= (p - i\omega)\,\hat{\psi}_+, \\ \hat{Q}_- &= (p + i\omega)\,\hat{\psi}_-, \end{aligned} \tag{5.2.16}$$

and satisfying

$$\hat{Q}_\pm^\dagger = \hat{Q}_\mp. \tag{5.2.17}$$

In matrix notation, they can be written as

$$\hat{Q}_+ = \begin{pmatrix} 0 & -i\left(\partial_q + \omega(q)\right) \\ 0 & 0 \end{pmatrix}, \quad \hat{Q}_- = \begin{pmatrix} 0 & 0 \\ -i\left(\partial_q - \omega(q)\right) & 0 \end{pmatrix}. \tag{5.2.18}$$

Indeed, one finds that

$$[\hat{H}, \hat{Q}_\pm] = 0 \tag{5.2.19}$$

if and only if

$$Y(q) = \omega'(q), \qquad V'(q) = Y(q)\omega(q), \tag{5.2.20}$$

which lead to the conditions (5.2.15). In addition, one has

$$H = \frac{1}{2}\{\hat{Q}_+, \hat{Q}_-\}. \tag{5.2.21}$$

Alternatively, we can deduce the existence of conserved fermionic charges from the existence of fermionic symmetries in the Lagrangian describing the theory. The Lagrangian is given by,

$$L = \frac{1}{2}\dot{q}^2 + \frac{i}{2}(\psi_-\dot{\psi}_+ - \dot{\psi}_-\psi_+) + \frac{1}{2}D^2 + Df'(q) + \frac{[\psi_-, \psi_+]}{2}f''(q), \quad (5.2.22)$$

and it is invariant under the following infinitesimal transformation,

$$i\delta q = \epsilon_-\psi_- - \psi_+\epsilon_+,$$
$$\delta\psi_\pm = \mp i\epsilon_\mp D + \epsilon_\mp\dot{q}, \quad (5.2.23)$$
$$\delta D = \epsilon\dot{\psi}_+ + \dot{\psi}_-\epsilon_-,$$

where $\epsilon_\pm$ are fermionic generators for the two symmetries. Since D is an auxiliary field, we can integrate it out to obtain $D = -f'(q)$. It is easy to verify that the above Lagrangian leads to the Hamiltonian (5.2.11), and that the above fermionic symmetries lead to the charges $\hat{Q}_\pm$, after setting

$$f'(q) = \omega(q). \quad (5.2.24)$$

Note that in the supersymmetric theory, with Hamiltonian

$$H = \frac{1}{2}p^2 + \frac{1}{2}\omega^2(q) + \frac{1}{2}\omega'(q)\sigma_3, \quad (5.2.25)$$

the fermionic sectors with σ_3 eigenvalues ± 1 have different potentials,

$$W_\pm(q) = \frac{1}{2}\omega^2(q) \pm \frac{1}{2}\omega'(q). \quad (5.2.26)$$

Supersymmetry breaking

Theories with supersymmetry have peculiar properties. First of all, the energy of any eigenstate of the Hamiltonian is positive or zero. Moreover, a state can have zero energy if and only if it is annihilated by all supercharges. Indeed, if

$$H|\psi\rangle = E|\psi\rangle, \quad (5.2.27)$$

then

$$E = \langle\psi|H|\psi\rangle = \frac{1}{2}\langle\psi|\left(\hat{Q}_+\hat{Q}_- + \hat{Q}_+\hat{Q}_-\right)|\psi\rangle$$
$$= \frac{1}{2}\left(\|\hat{Q}_-|0\rangle\|^2 + \|\hat{Q}_+|0\rangle\|^2\right), \quad (5.2.28)$$

where we used (5.2.17) and (5.2.21). Since the last term is a sum of positive-definite quantities, we must have $E \geq 0$. Moreover, $E = 0$ if and only if

$$\hat{Q}_+|\psi\rangle = \hat{Q}_-|\psi\rangle = 0. \quad (5.2.29)$$

This means that a state with zero energy must be annihilated by the supercharges. Conversely, a state which is not annihilated by both supercharges must have a non-zero energy.

We can now ask what are the conditions to have spontaneous breaking of supersymmetry. As with other symmetries, a supersymmetric theory is spontaneously broken if the ground state is not invariant under the action of at least one of the supercharges. Therefore, such a state must have a non-zero energy E_0. If this is the case, there are actually *two* states with the same energy E_0 (but different fermion number). Indeed, let $|\psi\rangle$ be a non-supersymmetric ground state:

$$\hat{H}|\psi\rangle = E_0|\psi\rangle. \tag{5.2.30}$$

Since we can diagonalize $\hat{H}$ and σ_3 simultaneously, we can assume that $|\psi\rangle$ has $\sigma_3 = \pm 1$, and in particular that $\hat{Q}_\pm|\psi\rangle = 0$. Then, if

$$|\psi_\mp\rangle = \hat{Q}_\mp|\psi\rangle \neq 0 \tag{5.2.31}$$

we have

$$\hat{H}|\psi_\mp\rangle = \hat{H}\hat{Q}_\mp|\psi\rangle = \hat{Q}_\mp\hat{H}|\psi\rangle = E_0|\psi_\mp\rangle \tag{5.2.32}$$

and the ground state is doubly degenerate.

Witten has pointed out a very simple criterion to decide whether SUSY is broken, in supersymmetric QM, for a given superpotential $\omega(q)$. A SUSY ground state must be annilihated by both supercharges, i.e. it has to satisfy the equations

$$\left(\frac{\partial}{\partial q} - \omega(q)\right)\phi_1(q) = 0, \qquad \left(\frac{\partial}{\partial q} + \omega(q)\right)\phi_2(q) = 0, \tag{5.2.33}$$

with the immediate solution

$$\phi_1(q) = \phi_1(q_0)\exp\left(\int_{q_0}^q \omega(q')\mathrm{d}q'\right),$$
$$\phi_2(q) = \phi_2(q_0)\exp\left(-\int_{q_0}^q \omega(q')\mathrm{d}q'\right). \tag{5.2.34}$$

Let us assume that $\omega(q)$ is a polynomial. Then, there are two cases to consider.

1. If the highest power of $\omega(q)$ is odd, the highest power of $\int \omega(q')\mathrm{d}q'$ will be even, and we will have one normalizable ground state with $\phi_1 = 0$ or $\phi_2 = 0$ (depending on the sign of the highest power). In this case supersymmetry is unbroken.
2. If the highest power of $\omega(q)$ is even, the highest power of $\int \omega(q')\mathrm{d}q'$ will be odd, and none of the above functions is normalizable. In this case SUSY is broken, and we expect a degenerate ground state of non-zero energy.

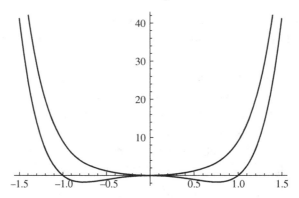

Figure 5.1 The potentials $W_\pm(q)$ (top and bottom line, respectively) for the Hamiltonian (5.2.36), represented here for $g = 3$.

As an example of the first situation, let us take

$$\omega(q) = gq^3, \tag{5.2.35}$$

so that

$$H = \frac{1}{2}p^2 + \frac{g^2 q^6}{2} + \frac{3gq^2}{2}\sigma_3. \tag{5.2.36}$$

The ground state in this case is

$$\Phi = \begin{pmatrix} 0 \\ \phi_2(q) \end{pmatrix}, \qquad \phi_2(q) = Ce^{-gq^4/4}, \tag{5.2.37}$$

which is normalizable. Notice that the potentials $W_\pm(q)$ are quite different in this case (see Fig. 5.1), and the ground state is an eigenstate for $W_-(q)$.

An example of the second situation is given by the quadratic superpotential

$$\omega(q) = \lambda q^2 - \mu^2. \tag{5.2.38}$$

In this case, the Hamiltonian is

$$H = \frac{1}{2}p^2 + \frac{\lambda^2}{2}\left(q^2 - \frac{\mu^2}{\lambda}\right)^2 + \lambda q\sigma_3. \tag{5.2.39}$$

The potentials $W_\pm(q)$ are shown in Fig. 5.2.

The above result on spontaneous supersymmetry breaking is an exact result. What would we obtain by using perturbation theory? In order to see this, it is useful to restore $\hbar$ by using dimensional analysis. It is easy to see that there is a relative factor of $\hbar$ in the expression for the potentials $W_\pm(q)$ in (5.2.26),

$$W_\pm(q) = \frac{1}{2}\omega^2(q) \pm \frac{\hbar}{2}\omega'(q). \tag{5.2.40}$$

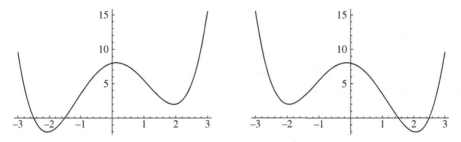

Figure 5.2 The potentials $W_\pm(q)$ (left and right, respectively) for the Hamiltonian (5.2.39), represented here for $\lambda = 1$, $\mu = 2$.

Classically, when $\hbar \to 0$, the number of supersymmetric ground states corresponds to the number of zeros of $\omega(q)^2/2$, i.e. to the zeros of $\omega(q)$, since each of them leads to a state of zero energy. This can be seen at first order in perturbation theory in $\hbar$: if the superpotential behaves near the zero at $q = q_0$ as

$$\omega(q) = c(q - q_0) + \mathcal{O}\left((q - q_0)^2\right), \tag{5.2.41}$$

then the Hamiltonian is given, at leading order, by

$$H \approx \frac{p^2}{2} + \frac{c^2}{2}(q - q_0)^2 + \frac{\hbar c}{2}\sigma_3. \tag{5.2.42}$$

The ground state of this approximate Hamiltonian has zero energy, since the first term is a harmonic oscillator Hamiltonian with energy $\hbar|c|/2$ for its ground state, while the last term has energies $\pm\hbar|c|/2$, where the sign depends on the eigenvalue of σ_3 and the sign of c. Therefore, the state with eigenvalue $-\text{sign}(c)$ for σ_3 has zero energy at first order in perturbation theory in $\hbar$.

However, the consequences of this semiclassical analysis are at odds with the exact result found before: when the superpotential is even, the perturbative analysis predicts that the number of supersymmetric vacua is equal to the number of zeros of $\omega(q)$, while we have just seen that in an exact analysis there are none. For example, when $\omega(q)$ is given by (5.2.38), we find two ground states at leading order in perturbation theory, and located at

$$q_\pm = \pm\frac{\mu}{\sqrt{\lambda}}. \tag{5.2.43}$$

We might think that higher orders of perturbation theory will give a non-zero energy to these states, but this is not the case. The reason is that, in perturbation theory, we expand around a given zero of the superpotential. This procedure is not sensitive to the parity of the total number of zeros of $\omega(q)$, which is the crucial ingredient to determine whether supersymmetry is unbroken or not. Therefore, the ground state energy will vanish at all orders in perturbation theory if it vanishes at leading order.

We conclude that, when $\omega(q)$ is given by (5.2.38), the spontaneous breaking of supersymmetry must be due to non-perturbative effects. Let us study this issue in more detail. If supersymmetry is spontaneously broken, there must be two degenerate ground states, of the form

$$\langle q|\Psi\rangle = \begin{pmatrix} \phi_1(q) \\ 0 \end{pmatrix}, \qquad \langle q|\chi\rangle = \begin{pmatrix} 0 \\ \phi_2(q) \end{pmatrix}. \tag{5.2.44}$$

These two states are related by the supersymmetric charges. If we assume that they are both normalized, we find

$$\hat{Q}_-|\Psi\rangle = i\sqrt{2E}|\chi\rangle, \qquad \hat{Q}_+|\chi\rangle = -i\sqrt{2E}|\Psi\rangle, \tag{5.2.45}$$

where we have chosen the phase on the right hand side to simplify our analysis below. These relations can be written, in the q representation, as

$$\begin{aligned} \left(\frac{\partial}{\partial q} - \omega(q)\right)\phi_1(q) &= -\sqrt{2E}\phi_2(q), \\ \left(\frac{\partial}{\partial q} + \omega(q)\right)\phi_2(q) &= \sqrt{2E}\phi_1(q). \end{aligned} \tag{5.2.46}$$

When $\hbar$ is small, and due to our perturbative analysis above, we expect the ground states $|\Psi\rangle$, $|\chi\rangle$ to be localized around $q_{\mp}$, respectively (the values of $q_{\pm}$ are given in (5.2.43)). This is because the coefficient c appearing in (5.2.41) is negative for q_- and positive for q_+, and they correspond to the eigenvalues ± 1 of σ_3, respectively. We can find a solution to the equations (5.2.46) by starting with a localized ansatz and then iterating. We take as our starting point

$$\phi_1^{(0)}(q) = \delta(q - q_-), \qquad \phi_2^{(0)}(q) = \delta(q - q_+). \tag{5.2.47}$$

To construct the iterations, it is useful to consider the function $f(q)$, defined by (5.2.24) up to an integration constant. It is chosen in such a way that $f(q)$ is odd. In the case of (5.2.38), $f(q)$ is given by

$$f(q) = \frac{\lambda}{3}q^3 - \mu^2 q. \tag{5.2.48}$$

It has a local maximum at $q = q_-$, and a local minimum at $q = q_+$. It is easy to see that the first iteration of (5.2.46) gives

$$\phi_1^{(1)}(q) = \frac{1}{N_1}\theta(q_+ - q)e^{f(q)}, \qquad \phi_2^{(2)}(q) = \frac{1}{N_2}\theta(q - q_-)e^{-f(q)}. \tag{5.2.49}$$

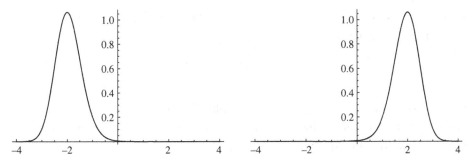

Figure 5.3 The functions (5.2.49), for $\lambda = 1$, $\mu = 2$. The figure on the left represents the function $\phi_1^{(1)}(q)$, centered around $q_- = -2$, while the figure on the right represents the function $\phi_2^{(1)}(q)$, centered around $q_+ = 2$.

After including $\hbar$ factors explicitly, we see that these wavefunctions are proportional to

$$\exp\left(\pm f(q)/\hbar\right). \tag{5.2.50}$$

In the limit of $\hbar \to 0$, they are very peaked around $q_\mp$, respectively, as shown in Fig. 5.3. In fact, (5.2.49) can be taken to be the leading normalizable approximation to the wavefunctions in the semiclassical limit, as can be seen by looking at the next iteration. The coefficients $N_{1,2}$ can be fixed by normalization. We find:

$$N_1^2 = \int_{-\infty}^{q_+} e^{2f(q)} \, dq, \qquad N_2^2 = \int_{q_-}^{\infty} e^{-2f(q)} \, dq. \tag{5.2.51}$$

Again, when $\hbar \to 0$, these integrals can be evaluated by a Gaussian integral around the peaks at $q = q_\mp$, respectively, and one finds

$$N_1^2 \approx N_2^2 \approx N^2 = e^{\Delta f} \sqrt{\frac{\pi}{|f''(q_+)|}}, \tag{5.2.52}$$

where

$$\Delta f = 2f(q_-) = f(q_-) - f(q_+). \tag{5.2.53}$$

The approximate solutions (5.2.49) can be used to calculate the approximate values of some correlators. We want to use this approximation to show that the spontaneous breaking of supersymmetry is a non-perturbative phenomenon. To do this, we will calculate a simple order parameter for supersymmetry breaking, as well as the energy of the ground state. Let us consider the vev

$$F = i\langle\Psi|\{\widehat{Q}_+, \widehat{\psi}_-\}|\Psi\rangle, \tag{5.2.54}$$

where $|\Psi\rangle$ is the state in (5.2.44). Clearly, if supersymmetry were unbroken and $|\Psi\rangle$ were a ground state, the above vev would vanish. A non-zero value for this vev is therefore a signature of supersymmetry breaking. In order to evaluate the vev, we insert a complete set of states,

$$F = i \sum_{\phi} \langle \Psi | \widehat{Q}_+ | \phi \rangle \langle \phi | \hat{\psi}_- | \Psi \rangle. \tag{5.2.55}$$

It is easy to see that $\langle \Psi | \widehat{Q}_+ | \phi \rangle$ is non-zero unless $|\phi\rangle$ has the same energy as Ψ (but different fermion number): indeed, since $\widehat{Q}_+$ is a conserved charge and commutes with the Hamiltonian, one has,

$$\Psi | [H, \widehat{Q}_+] | \phi \rangle = (E_\Psi - E_\phi) \langle \Psi | \widehat{Q}_+ | \phi \rangle = 0. \tag{5.2.56}$$

Therefore, the only state contributing to the sum in (5.2.55) is the other state in (5.2.44), $|\chi\rangle$, and we find

$$F = i \langle \Psi | \widehat{Q}_+ | \chi \rangle \langle \chi | \hat{\psi}_- | \Psi \rangle. \tag{5.2.57}$$

We can now calculate these products at leading order as $\hbar \to 0$ by using (5.2.52). We find

$$\langle \chi | \hat{\psi}_- | \Psi \rangle = \int \phi_1(q) \phi_2(q) \mathrm{d}q \approx \frac{1}{N^2} \Delta q, \tag{5.2.58}$$

where

$$\Delta q = q_+ - q_-. \tag{5.2.59}$$

Similarly, we can calculate

$$i \langle \Psi | \widehat{Q}_+ | \chi \rangle \approx \frac{1}{N^2}. \tag{5.2.60}$$

Using the value of N^2 calculated above, we find

$$\langle \chi | \hat{\psi}_- | \Psi \rangle \approx \sqrt{\frac{|f''(q_+)|}{\pi}} e^{-\Delta f} \Delta q,$$

$$i \langle \Psi | \widehat{Q}_+ | \chi \rangle \approx \sqrt{\frac{|f''(q_+)|}{\pi}} e^{-\Delta f}, \tag{5.2.61}$$

and we conclude that

$$F = \frac{|f''(q_+)|}{\pi} e^{-2\Delta f} \Delta q. \tag{5.2.62}$$

If we remember that, after restoring units, Δf should be multiplied by a factor of $1/\hbar$, we conclude that this quantity is *exponentially small* when $\hbar \to 0$. We have then shown that spontaneous supersymmetry breaking is, in these models, a non-perturbative effect. To verify this, we calculate the energy of the ground state. This is given by

$$E = \langle \Psi | \widehat{H} | \Psi \rangle. \tag{5.2.63}$$

We can evaluate this vev by inserting a complete set of states, as in (5.2.55), and we find, by using the second vev in (5.2.61),

$$E = \frac{1}{2}\langle\Psi|\widehat{Q}_+|\chi\rangle\langle\chi|\widehat{Q}_-|\Psi\rangle = \frac{1}{2}\|\langle\Psi|\widehat{Q}_+|\chi\rangle\|^2 \approx \frac{|f''(q_+)|}{2\pi}e^{-2\Delta f}. \tag{5.2.64}$$

This is also exponentially small in $\hbar$. This explains why we could not detect a non-zero ground state energy by doing perturbation theory in $\hbar$.

Instantons and fermionic zero modes

In the previous section, we showed that, in supersymmetric QM with a super-potential ω given by (5.2.38), supersymmetry is broken, and SUSY breaking is non-perturbative in $\hbar$. We will now show explicitly that this is an instanton effect, i.e. we will show that it is due to a non-trivial solution to the EOM in the Euclidean path integral. To do this, we will compute the vev (5.2.58) in the path integral formalism. Since the wavefunction χ is concentrated around q_+, and Ψ is concentrated around q_-, this is equivalent to calculating the propagator from q_- to q_+ with an insertion of the fermion field $\psi_-(t)$. In the path integral formalism, this is given by

$$\langle\chi|\widehat{\psi}_-|\Psi\rangle = \lim_{\beta\to\infty} \int \mathcal{D}q\,\mathcal{D}\psi_+\,\mathcal{D}\psi_-\,\psi_-(t)\,e^{-S}, \tag{5.2.65}$$

where the integration is over paths satisfying

$$q(-\beta/2) = q_-, \qquad q(\beta/2) = q_+. \tag{5.2.66}$$

In the semiclassical limit, the leading contribution to the integration over paths comes from the instanton of the double-well potential. As we will see, for this path integral not to be zero, the field ψ_- must have a zero mode in the background of this instanton, which can be used to "soak up" the insertion of the Grasmannian operator ψ_- in the path integral. This is a basic aspect of instanton calculus in the presence of fermions, which is very well illustrated by this simple example.

To do the instanton calculation, we need the Euclidean version of the theory with Lagrangian (5.2.22). After a Wick rotation, we find

$$L_E = \frac{1}{2}\dot{q}^2 + \frac{1}{2}\omega^2(q) - \frac{1}{2}\psi_-\dot{\psi}_+ - \frac{1}{2}\psi_+\dot{\psi}_- + \psi_-\psi_+\omega'(q). \tag{5.2.67}$$

The Euclidean EOMs are

$$\ddot{q} - \omega'(q)\omega(q) - \psi_-\psi_+\omega''(q) = 0,$$
$$\dot{\psi}_\pm \mp \omega'(q)\psi_\pm = 0. \tag{5.2.68}$$

A solution to the EOM can be obtained by setting to zero the fermionic fields:

$$\psi_\pm^c = 0. \tag{5.2.69}$$

The bosonic equation can then be integrated once, and the EOM, which was originally a second order ODE, becomes a first order ODE,

$$\dot{q}_c(t) = \pm \omega(q_c(t)). \tag{5.2.70}$$

We have fixed the integration constant in such a way that the resulting solution has zero energy, since these are the solutions that are relevant in the limit $\beta \to \infty$. The equation (5.2.70) can be integrated immediately:

$$t - t_0 = \pm \int_{q_c(t_0)}^{q_c(t)} \frac{dq'}{\omega(q')}. \tag{5.2.71}$$

It is interesting to note that (5.2.70) is equivalent to the conditions

$$\delta\psi_\pm = 0. \tag{5.2.72}$$

Indeed, as we can see from (5.2.23), in the Euclidean theory we have

$$\delta\psi_\pm = i\epsilon_\mp (\dot{q} \pm \omega(q)). \tag{5.2.73}$$

The condition (5.2.72) defines *supersymmetric configurations*, i.e. configurations which preserve supersymmetry, and it leads automatically to first order equations which solve the EOM. We saw a similar reduction to first order equations in the analysis of the instantons of YM theory, which could be obtained from the (anti)self-duality equations. As in that case, we also have here two types of solutions depending on the choice of sign in this equation. They go from one of the zeros of $\omega(q)$ in the infinite past, to the other zero in the infinite future, i.e. they interpolate between q_- and q_+. The solution with the $+$ sign in (5.2.70) goes from q_+ in the infinite past to q_- in the infinite future, while the solution with the $-$ sign goes the other way around. The classical action evaluated on these trajectories is

$$S_c = \int_{-\infty}^{\infty} \dot{q}_c^2(t)dt = \int_{q_+}^{q_-} \omega(q)dq = \Delta f, \tag{5.2.74}$$

where we used that $f(q)$ is a primitive of $\omega(q)$, and Δf was defined in (5.2.53). For the superpotential (5.2.38) the solutions to (5.2.70) are

$$q_c(t) = \mp \frac{\mu}{\sqrt{\lambda}} \tanh\left(\mu\sqrt{\lambda}(t - t_0)\right), \tag{5.2.75}$$

corresponding to the $\pm$ sign in (5.2.70). These are precisely the (anti)instantons of the double-well potential (1.8.13), and they are depicted in Fig. 1.12.

As in the bosonic case, in order to do the calculation at one-loop, we should consider the quadratic fluctuations around the classical solution (5.2.70). If we denote

$$q(t) = q_c(t) + r(t), \qquad \psi_{\pm}(t) = \psi_{\pm}^c(t) + \eta_{\pm}(t), \tag{5.2.76}$$

we find that the Euclidean action is given by

$$S = S_c + \int r(t) \, (\mathbf{M}_B r)(t) dt$$

$$- \frac{1}{2} \int dt \, (\eta_{+}(t) \quad \eta_{-}(t)) \, \mathbf{M}_F \begin{pmatrix} \eta_{+}(t) \\ \eta_{-}(t) \end{pmatrix}. \tag{5.2.77}$$

Here, the bosonic operator $\mathbf{M}_B$ is just a particular case of (1.4.23) for $V(q) = -\omega^2(q)/2$, i.e.

$$\mathbf{M}_B = -\partial_t^2 + \omega''(q_c(t))\omega(q_c(t)) + (\omega'(q_c(t)))^2. \tag{5.2.78}$$

The fermionic operator is of the form

$$\mathbf{M}_F = \begin{pmatrix} 0 & -M_F^{\dagger} \\ M_F & 0 \end{pmatrix}, \tag{5.2.79}$$

with

$$M_F = \partial_t - \omega'(q_c(t)), \qquad M_F^{\dagger} = -\partial_t - \omega'(q_c(t)). \tag{5.2.80}$$

An important property of the bosonic operator is that it factorizes: when $\dot{q}_c(t) = \pm\omega(q_c(t))$, we have

$$\mathbf{M}_B = -\left(\partial_t \pm \omega'(q_c(t))\right)\left(\partial_t \mp \omega'(q_c(t))\right)$$

$$= \begin{cases} M_F^{\dagger} M_F, & \text{if } \dot{q}_c(t) = \omega(q_c(t)), \\ M_F M_F^{\dagger}, & \text{if } \dot{q}_c(t) = -\omega(q_c(t)). \end{cases} \tag{5.2.81}$$

In the case of the quadratic superpotential (5.2.38) we are considering, the bosonic operator is given by

$$- \partial_t^2 + 2\lambda\mu^2 \left(2 - \frac{3}{\cosh\left(\mu\sqrt{\lambda}(t - t_0)\right)} \right) = \lambda\mu^2 \mathbf{M}_{2,2}, \tag{5.2.82}$$

where $\mathbf{M}_{2,2}$ is the Pöschl–Teller operator (1.5.1), and its variable is $\mu\sqrt{\lambda}(t - t_0)$. In this case, the factorization (5.2.81) is nothing but the factorization of the Hamiltonian in (1.5.3). If we take for example the case with $\dot{q}_c(t) = -\omega(q_c(t))$, we find that

$$M_F^{\dagger} = -\partial_t - 2\mu\sqrt{\lambda}\tanh\left(\mu\sqrt{\lambda}(t - t_0)\right) \tag{5.2.83}$$

which is (minus) the operator A_2, up to a rescaling of t. Similarly, M_F is minus $A_2^\dagger$, and the factorization (5.2.81) can be written as

$$\mathbf{M}_{2,2} = A_2^\dagger A_2. \tag{5.2.84}$$

This is the case $\ell = m = 2$ of the first factorization property in (1.5.3). The solvability of the spectral problem for the Pöschl–Teller operator can then be understood as a consequence of this hidden supersymmetric structure.

We now study the spectrum of the operators $\mathbf{M}_{B,F}$. As in our analysis of instantons in ordinary QM, we have a bosonic zero mode corresponding to $\dot{q}_c(t)$, i.e.

$$q_0(t) \propto \omega(q_c(t)), \tag{5.2.85}$$

as well as non-zero modes. The eigenvalue problem for the operators M_F, $M_F^\dagger$ is given by:

$$\begin{aligned} M_F \psi_+^n &= \lambda_F^n \psi_-^n, \\ M_F^\dagger \psi_-^n &= \lambda_F^n \psi_+^n. \end{aligned} \tag{5.2.86}$$

The zero modes are then defined by

$$M_F \psi_+^0 = 0, \qquad M_F^\dagger \psi_-^0 = 0. \tag{5.2.87}$$

The solution to the zero mode equations is, as in (5.2.34),

$$\psi_\pm^0(t) \propto \exp\left(\pm \int^t \omega'(q_c(t')) \, dt' \right). \tag{5.2.88}$$

However, they cannot both be normalizable: when $\dot{q}_c(t) = \pm\omega(q_c(t))$, there is a normalizable zero mode

$$\psi_\pm^0(t) \propto \omega(q_c(t)), \tag{5.2.89}$$

i.e. there are fermionic zero modes for the fluctuations η_+ *or* for η_-, but not for both. Which of the two fluctuation fields has a zero mode depends on the choice of instanton or anti-instanton background in (5.2.70). Another way to state this result is to introduce an "instanton number" ν which is $+1$ (-1) for instantons (anti-instantons, respectively). Then, if we denote by $N_\pm$ the number of zero modes of $\eta_\pm$, we have the equality

$$N_+ - N_- = -\nu. \tag{5.2.90}$$

This is a toy example of an "index theorem," relating the number of zero modes of fermions to a topological index characterizing the bosonic instanton background.

In our calculation of (5.2.65), we have to consider an instanton satisfying $\dot{q}_c(t) = -\omega(q_c(t))$, therefore $\eta_-(t)$ has a zero mode but $\eta_+(t)$ does not. By acting with M_F on the second equation in (5.2.86) we find that ψ_-^n satisfies

$$\mathbf{M}_B \psi_-^n = \left(\lambda_F^n\right)^2 \psi_-^n, \tag{5.2.91}$$

i.e. it is an eigenvector of the bosonic operator $\mathbf{M}_B$. Therefore, for any solution of the bosonic eigenvalue equation

$$\mathbf{M}_B q_n = (\lambda_B^n)^2 q_n \tag{5.2.92}$$

we can obtain a properly normalized solution of the fermionic eigenvalue equations by setting

$$\psi_-^n(t) = q_n(t), \qquad \lambda_F^n = \lambda_B^n, \qquad n \geq 0. \tag{5.2.93}$$

This pairing of the bosonic and fermionic modes is typical of supersymmetric systems. The eigenfunctions ψ_+^n are then given by

$$\psi_+^n(t) = \frac{1}{\lambda_F^n} M_F^\dagger q_n(t), \qquad n \geq 1, \tag{5.2.94}$$

since the zero mode q_0 is annihilated by $M_F^\dagger$.

Example 5.1 For the quadratic superpotential (5.2.38) we can write the modes $\psi_\pm^n$ very explicitly by using the results for the Pöschl–Teller operator with $\ell = 2$ (for simplicity we set $\mu = \lambda = 1$). For $\psi_-^n(t) = q_n(t)$ we have two bound states, given in (1.5.15), (1.5.14), and a continuum of scattering states (1.5.13) with the explicit expression

$$\psi_-^{(k)}(t) \propto -\left(k^2 - 2 + 3\mathrm{sech}^2(t) + 3ik\tanh(t)\right) e^{ikt}. \tag{5.2.95}$$

For the states $\psi_+^n(t)$ we have a bound state

$$\psi_+^1(t) \propto \frac{1}{\cosh(t)} \tag{5.2.96}$$

and a continuum given by (1.5.11), or explicitly,

$$\psi_+^{(k)}(t) \propto (-ik + \tanh(t)) e^{ikt}. \tag{5.2.97}$$

The fermionic eigenvalue for the scattering states is given by the square root of the energy (1.5.17) with $\ell = m = 2$,

$$\omega_F^{(k)} = \sqrt{k^2 + 4}. \tag{5.2.98}$$

$\square$

We are now ready to evaluate the path integral (5.2.65) at one-loop. The expansion of the fermionic fluctuation $\eta_-(t)$ is given by

$$\eta_-(t) = \zeta_0^- \frac{\dot{q}_c^{t_0}(t)}{\|\dot{q}_c\|} + \sum_{n \geq 1} \zeta_n^- q_n(t), \qquad (5.2.99)$$

where the first term corresponds to the zero mode $\dot{q}_c(t)$, while the fermionic fluctuation $\eta_+(t)$ has the expansion

$$\eta_+(t) = \sum_{n \geq 1} \zeta_n^+ \psi_+^n(t). \qquad (5.2.100)$$

Using the orthonormality of the q_n and the ψ_+^n, we find that the second term in (5.2.77) is

$$\sum_{n \geq 1} \lambda_F^n \zeta_n^+ \zeta_n^- . \qquad (5.2.101)$$

The measure for the zero modes is given by,

$$\int \frac{dc_0}{\sqrt{2\pi}} \int d\zeta_0^- = \frac{S_c^{1/2}}{\sqrt{2\pi}} \int dt_0 \int d\zeta_0^- . \qquad (5.2.102)$$

The integration over ζ_0^- picks up the first term in (5.2.99). Notice that, if there were no insertion of the fermionic field $\psi_-(t)$, the path integral would *vanish* due to the presence of the fermionic zero mode. The Gaussian integration over the fermionic non-zero modes is simply

$$\int \prod_{n \geq 1} d\zeta_n^+ d\zeta_n^- \exp\left[-\sum_{n \geq 1} \lambda_F^n \zeta_n^+ \zeta_n^- \right] = \prod_{n \geq 1} \lambda_F^n . \qquad (5.2.103)$$

Putting all these results together we find that, at one-loop,

$$\lim_{\beta \to \infty} \int \mathcal{D}q \, \mathcal{D}\psi_+ \, \mathcal{D}\psi_- \, \psi_-(t) \, e^{-S} \approx -\frac{e^{-S_c}}{\sqrt{2\pi}} \left[\frac{\det' \mathbf{M}_F}{\det' \mathbf{M}_B} \right]^{1/2} \int_{-\infty}^{\infty} dt_0 \frac{d}{dt_0} q_c^{t_0}(t)$$

$$= \frac{e^{-\Delta f}}{\sqrt{2\pi}} \left[\frac{\det' \mathbf{M}_F}{\det' \mathbf{M}_B} \right]^{1/2} \Delta q. \qquad (5.2.104)$$

In writing this result, we used that $\dot{q}_c^{t_0}(t) = -dq_c^{t_0}(t)/dt_0$, as well as the value of the classical action given in (5.2.74). Δq was defined in (5.2.59). The quotient of determinants is given by,

$$\left[\frac{\det' \mathbf{M}_F}{\det' \mathbf{M}_B} \right]^{1/2} = \prod_{n \geq 1} \frac{\lambda_F^n}{\lambda_B^n} . \qquad (5.2.105)$$

Due to (5.2.93), one could think that this is just one. Indeed, this would be the case if the spectrum were discrete, as we assumed in the above simplified derivation. However, in the presence of a continuous spectrum this "cancellation between fermionic and bosonic degrees of freedom" does not take place. To see this, we have first to make sense of the formal expression (5.2.105). For the bosonic determinant, we can use the results in Chapter 1. It remains to give an appropriate definition of the fermionic determinant. One way to proceed is to define it in terms of bosonic determinants, as the square root of

$$\det'(\mathbf{M}_F\mathbf{M}_F^\dagger) = \det' \begin{pmatrix} M_F^\dagger M_F & 0 \\ 0 & M_F M_F^\dagger \end{pmatrix} = \det\left(M_F^\dagger M_F\right) \det'\left(M_F M_F^\dagger\right),$$

(5.2.106)

where the operator

$$M_F^\dagger M_F = -\partial_t^2 + \left(\omega'(q_c(t))\right)^2 - \omega''(q_c(t))\omega(q_c(t))$$

(5.2.107)

has no zero modes, and $M_F M_F^\dagger = \mathbf{M}_B$. Therefore, we find,

$$\frac{\det' \mathbf{M}_F}{\det' \mathbf{M}_B} = \left(\frac{\det\left(-\partial_t^2 + \left(\omega'(q_c(t))\right)^2 - \omega''(q_c(t))\omega(q_c(t))\right)}{\det'\left(-\partial_t^2 + \left(\omega'(q_c(t))\right)^2 + \omega''(q_c(t))\omega(q_c(t))\right)}\right)^{1/2}.$$

(5.2.108)

In the case of the quadratic superpotential, we can compute (5.2.108) by using the results on the spectrum of Pöschl–Teller operators. Indeed, for (5.2.38), we find

$$-\partial_t^2 + \left(\omega'(q_c(t))\right)^2 - \omega''(q_c(t))\omega(q_c(t))$$

$$= -\partial_t^2 + \mu^2\lambda\left(4 - \frac{2}{\cosh^2\left(\mu\sqrt{\lambda}(t - t_0)\right)}\right).$$

(5.2.109)

The resulting operator is, up to a rescaling of t, the Pöschl–Teller operator $\mathbf{M}_{1,2}$:

$$-\partial_t^2 + \mu^2\lambda\left(4 - \frac{2}{\cosh^2\left(\mu\sqrt{\lambda}(t - t_0)\right)}\right) = \mu^2\lambda\mathbf{M}_{1,2}.$$

(5.2.110)

As expected, $\mathbf{M}_{1,2}$ has no zero modes, and one has, by using (1.5.31), that

$$\frac{\det \mathbf{M}_{1,2}}{\det \mathbf{M}_{0,2}} = \frac{1}{3}.$$

(5.2.111)

Therefore,

$$\frac{\det \mathbf{M}_{1,2}}{\det' \mathbf{M}_{2,2}} = 16.$$

(5.2.112)

In order to take into account the effect of the rescaling, we proceed as in (1.7.23). By using the density of states, one computes the difference $N_{1,2} - N'_{2,2} = 1$, and we find

$$\frac{\det' \mathbf{M}_F}{\det' \mathbf{M}_B} = 4\mu\sqrt{\lambda}. \tag{5.2.113}$$

A more physical way to understand the lack of cancellation between the fermionic and the bosonic degrees of freedom is to look carefully at the densities of states of the fields $\eta_{\pm}$. Since the modes ψ''_- satisfy the same equation as the bosonic modes, the continuous eigenvectors have the same phase shifts that we found in (1.5.23):

$$\psi_-^{(k)}(t) \sim \exp\left[i\left(kt \pm \frac{\theta(k)}{2}\right)\right], \qquad t \to \pm\infty. \tag{5.2.114}$$

On the other hand, the asymptotic behavior of $\psi_+^{(k)}(t)$ can be deduced from the second equation in (5.2.86):

$$\psi_+^{(k)}(t) \sim \left(k \mp 2i\mu\sqrt{\lambda}\right) \exp\left[i\left(kt \pm \frac{\theta(k)}{2}\right)\right], \qquad t \to \pm\infty. \tag{5.2.115}$$

This means that there is an extra phase shift $\delta(k)/2$ given by

$$\tan\left(\frac{\delta(k)}{2}\right) = -\frac{2\mu\sqrt{\lambda}}{k}. \tag{5.2.116}$$

The densities of states are then given by

$$\rho_-(k) = \rho(k), \qquad \rho_+(k) = \rho(k) + \frac{1}{2\pi}\delta'(k), \tag{5.2.117}$$

where $\rho(k)$ is the density of states determined in (1.5.26), for $\ell = 2$ in this case. We then see that the eigenstates of the one-dimensional Dirac operator $\mathbf{M}_F$ with different chiralities have different densities of states. This phenomenon is called *spectral asymmetry*.

We can now use this information to calculate the quotient of determinants, as we did in (1.5.29). The discrete spectrum cancels in (5.2.105). To calculate $\log \det \mathbf{M}_F$ we have to add the densities $\rho_-(k)$ and $\rho_+(k)$, with the eigenvalues $\lambda_F^{(k)}$, while in calculating $\log \det \mathbf{M}_B$ we use the density $\rho(k)$ with the eigenvalues $\left(\lambda_B^{(k)}\right)^2$. We then find,

$$\log\left[\frac{\det' \mathbf{M}_F}{\det' \mathbf{M}_B}\right]^{1/2} = \int_{-\infty}^{\infty} dk \, \log\left(\sqrt{k^2 + 4\lambda\mu^2}\right) \rho_t(k), \tag{5.2.118}$$

where

$$\rho_t(k) = \rho_+(k) + \rho_-(k) - 2\rho(k) = \frac{1}{2\pi}\delta'(k) = \frac{1}{\pi}\frac{\sqrt{\lambda}\mu}{k^2 + 4\lambda\mu^2}, \tag{5.2.119}$$

and we have restored the dependence on λ, μ in the energies (5.2.98). The integral in (5.2.118) is easily evaluated, and we obtain,

$$\int_{-\infty}^{\infty} dk \, \log\left(\sqrt{k^2 + 4\lambda\mu^2}\right) \rho_t(k) = \frac{1}{2} \log\left(4\sqrt{\lambda}\mu\right), \qquad (5.2.120)$$

in agreement with (5.2.113).

One can show, for example by using the Gelfand–Yaglom theorem, that for general $\omega(q)$ the quotient of determinants (5.2.108) is given by

$$\frac{\det' \mathbf{M}_F}{\det' \mathbf{M}_B} = 2\omega'(q_+), \qquad (5.2.121)$$

and of course our previous result (5.2.113) is a particular case of (5.2.121). We conclude that, in the one-loop approximation,

$$\lim_{\beta \to \infty} \int \mathcal{D}q \, \mathcal{D}\psi_+ \, \mathcal{D}\psi_- \, \psi_-(t) \, e^{-S} \approx \sqrt{\frac{\omega'(q_+)}{\pi}} e^{-\Delta f} \Delta q, \qquad (5.2.122)$$

in agreement with the result in the first line of (5.2.61), which we obtained by solving directly for the wavefunctions. It is possible to perform a path-integral derivation of the result for the second vev in (5.2.61), but this requires a two-loop analysis in the background of an instanton. The interested reader will find more details in the references at the end of the chapter.

5.3 Fermions and chiral symmetry in Quantum Chromodynamics

Quantum Chromodynamics (QCD) is the quantum theory of a YM connection coupled to matter spinor fields in the fundamental representation of the gauge group (which is usually taken to be $SU(N)$). The matter fields are represented by Dirac spinors or "quarks" ψ_f, where f is an index running from 1 to N_f (the number of "flavors"). The Lagrangian of QCD is

$$g_0^2 \mathcal{L}_{\text{QCD}} = -\frac{1}{4} F_{\mu\nu}^a F^{\mu\nu a} + \sum_{f=1}^{N_f} \overline{\psi}_f \left(i\slashed{D} - m_f\right) \psi_f, \qquad (5.3.1)$$

where as usual

$$\overline{\psi}_f = \psi_f^\dagger \gamma^0, \qquad \slashed{D} = \gamma^\mu D_\mu, \qquad (5.3.2)$$

D_μ is the covariant derivative (4.2.7) in the fundamental representation (i.e. the T^a are the generators of the Lie algebra in the fundamental representation), γ^μ are gamma matrices, and m_f are the masses of the quarks. Our conventions for the gamma matrices are as follows. Let us introduce the matrix-valued four-vectors,

$$\sigma^\mu = (\mathbf{1}, \boldsymbol{\sigma}), \qquad \overline{\sigma}^\mu = (\mathbf{1}, -\boldsymbol{\sigma}), \qquad (5.3.3)$$

where we used the same notation as in (4.4.11). Then, we will set

$$\gamma^\mu = \begin{pmatrix} 0 & \sigma^\mu \\ \bar{\sigma}^\mu & 0 \end{pmatrix}. \tag{5.3.4}$$

We will also define

$$\gamma_5 = i\gamma^0\gamma^1\gamma^2\gamma^3 = \begin{pmatrix} -1 & 0 \\ 0 & 1 \end{pmatrix}. \tag{5.3.5}$$

With this choice of gamma matrices, γ^0 is Hermitian, while γ^i is anti-Hermitian, i.e. $(\gamma^i)^\dagger = -\gamma^i$. When we use the rescaled field (4.2.30), we should also redefine

$$\hat{\psi}_f = \frac{1}{g_0}\psi_f. \tag{5.3.6}$$

The Euclidean version of the fermionic sector goes as follows. We Wick-rotate the gamma matrices as

$$\gamma_4^E = \gamma_0, \qquad \gamma_k^E = -i\gamma^k, \quad k = 1, 2, 3, \tag{5.3.7}$$

so that

$$\{\gamma_\mu^E, \gamma_\nu^E\} = 2\delta_{\mu\nu}. \tag{5.3.8}$$

With this convention, the Euclidean gamma matrices are Hermitian. We will also define the Euclidean conjugate spinors as

$$\overline{\psi}_f^E = i\overline{\psi}_f. \tag{5.3.9}$$

We also note that in the Euclidean version of the theory, the fields ψ_f^E, $\overline{\psi}_f^E$ should be regarded as independent, anticommuting variables in the path integral. The Euclidean version of the QCD Lagrangian is

$$g_0^2 \mathcal{L}_{QCD}^E = \frac{1}{4}F_{\mu\nu}^{Ea}F^{E\mu\nu a} - \sum_{f=1}^{N_f} \overline{\psi}_f^E \left(i\slashed{D} + im_f \right) \psi_f^E. \tag{5.3.10}$$

Finally, let us briefly recall that the beta function of QCD (which is still defined by (4.2.40)), has

$$\beta_0 = \frac{1}{(4\pi)^2} \left(\frac{11N}{3} - \frac{2}{3}N_f \right). \tag{5.3.11}$$

Therefore, if the number of flavors is small enough, the theory is still asymptotically free.

Let us now come back to the Lagrangian (5.3.1) in Minkowski space and study its global symmetries. In the rest of this section, we will use the fields $\hat{\psi}_f$ with the normalization in (5.3.6), and for simplicity of notation we will remove the hats. This different normalization has to be taken into account when comparing to other results in the book. It is easy to see that, if the quark masses vanish, i.e. if $m_f = 0$ for all $f = 1, \ldots, N_f$, one has a global symmetry group

$$SU(N_f)_V \times SU(N_f)_A, \tag{5.3.12}$$

with an action on the Dirac spinors

$$\psi \to e^{i\theta_V^a T_a} \psi, \qquad \psi \to e^{i\theta_A^a T_a \gamma_5} \psi \tag{5.3.13}$$

where we have denoted by T_a the generators of the group $SU(N_f)$. Here,

$$\psi = \begin{pmatrix} \psi_1 \\ \vdots \\ \psi_{N_f} \end{pmatrix} \tag{5.3.14}$$

is a vector of N_f components, in the fundamental representation of $SU(N_f)$, and $\theta_{V,A}^a$ parametrize $SU(N_f)_{V,A}$. The symmetry (5.3.13) is the *chiral symmetry* of massless QCD. The vectorial part of the symmetry $SU(N_f)_V$ is sometimes called the *isospin symmetry* of QCD.

Another way of understanding these symmetries is by introducing Weyl spinors,

$$\psi_{L,f} = \frac{1 - \gamma_5}{2} \psi_f, \qquad \psi_{R,f} = \frac{1 + \gamma_5}{2} \psi_f. \tag{5.3.15}$$

The fermionic part of the Lagrangian (5.3.1) reads, in terms of these,

$$i \sum_{f=1}^{N_f} \left(\overline{\psi}_{L,f} \slashed{D} \psi_{L,f} + \overline{\psi}_{R,f} \slashed{D} \psi_{R,f} \right) - \sum_{f=1}^{N_f} m_f \left(\overline{\psi}_{R,f} \psi_{L,f} + \overline{\psi}_{L,f} \psi_{R,f} \right). \tag{5.3.16}$$

We can now write the chiral symmetry group as

$$SU(N_f)_L \times SU(N_f)_R. \tag{5.3.17}$$

This acts on Weyl spinors as

$$\psi_{L,f} \to \sum_{f'=1}^{N_f} L_{ff'} \psi_{L,f'}, \qquad \psi_{R,f} \to \sum_{f'=1}^{N_f} R_{ff'} \psi_{R,f'}, \tag{5.3.18}$$

where the rotation matrices L, R are given by

$$L = e^{i\theta_L^a T_a}, \qquad R = e^{i\theta_R^a T_a}. \tag{5.3.19}$$

The left and right angles are related to the vectorial and axial angles (5.3.13) as,

$$\theta_V = \frac{1}{2}(\theta_L + \theta_R), \qquad \theta_A = \frac{1}{2}(\theta_R - \theta_L). \tag{5.3.20}$$

Therefore, the vectorial group in (5.3.12) corresponds to the diagonal part of (5.3.17), while the axial group corresponds to the anti-diagonal part.

The chiral symmetries lead to conserved currents:

$$J_L^{a\mu} = \sum_{f,f'} \overline{\psi}_{L,f} T^a_{ff'} \gamma^\mu \psi_{L,f'}, \qquad J_R^{a\mu} = \sum_{f,f'} \overline{\psi}_{R,f} T^a_{ff'} \gamma^\mu \psi_{R,f'}. \qquad (5.3.21)$$

Equivalently, we can consider vector and axial currents,

$$V^{a\mu} = \sum_{f,f'} V^\mu_{ff'} T^a_{ff'}, \qquad A^{a\mu} = \sum_{f,f'} A^\mu_{ff'} T^a_{ff'}. \qquad (5.3.22)$$

with

$$V^\mu_{ff'} = \overline{\psi}_f \gamma^\mu \psi_{f'}, \qquad A^\mu_{ff'} = \overline{\psi}_f \gamma^\mu \gamma_5 \psi_{f'}. \qquad (5.3.23)$$

There are two additional $U(1)$ symmetries of massless QCD, which in fact enhance (5.3.12) to

$$U(N_f)_V \times U(N_f)_A. \qquad (5.3.24)$$

The first one is a vectorial $U(1)_V$,

$$\psi_f \to e^{i\theta} \psi_f, \qquad f = 1, \ldots, N_f. \qquad (5.3.25)$$

The associated current

$$Q^\mu = \sum_{f=1}^{N_f} \overline{\psi}_f \gamma^\mu \psi_f \qquad (5.3.26)$$

is conserved quantum mechanically. This leads to a conserved quantum number which is just the number of quarks minus the number of antiquarks. The second classical symmetry is the axial $U(1)_A$

$$\psi_f \to e^{i\alpha\gamma_5} \psi_f, \qquad f = 1, \ldots, N_f, \qquad (5.3.27)$$

with current

$$J^\mu = \sum_{f=1}^{N_f} \overline{\psi}_f \gamma^\mu \gamma_5 \psi_f. \qquad (5.3.28)$$

What is the fate of these symmetries of the QCD Lagrangian? Let us first consider the chiral symmetry (5.3.12). In a world of massless quarks, this symmetry of the QCD Lagrangian is *spontaneously broken* by quantum effects. This means that the vacuum is not invariant under the full symmetry group, but only under a subgroup. The symmetry breaking pattern is,

$$SU(N_f)_V \times SU(N_f)_A \to SU(N_f)_V. \qquad (5.3.29)$$

In other words, the charges of the axial current

$$Q^{5a}(t) = \int d^3x \, A^{a0}(t, \mathbf{x}) \qquad (5.3.30)$$

do not leave the vacuum invariant, while the charges of the vectorial current do preserve the vacuum. This phenomenon is called *chiral symmetry breaking* (χSB). χSB is manifested in the fact that the vev,

$$\langle 0|\overline{\psi}_f \psi_f|0\rangle, \tag{5.3.31}$$

which mixes the L and the R sectors, does not vanish. Notice that, due to isospin symmetry, this vev is the same for any flavor $f = 1, \ldots, N_f$. The above vev is called the *quark condensate*.

Chiral symmetry breaking involves the dynamics of QCD theory at low energies, where the effective coupling is large, and therefore it cannot be established using perturbation theory. However, χSB explains to a large extent the observed properties of light mesons. Goldstone's theorem states that, for each charge that fails to annihilate the vacuum, there is a massless boson with the quantum numbers of this generator. In the case of chiral symmetry, the charges are the axial charges (5.3.30). They are constructed from the axial current, which involves the γ_5 matrix, therefore $Q^{5a}a$ changes sign under a parity transformation. We conclude that, as a consequence of χSB, there must be $N_f^2 - 1$ pseudo-scalar Goldstone bosons, which are identified with the light mesons. They are not really massless, but their masses can be understood as due to an explicit breaking of chiral symmetry due to the masses of the quarks, and for this reason they are sometimes called *pseudo-Goldstone bosons*. For this reason, it only makes sense to talk about Goldstone bosons if one considers light quarks, since for heavy quarks chiral symmetry is no longer an approximate symmetry of the QCD Lagrangian. Taking the quarks u, d, s as light, we have eight mesons. These are the three pions, the four kaons, and the η. We list them in Table 5.1, together with their quark content and masses. Notice that we have listed an extra meson, the η'. From the point of view of the quark model, the η and the η' should be regarded as different linear combinations of two states, the η_0 and the η_8, with quark content

$$\eta_8 = \frac{1}{\sqrt{6}}\left(d\bar{d} + u\bar{u} - 2s\bar{s}\right),$$

$$\eta_0 = \frac{1}{\sqrt{3}}\left(d\bar{d} + u\bar{u} + s\bar{s}\right). \tag{5.3.32}$$

Experimentally, the mixing angle θ appearing in Table 5.1 is approximately $\theta \approx -17°$. The role of the η' will be discussed in detail in this chapter, as well as in Chapter 7 in the context of the Witten–Veneziano formula.

The low-energy dynamics of these pseudo-Goldstone bosons can be understood by using an effective theory. The philosophy of effective theories is to write down the most general Lagrangian, not necessarily renormalizable, which is compatible with the symmetries of the theory. Of course, there are an infinite number of terms

Table 5.1 *The eight light mesons and the η', together*
with their quark content and their approximate mass

Meson	Quark content	Mass (MeV)
π^+	$u\bar{d}$	140
π^-	$d\bar{u}$	140
π^0	$(d\bar{d} - u\bar{u})/\sqrt{2}$	135
K^+	$u\bar{s}$	494
K^0	$d\bar{s}$	498
K^-	$s\bar{u}$	494
$\overline{K}^0$	$s\bar{d}$	498
η	$\cos\theta\,\eta_8 - \sin\theta\,\eta_0$	547
η'	$\sin\theta\,\eta_8 + \cos\theta\,\eta_0$	958

that can be included in such a Lagrangian, but one can organize them in terms of
their number of derivatives, or equivalently by powers of the momenta. The terms
with a low number of derivatives describe the physics at low energies. In order to
write a Lagrangian for the Goldstone bosons, we first note that under a combined
vectorial plus axial transformation, the quark fields transform as

$$\psi \to \exp\left\{ i\sum_a \left(\theta_V^a T_a + \theta_A^a T_a \gamma_5\right) \right\} \psi. \tag{5.3.33}$$

We want to "extract" from these fields the Goldstone bosons associated to the broken axial symmetry, and to construct a new field which only transforms under the
unbroken, vectorial symmetry. To do this, we write

$$\psi = \exp\left(-i\gamma_5 \sum_a \xi_a(x) T_a \right) \tilde{\psi}, \tag{5.3.34}$$

where $\xi_a(x)$ are fields which represent the Goldstone bosons The new quark fields
$\tilde{\psi}$, are required to transform only under $SU(N_f)_V$, i.e.

$$\tilde{\psi}' = \exp\left(i\sum_a \theta_a(x) T_a \right) \tilde{\psi}, \tag{5.3.35}$$

where $\theta_a(x)$ depends on θ_V, θ_A and $\xi(x)$. This imposes on the fields $\xi_a(x)$ the
transformation rule

$$\exp\left\{ i\sum_a \left(\theta_V^a T_a + \theta_A^a T_a \gamma_5\right) \right\} \exp\left(-i\gamma_5 \sum_a \xi_a T_a \right)$$

$$= \exp\left(-i\gamma_5 \sum_a \xi_a' T_a \right) \exp\left(i\sum_a \theta_a T_a \right). \tag{5.3.36}$$

We can decompose this equation in terms of right and left parts, by acting on Weyl spinors, and we find

$$\exp\left(i\sum_a \theta_L^a T_a\right)\exp\left(i\sum_a \xi_a T_a\right) = \exp\left(i\sum_a \xi_a' T_a\right)\exp\left(i\sum_a \theta_a T_a\right),$$

$$\exp\left(i\sum_a \theta_R^a T_a\right)\exp\left(-i\sum_a \xi_a T_a\right) = \exp\left(-i\sum_a \xi_a' T_a\right)\exp\left(i\sum_a \theta_a T_a\right),$$

$$(5.3.37)$$

where we have used (5.3.20). Let us now consider the inverse of the second equation, and multiply it by the first equation. We find that the field,

$$U = \exp\left(2i\sum_a \xi_a T_a\right),\qquad (5.3.38)$$

which contains the Goldstone boson fields, transforms as

$$U' = \exp\left(i\sum_a \theta_L^a T_a\right)U\exp\left(-i\sum_a \theta_R^a T_a\right),\qquad (5.3.39)$$

i.e. it belongs to the representation $(N_f, \overline{N}_f)$ of $SU(N_f)_L \times SU(N_f)_R$. We now want to write down an effective Lagrangian for the Goldstone bosons, in terms of U, invariant under the Lorentz symmetry and the chiral symmetry group. Up to two derivatives, the only choice is

$$\mathcal{L}_{\text{eff}} = \frac{F_\pi^2}{4}\text{Tr}\left[\partial_\mu U \partial^\mu U^\dagger\right].\qquad (5.3.40)$$

The constant F_π is called the *pion decay constant*. It can be determined by comparing the predictions of the effective Lagrangian (5.3.40) to experimental results, and its value is

$$F_\pi \approx 93\,\text{MeV}.\qquad (5.3.41)$$

To understand the field content of (5.3.40), let us consider the case $N_f = 3$. We will now rewrite the Goldstone boson fields ξ_a in terms of canonically normalized fields. To do that, we set

$$U = \exp\left(i\frac{\sqrt{2}}{F_\pi}B\right),\qquad (5.3.42)$$

where the matrix $\mathcal{B}$ is given by

$$\mathcal{B} = \sqrt{2\pi} \cdot \mathbf{T} = \begin{pmatrix} \frac{1}{\sqrt{2}}\pi^0 + \frac{1}{\sqrt{6}}\eta_8 & \pi^+ & K^+ \\ \pi^- & -\frac{1}{\sqrt{2}}\pi^0 + \frac{1}{\sqrt{6}}\eta_8 & K^0 \\ K^- & \overline{K}^0 & -\sqrt{\frac{2}{3}}\eta_8 \end{pmatrix}. \tag{5.3.43}$$

Here, we have chosen an explicit basis T_a for the Lie algebra of $SU(3)$, $a = 1, \ldots, 8$, satisfying the normalization (4.2.5):

$$T_1 = \frac{1}{2}\begin{pmatrix} 0 & 1 & 0 \\ 1 & 0 & 0 \\ 0 & 0 & 0 \end{pmatrix}, \quad T_2 = \frac{1}{2}\begin{pmatrix} 0 & -i & 0 \\ i & 0 & 0 \\ 0 & 0 & 0 \end{pmatrix}, \quad T_3 = \frac{1}{2}\begin{pmatrix} 1 & 0 & 0 \\ 0 & -1 & 0 \\ 0 & 0 & 0 \end{pmatrix},$$

$$T_4 = \frac{1}{2}\begin{pmatrix} 0 & 0 & 1 \\ 0 & 0 & 0 \\ 1 & 0 & 0 \end{pmatrix}, \quad T_5 = \frac{1}{2}\begin{pmatrix} 0 & 0 & -i \\ 0 & 0 & 0 \\ i & 0 & 0 \end{pmatrix}, \quad T_6 = \frac{1}{2}\begin{pmatrix} 0 & 0 & 0 \\ 0 & 0 & 1 \\ 0 & 1 & 0 \end{pmatrix},$$

$$T_7 = \frac{1}{2}\begin{pmatrix} 0 & 0 & 0 \\ 0 & 0 & -i \\ 0 & i & 0 \end{pmatrix}, \quad T_8 = \frac{1}{2\sqrt{3}}\begin{pmatrix} 1 & 0 & 0 \\ 0 & 1 & 0 \\ 0 & 0 & -2 \end{pmatrix}, \tag{5.3.44}$$

and we have put them together inside a single vector $\mathbf{T}$. The Goldstone bosons have been gathered in the vector

$$\boldsymbol{\pi} = \left(\pi^1, -\pi^2, \pi^0, K^1, -K^2, \mathrm{Re}(K^0), -\mathrm{Im}(K^0), \eta_8\right), \tag{5.3.45}$$

and

$$\pi^\pm = \frac{1}{\sqrt{2}}\left(\pi^1 \pm i\pi^2\right), \quad K^\pm = \frac{1}{\sqrt{2}}\left(K^1 \pm iK^2\right). \tag{5.3.46}$$

We can now write the Lagrangian in terms of the fields appearing in (5.3.45), by expanding U around the vacuum $U = \mathbf{1}$,

$$\frac{F_\pi^2}{4}\mathrm{Tr}(\partial_\mu U \partial^\mu U^\dagger)$$

$$\approx \frac{1}{2}\partial_\mu\pi^0\partial^\mu\pi^0 + \partial_\mu\pi^+\partial_\mu\pi^- + \partial_\mu K^+\partial_\mu K^- + \partial_\mu K^0\partial_\mu\overline{K}^0 + \frac{1}{2}\partial_\mu\eta_8\partial^\mu\eta_8. \tag{5.3.47}$$

Here we have only kept the quadratic terms in the light meson fields. The Lagrangian (5.3.40) can be taken as the starting point to calculate matrix elements of currents of the underlying, microscopic QCD Lagrangian. To do this, we have to add sources to the QCD Lagrangian:

$$-\overline{\psi}_L(s+ip)\psi_R - \overline{\psi}_R(s+ip)^\dagger\psi_L + \overline{\psi}_L\gamma_\mu\ell_\mu\psi_L + \overline{\psi}_R\gamma_\mu r_\mu\psi_R, \tag{5.3.48}$$

where s, p, ℓ_μ, r_μ are $N_f \times N_f$ matrices. Here, $\psi_{L,R}$ are column vectors of N_f components, while $\overline{\psi}_{L,R}$ are row vectors. The standard QCD Lagrangian for $N_f = 3$ is obtained by setting

$$\ell_\mu = r_\mu = p = 0, \qquad s = \mathcal{M}, \tag{5.3.49}$$

where

$$\mathcal{M} = \begin{pmatrix} m_u & 0 & 0 \\ 0 & m_d & 0 \\ 0 & 0 & m_s \end{pmatrix} \tag{5.3.50}$$

is the quark mass matrix. In this way, vevs of different operators can be computed by taking derivatives of the generating functional

$$\mathcal{L}_{\text{eff}} = -i \log Z[\ell, r, s, p], \tag{5.3.51}$$

with respect to the sources. Here, as usual, $Z[\ell, r, s, p]$ is the partition function in the presence of the sources ℓ, r, s, p. For example, the vev of $\overline{\psi}_f \psi_{f'}$ is obtained by setting s to be a Hermitian matrix, and taking a derivative with respect to $s_{ff'}$:

$$\langle \overline{\psi}_f \psi_{f'} \rangle = -\frac{\delta \mathcal{L}_{\text{eff}}}{\delta s_{ff'}}. \tag{5.3.52}$$

To see how the sources appear in the effective Lagrangian (5.3.40), we gauge the chiral symmetry and we promote the sources to gauge fields. Under the transformations

$$\psi_L \to L(x)\psi_L, \qquad \psi_R \to R(x)\psi_R, \tag{5.3.53}$$

we have, as we have seen in (5.3.39),

$$U \to L(x)U R^\dagger(x). \tag{5.3.54}$$

The sources ℓ_μ, r_μ behave as gauge potentials for L, R, respectively, therefore they transform as in (4.2.23),

$$\begin{aligned} \ell_\mu &\to L(x)\ell_\mu L^\dagger(x) + iL(x)\left(\partial_\mu L^\dagger\right)(x), \\ r_\mu &\to R(x)r_\mu R^\dagger(x) + iR(x)\left(\partial_\mu R^\dagger\right)(x). \end{aligned} \tag{5.3.55}$$

The coupling $s + ip$ transforms as

$$s + ip \to L(x)\,(s + ip)\,R^\dagger(x). \tag{5.3.56}$$

The Lagrangian (5.3.48) is now invariant under the gauged chiral symmetry. The goal is to construct a gauge invariant low-energy Lagrangian. The gauge covariant derivative acting on U is

$$D_\mu U = \partial_\mu U - i\ell_\mu U + iU r_\mu, \tag{5.3.57}$$

and transforms covariantly

$$D_\mu U \rightarrow L(x) D_\mu U R^\dagger(x). \tag{5.3.58}$$

The effective Lagrangian, in the presence of the gauged global symmetry, can now be determined by using gauge invariance, Lorentz invariance, and parity invariance. At leading order in the derivative expansion it is just given by

$$\mathcal{L}_{\text{eff}} = \frac{F_\pi^2}{4} \text{Tr}(D_\mu U D^\mu U^\dagger) + \frac{F_\pi^2}{4} \text{Tr}(\chi U^\dagger + U\chi^\dagger), \tag{5.3.59}$$

where

$$\chi = 2B_0(s + ip), \tag{5.3.60}$$

and B_0 is a constant to be determined. This tells us how the different sources in the microscopic QCD Lagrangian appear in the effective Lagrangian. In particular, it tells us how the effective Lagrangian depends on the masses of the quarks, since we know that these masses enter through s: we can evaluate the effective Lagrangian (5.3.59) on the configuration (5.3.49), and read off the masses of the light mesons. Since

$$U + U^\dagger = 2 - \frac{2}{F_\pi^2} B^2 + \cdots, \tag{5.3.61}$$

the mass term in the effective Lagrangian is

$$\sum_{a,b=1}^{3} M_{(ab)}^2 \pi_a \pi_b = 2B_0 \text{Tr}(B^2 \mathcal{M}), \tag{5.3.62}$$

where $M_{(ab)}^2$ is the symmetrization of the matrix

$$M_{ab}^2 = 4B_0 \text{Tr}\left[T^a T^b \mathcal{M}\right]. \tag{5.3.63}$$

From this we can easily obtain the masses of the mesons:

$$\begin{aligned}
m_\pi^2 &= B_0(m_u + m_d), \\
m_{K^\pm}^2 &= B_0(m_u + m_s), \\
m_{K^0}^2 &= B_0(m_d + m_s), \\
m_{\eta_8}^2 &= B_0 \frac{m_u + m_d + 4m_s}{3}.
\end{aligned} \tag{5.3.64}$$

There is also a mixing term between the η_8 and the π^0,

$$m_{\pi\eta}^2 = B_0 \frac{m_u - m_d}{\sqrt{3}}. \tag{5.3.65}$$

One particular prediction of chiral symmetry, following from (5.3.64), is that

$$m_{\eta_8}^2 = \frac{1}{3}(2m_{K^\pm}^2 + 2m_{K^0}^2 - m_\pi^2) \qquad (5.3.66)$$

which is called the *Gell-Mann–Okubo mass formula*. Taking as data the masses of the kaons and the pions, it predicts

$$m_{\eta_8}^2 = 566 \text{ MeV}, \qquad (5.3.67)$$

which is not far from the experimental value of the squared mass of the η meson. The constant B_0 appearing in the above formulae can be related to the quark condensate in the massless theory by setting $\chi = 2B_0 s$, where s is a Hermitian matrix, and using the relation (5.3.52). The derivative is simply,

$$-\frac{\partial \mathcal{L}_{\text{eff}}}{\partial s_{ff'}} = -\frac{F_\pi^2}{4} \frac{\partial}{\partial s_{ff'}} \text{Tr}\left[\chi\left(U^\dagger + U\right)\right] = -\frac{F_\pi^2 B_0}{2}(U^\dagger + U)_{f'f}. \qquad (5.3.68)$$

Evaluating this in the vacuum $U = 1$, we find that the quark condensate is proportional to B_0,

$$\langle 0 | \overline{\psi}_f \psi_{f'} | 0 \rangle = -F_\pi^2 B_0 \delta_{ff'}. \qquad (5.3.69)$$

Expressing B_0 in terms of the quark condensate, we find

$$m_\pi^2 = -\frac{m_u + m_d}{F_\pi^2} \langle 0 | \overline{\psi}_f \psi_f | 0 \rangle. \qquad (5.3.70)$$

Note that, as the masses of the quarks go to zero, all mesons become massless, as expected.

The axial current in (5.3.22) creates light meson states out of the vacuum. To see this, we first note that, in terms of the low-energy degrees of freedom appearing in the effective Lagrangian (5.3.40), this current is given by

$$A_\mu^a = i\frac{F_\pi^2}{2} \text{Tr}\left[T^a\left(U^\dagger \partial_\mu U - U \partial_\mu U^\dagger\right)\right]. \qquad (5.3.71)$$

This current is classically conserved when the quarks are massless, but for massive quarks its divergence does not vanish. To calculate this divergence, we consider the EOM derived from the Lagrangian (5.3.59) with $\chi = 2B_0 \mathcal{M}$. A simple calculation leads to,

$$\frac{F_\pi^2}{2} \partial^\mu \left(U^\dagger \partial_\mu U\right) + \frac{B_0 F_\pi^2}{2} \mathcal{M}\left(U - U^\dagger\right) = 0, \qquad (5.3.72)$$

and we deduce that

$$\partial^\mu A_\mu^a = -i B_0 F_\pi^2 \text{Tr}\left[T^a \mathcal{M}\left(U - U^\dagger\right)\right]. \qquad (5.3.73)$$

If we use the expression for U in (5.3.42), in terms of the canonically normalized fields π^a in the vector (5.3.45), we find, at leading order,

$$\partial^\mu A^a_\mu \approx F_\pi \pi^b M^2_{ba},\tag{5.3.74}$$

where M^2_{ab} is the mass matrix (5.3.63). We conclude that

$$\langle 0|A^a_\mu|\pi^b(p)\rangle = ip^\mu C_\pi e^{-ip\cdot x}M^2_{ba}.\tag{5.3.75}$$

In this equation, C_π is given by

$$C_\pi = \frac{F_\pi}{(2\pi)^{3/2}\sqrt{2E}},\tag{5.3.76}$$

and E is the energy of the state $|\pi^b(p)\rangle$. The fact that the axial current is proportional to the meson field, up to an overall constant, is the basis of methods for understanding the physics of mesons based on current algebra techniques. The equation (5.3.75) relates the axial current (which in the microscopic theory is a bilinear in the quark fields) to the meson states and the pion decay constant. It will be used later on to understand some of the large N properties of mesons in QCD.

5.4 The $U(1)$ problem and the axial anomaly

As we mentioned in the previous section, the axial current (5.3.28) is classically conserved if the quarks are massless. What is the quantum mechanical fate of this classical symmetry? If this current were conserved quantum mechanically, there would be an extra conserved quantum number. However, this is not what is seen in hadronic physics. If it were spontaneously broken, there would be a ninth Goldstone boson, in addition to the other mesons. In that case, the pattern of chiral symmetry breaking would rather be

$$U(N_f)_A \times U(N_f)_V \to U(N_f)_V.\tag{5.4.1}$$

It is possible to extend the construction of the effective Lagrangian to realize this pattern. In (5.3.42), we wrote U as the exponential of a field living in the Lie algebra of $SU(N_f)$. To include the ninth Goldstone boson, we should promote it to a field living in the Lie algebra $U(N_f)$. We then write

$$U = \exp\left(\frac{2i}{F_\pi}\pi\cdot\mathbf{T} + \frac{2i}{F_{\eta_0}}\eta_0 T_0\right),\tag{5.4.2}$$

where

$$T_0 = \frac{1}{\sqrt{2N_f}}\mathbf{1}\tag{5.4.3}$$

in order to preserve the normalization condition (4.2.5). The new Goldstone boson is denoted by η_0, and we have introduced a new coupling constant F_{η_0}. The field η_0 encodes the information about the determinant of U, and we have obviously

$$\log \det U = \frac{i\sqrt{2N_f}}{F_{\eta_0}} \eta_0, \qquad (5.4.4)$$

where we have used that

$$\log \det A = \operatorname{tr} \log A. \qquad (5.4.5)$$

Under an infinitesimal $U(N_f)_L \times U(N_f)_R$ transformation of the form

$$L = 1 + i\theta_L, \qquad R = 1 + i\theta_R, \qquad (5.4.6)$$

where θ_L, θ_R are Hermitian $N_f \times N_f$ matrices, we have that

$$\operatorname{Tr} \log U \rightarrow \operatorname{Tr} \log U + i\operatorname{Tr}(\theta_L - \theta_R). \qquad (5.4.7)$$

In particular, η_0 does not change under the vectorial $U(1)_V$ symmetry (5.3.25),

$$\theta_L = \theta_R = \theta \mathbf{1}, \qquad (5.4.8)$$

but it transforms under the axial $U(1)_A$ symmetry

$$\theta_L = -\theta_R = -\alpha \mathbf{1}. \qquad (5.4.9)$$

The matrix $\mathcal{B}$ introduced in (5.3.43) now reads

$$\mathcal{B} = \sqrt{2\pi} \cdot \mathbf{T} + \frac{F_\pi}{\sqrt{N_f} F_{\eta_0}} \eta_0 \mathbf{1}. \qquad (5.4.10)$$

We will now show that, if (5.4.1) is the pattern of chiral symmetry breaking realized in Nature, with $N_f = 3$, there will be a ninth light Goldstone boson whose mass will not exceed $\sqrt{3} m_\pi^2$. To see this, we calculate the new mass matrix, which is again given by (5.3.62). The mass matrix for the strange mesons remains unchanged, and we find a mass matrix for π^0, η_8, η_0 of the form

$$M^2 = B_0 \begin{pmatrix} m_u + m_d & \dfrac{m_u - m_d}{\sqrt{3}} & \sqrt{\dfrac{2}{3}} \xi (m_u - m_d) \\[2mm] \dfrac{m_u - m_d}{\sqrt{3}} & \dfrac{m_u + m_d + 4m_s}{3} & \dfrac{\sqrt{2}}{3} \xi (m_u + m_d - 2m_s) \\[2mm] \sqrt{\dfrac{2}{3}} \xi (m_u - m_d) & \dfrac{\sqrt{2}}{3} \xi (m_u + m_d - 2m_s) & \dfrac{2\xi^2}{3} (m_u + m_d + m_s) \end{pmatrix},$$

$$(5.4.11)$$

where

$$\xi = \frac{F_\pi}{F_{\eta_0}}. \qquad (5.4.12)$$

Since m_u, m_d are small, we can calculate the eigenvalues of this matrix perturbatively in m_u, m_d. When $m_u = m_d = 0$, we obtain the matrix

$$M_0^2 = m_s B_0 \begin{pmatrix} 0 & 0 & 0 \\ 0 & \dfrac{4}{3} & -\dfrac{2\sqrt{2}\xi}{3} \\ 0 & -\dfrac{2\sqrt{2}\xi}{3} & \dfrac{2\xi^2}{3} \end{pmatrix}, \tag{5.4.13}$$

which has two eigenvectors with zero eigenvalues:

$$v_1 = \begin{pmatrix} 1 \\ 0 \\ 0 \end{pmatrix}, \qquad v_2 = \frac{1}{\sqrt{\xi^2 + 2}} \begin{pmatrix} 0 \\ \xi \\ \sqrt{2} \end{pmatrix}. \tag{5.4.14}$$

The third eigenvalue is

$$\frac{2 B_0 m_s}{3} (2 + \xi^2). \tag{5.4.15}$$

We then have two light pseudo-Goldstone bosons, whose mass squared is linear in m_u, m_d, and a heavier pseudo-Goldstone boson whose mass (5.4.15) is proportional to the mass of the strange quark, and which corresponds to the η_8 in the limit $\xi \to 0$. In order to calculate the masses of the lightest pseudo-Goldstone bosons, we use degenerate perturbation theory. To linear order in m_u, m_d, their masses are obtained by diagonalizing the matrix with entries

$$^t v_i \left(M^2 - M_0^2 \right) v_j, \qquad i, j = 1, 2, \tag{5.4.16}$$

which is given by

$$B_0 \begin{pmatrix} m_u + m_d & \dfrac{\xi\sqrt{3}(m_u - m_d)}{\sqrt{2 + \xi^2}} \\ \dfrac{\xi\sqrt{3}(m_u - m_d)}{\sqrt{2 + \xi^2}} & \dfrac{3\xi^2(m_u + m_d)}{2 + \xi^2} \end{pmatrix}. \tag{5.4.17}$$

If we neglect $m_u - m_d$ in comparison with m_u, m_d, we find that the light pseudo-Goldstone bosons should have square masses

$$B_0(m_u + m_d), \qquad \frac{3 B_0 \xi^2 (m_u + m_d)}{2 + \xi^2}. \tag{5.4.18}$$

The first one is of course the pion, whose mass has been computed in (5.3.64). The second one has a mass

$$\frac{\sqrt{3}\xi m_\pi}{\sqrt{2 + \xi^2}} \le \sqrt{3} m_\pi. \tag{5.4.19}$$

Therefore, if the $U(1)$ anomaly is spontaneously broken, there will be an extra, light pseudo-Goldstone boson whose mass will not exceed $\sqrt{3}m_\pi$. However, such a particle does not exist. There is an extra pseudo-scalar meson, the η', but its mass is too large. This is the so-called $U(1)$ *problem*, or the problem of the missing Goldstone boson.

The solution to this puzzle is that the axial current is neither classically conserved nor spontaneously broken, but rather *anomalous*. This means that, if one takes into account quantum mechanical effects, the axial current (5.3.28) is not conserved, even when quarks are massless. We will show this by using an elegant method based on the analysis of the measure of the path integral, due to Fujikawa.

It is convenient to do the calculation in Euclidean signature, and then go back to Minkowski spacetime. The reason is that, in Euclidean signature, the Dirac operator $i\not{D}$ is Hermitian, and this facilitates the argument. We will assume that we have a single fermion, i.e. $N_f = 1$. The generalization to arbitrary N_f is straightforward. The first step is to define, at least formally, the path integral measure for the fermions. To do this we first consider a complete set of eigenfunctions for the Dirac operator:

$$i\not{D}\psi_n = \lambda_n \psi_n. \tag{5.4.20}$$

We assume that these eigenfunctions are orthonormal,

$$\langle \psi_n | \psi_m \rangle = \int d^4x \, \psi_n^\dagger(x)\psi_m(x) = \delta_{nm}. \tag{5.4.21}$$

We expand the fields in modes of the Dirac operator as:

$$\psi = \sum_n a_n \psi_n, \qquad \overline{\psi} = \sum_n \bar{b}_n \psi_n^\dagger. \tag{5.4.22}$$

We now define the path integral measure for the Dirac fermions as:

$$\mathcal{D}\psi\mathcal{D}\overline{\psi} = \prod_n da_n \prod_n d\bar{b}_n. \tag{5.4.23}$$

Notice that a_n, $\bar{b}_n$ are Grassmann variables. Under a space dependent, infinitesimal axial rotation, the Dirac fields transform as:

$$\psi(x) \to \psi'(x) = \psi(x) + i\alpha(x)\gamma_5\psi(x),$$
$$\overline{\psi}(x) \to \overline{\psi}'(x) = \overline{\psi}(x) + \overline{\psi}(x)i\gamma_5\alpha(x). \tag{5.4.24}$$

Let us now analyze the effects of this rotation on the path integral,

$$Z = \int \mathcal{D}\psi\mathcal{D}\overline{\psi} \, \exp\left\{\frac{1}{g_0^2}\int d^4x \, \overline{\psi}\left(i\not{D}\psi + im\right)\psi\right\}. \tag{5.4.25}$$

Here, we use the normalization of the fields determined by the Lagrangian (5.3.1). The action changes under this transformation as:

$$\frac{1}{g_0^2} \int d^4x \, \overline{\psi} \left(i \slashed{D} \psi + iM \right) \psi \rightarrow \frac{1}{g_0^2} \int d^4x \, \overline{\psi} \left(i \slashed{D} \psi + im \right) \psi$$

$$+ \int d^4x \, \alpha(x) \partial_\mu J^\mu(x) - \frac{2m}{g_0^2} \int d^4x \, \alpha(x) \overline{\psi} \gamma^5 \psi,$$

(5.4.26)

where

$$J^\mu = \frac{1}{g_0^2} \overline{\psi} \gamma^\mu \gamma^5 \psi \tag{5.4.27}$$

is the axial current for a single flavor. In order to obtain the *quantum* conservation law, we must take into account the change in the measure. Let us decompose the rotated fields as

$$\psi'(x) = \sum_n a'_n \psi_n(x), \qquad \overline{\psi}'(x) = \sum_n \overline{b}'_n \psi_n^\dagger. \tag{5.4.28}$$

The coefficients a'_n, $\overline{b}'_n$ are computed by using the inner product (5.4.21):

$$a'_n = \langle \psi_n | \psi' \rangle = \langle \psi_n | 1 + i\alpha(x)\gamma_5 | \psi \rangle$$

$$= \sum_m a_m \langle \psi_n | 1 + i\alpha(x)\gamma_5 | \psi_m \rangle = \sum_m C_{nm} a_m, \tag{5.4.29}$$

where

$$C_{nm} = \langle \psi_n | 1 + i\alpha(x)\gamma_5 | \psi_m \rangle = \delta_{nm} + i\langle \psi_n | \alpha(x)\gamma_5 | \psi_m \rangle. \tag{5.4.30}$$

Let us now consider the measure for the rotated fields. Since we are dealing with Grassmann quantities the change of variables involves the inverse of the Jacobian:

$$\prod_n da'_n = [\det C_{nm}]^{-1} \prod_m da_m = e^{-\mathrm{Tr} \log C_{mn}} \prod_m da_m$$

$$\approx \exp\left[-i \sum_n \langle \psi_n | \alpha(x)\gamma_5 | \psi_n \rangle \right] \prod_m da_m. \tag{5.4.31}$$

A similar analysis can be done for the $\overline{b}$ variables, which give the same contribution. Performing the inner product in (5.4.31) we obtain, for the change in the measure,

$$\prod_n da_n \prod_n d\overline{b}_n \rightarrow \prod_n da_n \prod_n d\overline{b}_n \exp\left(-2i \int d^4x \, \alpha(x) A(x) \right), \tag{5.4.32}$$

where

$$A(x) = \sum_n \psi_n^\dagger(x) \gamma_5 \psi_n(x) \tag{5.4.33}$$

is called the anomaly density. Putting this together with (5.4.26), we find that the axial current satisfies

$$\partial_\mu J^\mu(x) = 2iA(x) + \frac{2m}{g_0^2}\bar{\psi}(x)\gamma^5\psi(x).$$

(5.4.34)

It remains to evaluate $A(x)$. Since it involves a sum over an infinite number of modes, we should regulate it. In Euclidean space, a natural regularization is to weight each mode by the eigenvalue λ_n^2 of the second order operator $(i\slashed{D})^2$. Therefore, we define

$$A(x) = \lim_{\tau \to 0} \sum_n \psi_n^\dagger(x)\gamma_5 e^{-\tau\lambda_n^2}\psi_n(x) = \lim_{\tau \to 0} \text{tr}\left(\gamma^5 K_{(i\slashed{D})^2}(x, x; \tau)\right),$$

(5.4.35)

where $K_{(i\slashed{D})^2}(x, y; \tau)$ is the heat kernel of the operator $(i\slashed{D})^2$, as defined in (B.1) and (B.5). It is a matrix-valued function with both color indices and spinor indices, since it acts on the space of Dirac spinors in a given representation of the gauge group, and tr denotes the trace with respect to these internal indices. The quantity (5.4.35) can be evaluated by using the heat kernel expansion. Let us first calculate $(i\slashed{D})^2$. We multiply both sides of the equation (4.2.17) by $\gamma^\mu\gamma^\nu$:

$$\gamma^\mu\gamma^\nu\left(D_\mu D_\nu - D_\nu D_\mu\right) = -\frac{i}{2}[\gamma^\mu, \gamma^\nu]F_{\mu\nu},$$

(5.4.36)

where on the right hand side we have antisymmetrized the product of gammas. The left hand side can now be written, after using $\{\gamma^\mu, \gamma^\nu\} = 2g^{\mu\nu}$, as

$$2\left(\gamma^\mu D_\mu\right)^2 - 2g^{\mu\nu}D_\mu D_\nu.$$

(5.4.37)

We conclude that

$$(i\slashed{D})^2 = -\Box + \frac{i}{4}[\gamma^\mu, \gamma^\nu]F_{\mu\nu},$$

(5.4.38)

where

$$\Box = g^{\mu\nu}D_\mu D_\nu$$

(5.4.39)

is the scalar Laplacian in the presence of a gauge field. We now plug this result into the definition of the anomaly density. We find,

$$A(x) = \lim_{\tau \to 0}\langle x|\text{tr}\left(\gamma_5 \exp\left\{-\tau\left(-\Box + \frac{i}{4}[\gamma^\mu, \gamma^\nu]F_{\mu\nu}\right)\right\}\right)|x\rangle.$$

(5.4.40)

We can expand $e^{\tau\Box}$ by using the heat kernel expansion (B.11). Since we are in four dimensions, we find

$$\langle x|e^{\tau\Box}|x\rangle = \frac{1}{16\pi^2\tau^2}(1 + \mathcal{O}(\tau)).$$

(5.4.41)

On the other hand, if we expand

$$\exp\left\{-\frac{i\tau}{4}[\gamma^\mu, \gamma^\nu]F_{\mu\nu}\right\} \tag{5.4.42}$$

in powers of τ, the first non-vanishing term is already of order τ^2, since $\mathrm{Tr}(\gamma^5) = \mathrm{Tr}(\gamma^5\gamma^\mu\gamma^\nu) = 0$. We then conclude that

$$A(x) = -\frac{1}{128\pi^2}\mathrm{Tr}\left(F_{\mu\nu}F_{\rho\sigma}\right)\mathrm{Tr}\left(\gamma^5\gamma^\mu\gamma^\nu\gamma^\rho\gamma^\sigma\right). \tag{5.4.43}$$

In this equation, the traces are with respect to color and spinor indices, respectively. We note that the non-Abelian field strength takes values in the representation of the Lie algebra associated to the Dirac spinor, i.e.

$$F_{\mu\nu} = F_{\mu\nu}^a T_a^r. \tag{5.4.44}$$

With our conventions for Euclidean gamma matrices, one has

$$\mathrm{Tr}\left(\gamma^5\gamma^\mu\gamma^\nu\gamma^\rho\gamma^\sigma\right) = -4\epsilon^{\mu\nu\rho\sigma}. \tag{5.4.45}$$

We conclude that

$$A(x) = \frac{1}{16\pi^2}\mathrm{Tr}\left(F_{\mu\nu}\tilde{F}^{\mu\nu}\right), \tag{5.4.46}$$

where the dual field strength $\tilde{F}^{\mu\nu}$ is defined in (4.3.3). In particular, if the Dirac spinors are in the fundamental representation of the gauge group, as in QCD, we find

$$A(x) = q_{\mathrm{E}}(x), \tag{5.4.47}$$

where $q_{\mathrm{E}}(x)$ is the topological density in Euclidean signature (4.3.8). We then conclude that the axial current (5.3.28) satisfies the equation

$$\partial_\mu J^\mu(x) = 2iN_f q_{\mathrm{E}}(x) + \frac{2}{g_0^2}\sum_{f=1}^{N_f} m_f \overline{\psi}_f(x)\gamma^5\psi_f(x), \tag{5.4.48}$$

where we have already extended the result to N_f flavors.

The above result has been obtained in Euclidean signature. To obtain the corresponding result in Minkowski signature, we have to Wick-rotate back. Taking into account (4.3.7), as well as the rules for Euclidean continuation of the spinors and gamma functions, we find that, in Minkowski signature,

$$\partial_\mu J^\mu(x) = -2N_f q(x) + \frac{2i}{g_0^2}\sum_{f=1}^{N_f} m_f \overline{\psi}_f(x)\gamma^5\psi_f(x). \tag{5.4.49}$$

It is also possible to do the computation of the anomaly directly in Minkowski space, although in that case the regularization of the anomaly density involves the insertion of $\exp\{\tau(\mathrm{i}\slashed{D})^2\}$. Also, notice from (5.4.49) and the form of the action (5.3.1), that in Minkowski signature the change of fermionic measure under an axial transformation is given by

$$\exp\left\{-2\mathrm{i}N_f \int \mathrm{d}^4x\, \alpha(x)q(x)\right\}. \qquad (5.4.50)$$

The axial anomaly has many important consequences and it is crucial in the understanding of QCD. The first consequence is that, when the quarks are *massless*, there is no dependence on the theta angle. This just follows from considering an axial rotation (5.4.24) with a constant α. As we have just seen, the effect of such a rotation in Minkowski signature is to introduce the factor (5.4.50) in the path integral. In view of (4.3.5), this is equivalent to a shift of the theta angle by

$$\theta \rightarrow \theta + 2N_f\alpha. \qquad (5.4.51)$$

Since (5.4.24) with a constant α can be regarded as a change of variables in the path integral, we conclude that

$$Z(\theta) = Z(\theta + 2N_f\alpha), \qquad (5.4.52)$$

where α is an arbitrary angle. In other words, the partition function will be independent of θ. A similar reasoning can be applied to correlation functions. This implies, in particular, that the topological susceptibility in QCD with massless quarks vanishes.

Another remarkable consequence of the axial anomaly is that, in the Euclidean theory, and in the presence of a gauge field with non-zero topological charge, the Dirac operator will have zero modes. Let us suppose that $|\psi_n\rangle$ is an eigenfunction of $\mathrm{i}\slashed{D}$ with a non-zero eigenvalue λ_n. The state

$$|\psi_n\rangle^\chi = \gamma_5|\psi_n\rangle \qquad (5.4.53)$$

verifies:

$$\mathrm{i}\slashed{D}|\psi_n\rangle^\chi = \mathrm{i}\slashed{D}\gamma_5|\psi_n\rangle = -\gamma_5\mathrm{i}\slashed{D}\psi_n\rangle = -\lambda_n|\psi_n\rangle. \qquad (5.4.54)$$

Since $\mathrm{i}\slashed{D}$ is Hermitian, the eigenfunctions for different eigenvalues are orthogonal:

$$\langle\psi_n|\psi_n\rangle^\chi = \langle\psi_n|\gamma_5|\psi_n\rangle = 0. \qquad (5.4.55)$$

This means that, after integration, in the sum (5.4.33) defining $A(x)$ only zero modes give a non-zero contribution. We will denote these zero modes by $|\psi_{0,\alpha}^\pm\rangle$, where the index α runs from 1 to $\nu_\pm$, respectively. As our notation indicates, they can be classified according to their chirality:

$$\gamma_5|\psi_{0,\alpha}^\pm\rangle = \pm|\psi_{0,\alpha}^\pm\rangle. \qquad (5.4.56)$$

The integral of $A(x)$ is then given by

$$\int d^4x \, A(x) = \sum_{\alpha=1}^{\nu_+} \langle \psi^+_{0,\alpha} | \gamma_5 | \psi^+_{0,\alpha} \rangle + \sum_{\alpha=1}^{\nu_-} \langle \psi^-_{0,\alpha} | \gamma_5 | \psi^-_{0,\alpha} \rangle = \nu_+ - \nu_-. \qquad (5.4.57)$$

On the other hand, since $A(x)$ is given by the topological density, its integral is the topological charge Q of the background gauge field configuration. We conclude that

$$\nu_+ - \nu_- = Q. \qquad (5.4.58)$$

For example, if the gauge field is an instanton, $\nu_+ - \nu_-$ cannot be zero, and there will be zero modes for the fermions. The result (5.4.58) is a simple example of the index theorem of Atiyah and Singer for the Dirac operator. Note the similarity between (5.4.58) and the result (5.2.90) found in supersymmetric QM. Note as well that, as a consequence of the presence of fermionic zero modes, the vacuum-to-vacuum amplitude in the background of an instanton is always zero. This is very similar to our discussion in the context of supersymmetric QM: when there are zero modes, and due to the Grassmannian nature of the spinorial fields, we need to insert fermions in the path integral to "soak up" the integration over the zero modes, as we did in (5.2.104). There should be as many fermion fields inserted in the path integral as zero modes. This gives an important selection rule for the calculation of correlation functions in the background of an instanton.

Let us finish this section by pointing out that the axial anomaly opens the way to resolution of the $U(1)$ problem: since the axial current has a non-zero divergence due to the anomaly, even in a world of massless quarks, it is likely that the topological density appearing in (5.4.49) leads to a mass for the would-be Goldstone boson. This would allow us to identify this would-be Goldstone boson with the η'. Its large mass would then be explained as an effect of the topological density. We will see how to make these ideas precise in the framework of the large N expansion, and we will derive a result (the Witten–Veneziano formula) which gives a quantitative relationship between the mass of the η' and the topological density.

5.5 Bibliographical notes

The study of non-perturbative aspects of supersymmetric QM was initiated in Witten's seminal paper [193], which triggered a large literature. The paper [54] includes a superspace derivation of the supersymmetric Lagrangian (5.2.22). Our presentation follows the original arguments of Witten, and the explicit instanton calculation of [159]. A review of supersymmetric QM can be found in [55]. The peculiar spectral properties of the Hamiltonian (5.2.39) were pointed out in [106], and they are studied with conventional instanton techniques in [200, 201]. The fact that the bosonic and fermionic modes do not cancel in general in the quotient of functional determinants was first pointed out in [119].

The textbooks [74] and [186] provide excellent references on chiral symmetry and the effective Lagrangian for the light mesons. For calculation of the meson masses using current algebraic techniques, see [74, 197]. The chiral Lagrangian including the η' has been studied in many references, including [72, 92, 107, 192]. The $U(1)$ problem is summarized in [185] and reviewed in [47, 186]. The calculation of the axial anomaly with path integral techniques was pioneered by Fujikawa [90] and is reviewed, together with many applications, in [91].

Part II
Large N

6

Sigma models at large N

6.1 Introduction

As we have already seen in the previous chapters, developing a quantitative understanding of interacting models is a very difficult task, specially in the strong coupling regime. Usually, the only kind of approximation we have in an interacting system is perturbation theory, but often this does not capture the essential physics of the problem. In asymptotically free theories we are not always able to do perturbation theory, since the coupling depends on the energy scale, and at low energies we are always at strong coupling. Is there another type of approximation which sheds light on the strong coupling regime?

Such an approximation was first found in the context of Statistical Mechanics, and in particular in the study of models describing phase transitions, for example the ferromagnetic transition. These interacting models are also difficult to solve. A first approach to analyzing them quantitatively is to use mean-field theory, but it is known that its predictions for the critical exponents are incorrect at low dimension (for the standard Ising model with nearest-neighbor interactions, mean-field theory fails below dimension 4). This is simply because mean-field theory does not give an accurate treatment of the interactions. However, in models based on N-vector variables and with an $O(N)$ symmetry, one can regard N as a parameter and consider an approximation in which the number of components, N, becomes very large. In this regime, as first shown by Berlin, Kac and Stanley, one can obtain results for the critical exponents which go beyond the mean-field approximation and capture features of the strong coupling regime of the theory. Of course, this so-called large N limit is yet another approximation to the theory, but it is one which displays aspects of the interacting physics which are not visible in conventional treatments.

This strategy can be applied to various QFTs in various dimensions. The idea is to promote the original, finite number of fields to a large number thereof.

Surprisingly, although the resulting theory has more degrees of freedom, it often simplifies as the number of fields becomes large. This is similar to the thermodynamic limit of Statistical Mechanics, in which having a very large number of degrees of freedom leads to a simplified description. For example, one can take an interacting scalar QFT and promote the scalar field to an N-component vector, similarly to what was done in the theory of phase transitions. In these vector theories, the large N limit turns out to be simpler than the theory at finite N, and one can perform a systematic expansion of various relevant physical quantities in inverse powers of $1/N$.

As we will see in the next chapter, YM theory and QCD can also be analyzed in the large N limit, by considering the gauge group $SU(N)$ with N large. However, the large N limit of these theories remains poorly understood, and no analytic solutions are known. It is then important to consider toy models of YM theory which admit a large N expansion and can give us a hint of what we can expect for the large N limit of gauge theories. A particularly interesting family of models are two-dimensional, asymptotically free sigma models, which are in many respects similar to non-Abelian gauge theories in four dimensions. In some of these theories, like the $O(N)$ sigma model or the $\mathbb{CP}^{N-1}$ model, one can consider N real or complex fields, and the large N limit turns out to be exactly solvable. These models then constitute excellent models to test the methods of the $1/N$ expansion and they can be regarded as laboratories for QCD at large N. In this chapter we will study these models in some detail. In particular, the $\mathbb{CP}^{N-1}$ model has a topological charge and we will be able to study some general non-perturbative properties discussed in the previous chapter, in the framework of the large N limit.

6.2 The *O(N)* non-linear sigma model

In a non-linear sigma model, the basic fields are maps

$$\sigma : \Sigma \to X \tag{6.2.1}$$

from a two-dimensional spacetime Σ to a so-called target manifold X. We will often take $\Sigma = \mathbb{R}^2$ with Minkowski or Euclidean signature, but one could also consider other spacetimes. Sometimes it is convenient to describe the map in terms of local coordinates on X. In this case, we have d fields σ^I, $I = 1, \ldots, d$, where d is the dimension of the target space X. Some of the typical geometries for the target manifold X can be described as submanifolds of $\mathbb{R}^N$, defined by an algebraic equation. In this case, the field σ can be described by N fields representing the coordinates of $\mathbb{R}^N$, and satisfying the algebraic constraint which defines the target space X. An important example is the $O(N)$ non-linear sigma model. In this case, one takes $X = \mathbb{S}^{N-1} \subset \mathbb{R}^N$, the unit sphere in Euclidean N-dimensional space.

The field σ is then described in terms of N fields σ^a, $a = 1, \ldots, N$, satisfying the constraint that defines the $(N-1)$-dimensional sphere,

$$\sum_{a=1}^{N} \sigma_a^2 = 1. \tag{6.2.2}$$

The action of the theory is given by

$$S = \frac{1}{2g_0^2} \int d^2x \, \partial_\mu \sigma^a \partial^\mu \sigma^a, \tag{6.2.3}$$

where we use the convention that the repeated indices μ, a are summed over. The group $O(N)$ acts on the sphere by rotations, and this implies that the fields σ^a transform in the vector or fundamental representation of $O(N)$. Therefore, the geometric symmetry of the target sphere is inherited as a global symmetry of the field theory. g_0 is a bare coupling constant, which should be renormalized after taking into account quantum corrections.

The $O(N)$ non-linear sigma model in two dimensions is one of the best toy models in QFT. It shares two important properties with non-Abelian YM theory: it is asymptotically free and it has a mass gap. In addition, many of its properties can be obtained exactly, for example its S-matrix. However, this is a model in which, as in other models we have studied in this book, the information provided by perturbation theory is misleading. The theory described by the classical action (6.2.3) is a complicated non-linear theory of $N - 1$ independent fields, as can easily seen by solving the constraint (6.2.2). Perturbatively, we have a theory of $N - 1$ massless bosons in two dimensions. These can be regarded as the Goldstone bosons of a theory with spontaneously broken $O(N)$ symmetry. However, a famous theorem due to Coleman, Mermin and Wagner says that there are no Goldstone bosons in two dimensions, so the perturbative picture cannot be correct. Indeed, it is possible to see that the non-perturbative spectrum consists of N *massive* particles in a vector representation of $O(N)$.

Here we will look at the model in the limit in which N, the number of components, is very large. Although in this limit we have many more degrees of freedom, we will see that the theory simplifies and it can be solved exactly. More importantly, the large N solution contains information beyond perturbation theory: it confirms that there are no Goldstone bosons in the model, and it shows that the spectrum consists of massive particles. Moreover, their mass can be computed exactly in the large N limit.

To study the large N limit of the $O(N)$ sigma model, we first introduce the (bare) *'t Hooft parameter*,

$$t_0 = N g_0^2. \tag{6.2.4}$$

We will take the limit in which N is large and g_0^2 is small, in such a way that t_0 is fixed. The large N analysis proceeds as follows. First, we change the normalization of the fields in order to have a canonically normalized kinetic term,

$$\sigma^a \to \sqrt{\frac{t_0}{N}} \sigma^a. \tag{6.2.5}$$

We now impose the constraint (6.2.2) through an extra field α, which can be regarded as a Lagrange multiplier. The resulting action is

$$S = \frac{1}{2} \int d^d x \left\{ \partial_\mu \sigma^a \partial^\mu \sigma^a - i\alpha \left(\sigma^a \sigma^a - \frac{N}{t_0} \right) \right\}, \tag{6.2.6}$$

and the classical EOM for α imposes the constraint. We want to calculate the generating functional of Euclidean correlation functions,

$$Z[J] = \int \mathcal{D}\sigma \, \mathcal{D}\alpha \, \exp \left\{ -S + \int d^2 x \, J^a(x) \sigma^a(x) \right\}. \tag{6.2.7}$$

The path integral over α produces a delta function which imposes (6.2.2) on the space of field configurations for the σ^a. In this way one obtains a path integral over the constrained σ^a fields. However, *first* we can integrate the σ^a fields and produce an effective action for α. Since the action is quadratic in the σ^a, this is a Gaussian integral and we find,

$$Z[J] = \int \mathcal{D}\alpha \, \exp \left\{ -S_{\text{eff}}(\alpha) + \int d^2 x \, d^2 y \, J^a(x) G(x, y) J^a(y) \right\} \tag{6.2.8}$$

where

$$S_{\text{eff}}(\alpha) = \frac{N}{2} \text{Tr} \log \left(-\partial_x^2 - i\alpha(x) \right) + \frac{iN}{2t_0} \int d^2 x \, \alpha(x), \tag{6.2.9}$$

and $G(x, y)$ is the Green function defined by

$$G^{-1}(x, y) = \left(-\partial_x^2 - i\alpha(x) \right) \delta(x - y). \tag{6.2.10}$$

The first term in (6.2.9) is minus the logarithm of the functional determinant

$$\frac{1}{\left[\det \left(-\partial^2 - i\alpha(x) \right) \right]^{N/2}} \tag{6.2.11}$$

obtained by integrating out the N scalar fields σ^a. In the effective action (6.2.9), N plays the role of $1/\hbar$. Therefore, for large N, we can use the saddle-point method and expand the path integral around the configuration which extremizes the effective action,

$$\frac{\delta}{\delta \alpha(x)} \left[\frac{iN}{t_0} \int d^2 y \, \alpha(y) + N \text{Tr} \log G^{-1} \right] = 0. \tag{6.2.12}$$

We now notice that

$$\frac{\delta}{\delta\alpha(x)}\mathrm{Tr}\log G^{-1} = \mathrm{Tr}\left(G\frac{\delta G^{-1}}{\delta\alpha(x)}\right), \tag{6.2.13}$$

and since

$$\frac{\delta G^{-1}(z, w)}{\delta\alpha(x)} = -i\delta(x - z)\delta(z - w), \tag{6.2.14}$$

we find the equation

$$G(x, x) = \frac{1}{t_0}. \tag{6.2.15}$$

We will search for solutions characterized by a *constant* field $\alpha(x) = \alpha$. This is of course what one expects from a solution which respects translation invariance. In this case, the Green function is of the form $G(x, y) = G(x - y)$, and it satisfies

$$\left(-\partial_x^2 - i\alpha\right) G(x) = \delta(x), \tag{6.2.16}$$

which gives, after Fourier transform,

$$G(0) = \int \frac{d^2k}{(2\pi)^2} \frac{1}{k^2 - i\alpha}. \tag{6.2.17}$$

The EOM (6.2.15) then reads,

$$\int \frac{d^2k}{(2\pi)^2} \frac{1}{k^2 - i\alpha} = \frac{1}{t_0}. \tag{6.2.18}$$

We will assume that the solution to the equation (6.2.18) is of the form

$$\alpha = im^2, \qquad m^2 > 0, \tag{6.2.19}$$

and we will see in a moment that this is indeed the case. The integral over momentum space on the left hand side of (6.2.18) is divergent and needs to be regularized. We will use dimensional regularization and set $d = 2 + \epsilon$. We find,

$$\int \frac{d^d k}{(2\pi)^d} \frac{1}{k^2 + m^2} = \frac{1}{(4\pi)^{d/2}}\Gamma(1 - d/2)m^{d-2}$$

$$= \frac{1}{4\pi}\left\{-\frac{2}{\epsilon} - \gamma_E - \log\left(\frac{m^2}{4\pi}\right)\right\} + \mathcal{O}(\epsilon). \tag{6.2.20}$$

In order to proceed, we introduce the renormalized coupling constant $t = t(\mu)$ through the equation,

$$t_0 = \mu^{-\epsilon}t Z(\epsilon, t), \tag{6.2.21}$$

where μ is the renormalization scale and

$$Z(\epsilon, t) = 1 + \sum_{n=1}^{\infty} Z_n(\epsilon) t^n. \tag{6.2.22}$$

In the MS scheme, we require $Z_n(\epsilon)$ to be polynomials in ϵ^{-1} with no constant term. The equation (6.2.18) leads to

$$Z_1 = \frac{1}{2\pi\epsilon}, \qquad Z_2 = Z_1^2. \tag{6.2.23}$$

As usual, requiring the bare coupling t_0 to be independent of the renormalization scale, we find the equation determining the beta function,

$$- \epsilon t Z(\epsilon, t) + \beta(t) \frac{\partial}{\partial t} (t Z(\epsilon, t)) = 0, \tag{6.2.24}$$

where

$$\beta(t) = \mu \frac{\partial t}{\partial \mu}. \tag{6.2.25}$$

The above calculation gives

$$\beta(t) = \epsilon t - \frac{1}{2\pi} t^2. \tag{6.2.26}$$

For $\epsilon = 0$ we find, at large N, an asymptotically free theory. The dynamically generated scale is given by

$$\Lambda^2 = \mu^2 e^{-4\pi/t(\mu)}, \tag{6.2.27}$$

in analogy with (4.2.44). Finally, by comparing the finite part of (6.2.18), after the renormalization procedure, we find

$$m^2 = 4\pi e^{-\gamma_E} \Lambda^2, \tag{6.2.28}$$

which is a positive quantity proportional to the square of the dynamically generated scale, confirming our ansatz (6.2.19). The above computation of the β function gives the large N limit of the usual beta function of the non-linear sigma model, and for fixed 't Hooft parameter t. Indeed, at one-loop and fixed N, this function reads,

$$\beta(g) = - (N - 2) \frac{g^3}{4\pi}, \tag{6.2.29}$$

which agrees with (6.2.26) at $\epsilon = 0$ and $N \to \infty$, t fixed.

The physical interpretation of m^2 is now clear: it is a mass for the σ^a fields. This can be seen by looking at the original Lagrangian in (6.2.6), since a vev for α of the form $-im^2$ gives a mass term for these fields. Equivalently, the correlation function of two σ^a fields is obtained as

$$G^{ab}(x, y) = \langle \sigma^a(x)\sigma^b(y) \rangle = \frac{\delta^2 Z[J]}{\delta J^a(x)\delta J^b(y)}. \tag{6.2.30}$$

In the large N limit this is obtained by evaluating the Green function $G(x, y)$ at the saddle point. In momentum space, we have

$$\tilde{G}_{ab}(p) = \frac{\delta_{ab}}{p^2 + m^2}. \tag{6.2.31}$$

This mass term is the first dynamical effect that can be obtained at large N.

We have found a vacuum which is a constant field configuration for $\alpha(x)$ and gives the main non-perturbative properties of the theory. In order to go beyond the strict large N limit, we should expand the effective action around the vacuum. We then redefine,

$$\alpha(x) \rightarrow im^2 + \frac{\alpha(x)}{\sqrt{N}} \tag{6.2.32}$$

and we consider small fluctuations of the field $\alpha(x)$. The factor of $1/\sqrt{N}$ is introduced to have an appropriate normalization for the fluctuations at large N, as we will see. The effective action is now given by

$$\frac{N}{2}\text{Tr} \log \left(-\partial^2 + m^2 - i\frac{\alpha(x)}{\sqrt{N}} \right), \tag{6.2.33}$$

which can be written as

$$\frac{N}{2}\text{Tr} \log(-\partial^2 + m^2) + \frac{N}{2}\text{Tr} \log \left[1 + \Delta \left(-i\alpha(x)/\sqrt{N} \right) \right], \tag{6.2.34}$$

where

$$\Delta(x, y) = (-\partial^2 + m^2)^{-1}\delta(x - y). \tag{6.2.35}$$

We now expand in inverse powers of N. By the definition of saddle point, there is no linear term in α, so the next term in the effective action is quadratic in α and is given schematically by

$$\frac{1}{4}\text{Tr} (\Delta\alpha)^2. \tag{6.2.36}$$

The best way to compute this quantity is by using Feynman diagrams. The functional determinant (6.2.11) is obtained by integrating out the σ^a fields at one-loop, and it can be expressed as an even power series in the α fields. Each term in this

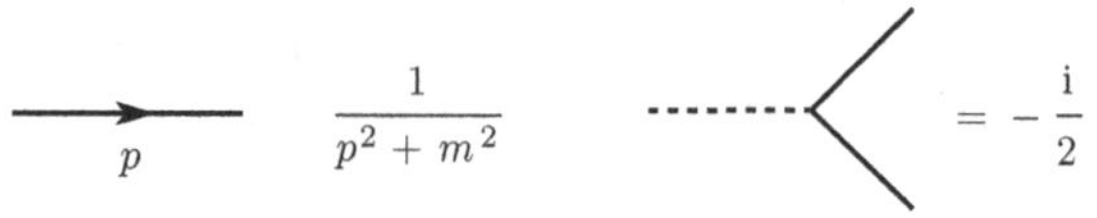

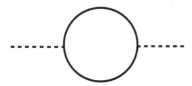

Figure 6.1 The Feynman rules for the computation of the one-loop determinant. The dashed lines correspond to the α field, while the thick lines correspond to the σ^a field.

Figure 6.2 The Feynman diagram computing the quadratic term in the α fluctuation.

series is given by a one-loop diagram with $2n$ external legs of the α fields, built out of the cubic vertex of the theory. The Feynman rules for constructing these diagrams are represented in Fig. 6.1. In terms of the Fourier-transformed field

$$\tilde{\alpha}(p) = \int d^2x \, e^{ipx} \alpha(x),$$ (6.2.37)

the quadratic term in S_{eff} in the α fluctuations reads,

$$\frac{1}{2} \int \frac{d^2p}{(2\pi)^2} \tilde{\alpha}(p) \tilde{\Gamma}^\alpha(p) \tilde{\alpha}(-p),$$ (6.2.38)

where $\tilde{\Gamma}^\alpha(p)$ is given by minus the Feynman diagram of Fig. 6.2:

$$\tilde{\Gamma}^\alpha(p) = \frac{1}{2} \int \frac{d^2q}{(2\pi)^2} \frac{1}{(q^2 + m^2)((p+q)^2 + m^2)} \equiv \frac{1}{2} f(p).$$ (6.2.39)

The computation of (6.2.39) is a standard exercise in Feynman diagram calculus. The details can be found in Appendix C, and the result is

$$f(p) = \frac{1}{2\pi \sqrt{p^2(p^2 + 4m^2)}} \log \frac{\sqrt{p^2 + 4m^2} + \sqrt{p^2}}{\sqrt{p^2 + 4m^2} - \sqrt{p^2}}.$$ (6.2.40)

It is now possible to compute the correlation functions in the original theory in terms of this effective theory, in which we have N massive particles with Green function (6.2.31) as well as an α particle with propagator

$$\tilde{G}^\alpha(p) = \frac{2}{f(p)}.$$ (6.2.41)

In addition, we have interactions suppressed at large N, and $1/\sqrt{N}$ becomes the effective coupling constant of the theory. Instead of working out other aspects of this model, let us move on and consider a closely related model, which has in addition many properties in common with YM theory: the $\mathbb{CP}^{N-1}$ model.

6.3 The $\mathbb{CP}^{N-1}$ model

Our second example of the uses of the $1/N$ expansion will be another two-dimensional non-linear sigma model: the $\mathbb{CP}^{N-1}$ model. As we have mentioned, this model is similar to YM theory, in the sense that it is asymptotically free and it has a topological charge. Therefore, one can introduce a theta angle and a two-dimensional analogue of the topological susceptibility. In addition, it admits instanton solutions, but instanton calculus at infinite volume is afflicted with the same IR divergences as in YM theory. We will see, however, that one can use the $1/N$ expansion to obtain non-perturbative results, and in particular a non-zero value for the topological susceptibility.

The model and its instantons

The $\mathbb{CP}^{N-1}$ model is a non-linear sigma model, in which the target is the complex projective space of $N - 1$ dimensions. Geometrically, this target can be described as a quotient of $\mathbb{C}^N$: given two vectors $z, z' \in \mathbb{C}^N$, we introduce an equivalence relation as follows:

$$z \sim z' \Leftrightarrow z = \lambda z', \qquad \lambda \in \mathbb{C}^*. \tag{6.3.1}$$

The $(N - 1)$-dimensional complex projective space is then defined as the quotient of $\mathbb{C}^N$ by this equivalence relation. It is very easy to see that, in order to define this quotient, we can restrict ourselves to complex vectors with unit norm, i.e. to the real subspace of $\mathbb{C}^N$ given by

$$\mathbb{S}^{2N-1} = \left\{ z \in \mathbb{C}^N : \|z\| = 1 \right\}, \tag{6.3.2}$$

where as usual

$$\|z\|^2 = \bar{z} \cdot z = \sum_{i=1}^{N} |z_i|^2, \tag{6.3.3}$$

and the dot denotes the standard Hermitian product in $\mathbb{C}^N$:

$$\bar{z} \cdot w = \sum_{i=1}^{N} \bar{z}_i w_i. \tag{6.3.4}$$

The equivalence relation is

$$z \sim z' \Leftrightarrow z = \lambda z', \qquad |\lambda| = 1, \quad z, z' \in \mathbb{S}^{2N-1}. \tag{6.3.5}$$

In order to construct the corresponding two-dimensional non-linear sigma model, we will use the second description of $\mathbb{CP}^{N-1}$: our basic field is defined on a two-dimensional spacetime Σ and takes values in the sphere (6.3.2). We can then describe it by N complex components $z_i(x)$ satisfying the constraint

$$\|z(x)\| = 1. \tag{6.3.6}$$

We also have to implement the equivalence relationship (6.3.5). Since the z_i are now fields depending on $x \in \Sigma$, the phase λ in (6.3.5) can also depend on x. This suggests that we introduce a $U(1)$ gauge symmetry

$$z_i(x) \rightarrow e^{i\alpha(x)} z_i(x). \tag{6.3.7}$$

We can cook up a gauge field out of the z_i, since the composite field

$$A_\mu = \frac{i}{2} \left(\bar{z} \cdot \partial_\mu z - (\partial_\mu \bar{z}) \cdot z \right), \tag{6.3.8}$$

which is real, transforms as

$$A_\mu \rightarrow A_\mu - \partial_\mu \alpha(x). \tag{6.3.9}$$

This is easy to check by using that

$$\partial_\mu z_i \rightarrow e^{i\alpha(x)} \left(i z_i \partial_\mu \alpha + \partial_\mu z_i \right). \tag{6.3.10}$$

Let us now suppose that $\Sigma = \mathbb{R}^2$, so that the sigma model is defined on a flat spacetime with Euclidean signature. In order to have a well-defined theory with target space $\mathbb{CP}^{N-1}$, fields which differ by a spacetime dependent phase, as in (6.3.5), have to lead to the same action. In other words, the action must be invariant under the gauge symmetry (6.3.7). Since we have a composite field A_μ which transforms as a gauge connection, we can use the gauge covariant derivative to write down the gauge invariant action

$$S = \frac{1}{g_0^2} \int d^2x \, \overline{D_\mu z} \cdot D^\mu z, \qquad D_\mu = \partial_\mu + i A_\mu. \tag{6.3.11}$$

The action (6.3.11) defines the $\mathbb{CP}^{N-1}$ model. We can keep the gauge field in the action as an auxiliary field, since it does not have a kinetic term and its EOM is purely algebraic. If we expand the Lagrangian of (6.3.11), we find

$$\mathcal{L} = \overline{D_\mu z} \cdot D^\mu z = \partial^\mu \bar{z} \cdot \partial_\mu z + A_\mu A^\mu - i A_\mu \left(\bar{z} \cdot \partial^\mu z - (\partial^\mu \bar{z}) \cdot z \right), \tag{6.3.12}$$

and the algebraic EOM for A_μ implies that it takes the value given in (6.3.8). After plugging this value into the Lagrangian, we obtain

$$\mathcal{L} = \partial^\mu \bar{z} \cdot \partial_\mu z + (\bar{z} \cdot \partial^\mu z)(\bar{z} \cdot \partial_\mu z). \qquad (6.3.13)$$

This shows that the $\mathbb{CP}^{N-1}$ model is a theory of N complex scalar fields z_i, subject to the constraint (6.3.6), and interacting through the quartic term in (6.3.13).

The $\mathbb{CP}^{N-1}$ model has instanton solutions, similar in many respects to the instantons of YM theory. These instantons are topologically non-trivial configurations with finite action. This requires that they decay sufficiently fast at infinity,

$$D_\mu z_i \to 0, \quad \text{at } |x| \to \infty, \quad i = 1, \dots, N. \qquad (6.3.14)$$

This means that

$$-iA_\mu \to \frac{\partial_\mu z_i}{z_i} = \frac{\partial_\mu |z_i|}{|z_i|} + i\partial_\mu \phi_i, \qquad (6.3.15)$$

where ϕ_i is the phase of z_i. Since iA_μ is pure imaginary and independent of the index i, we deduce that, at infinity,

$$\partial_\mu |z_i| = 0, \qquad \phi_i = -\sigma(x), \quad i = 1, \dots, N, \qquad (6.3.16)$$

therefore the field z_i is of the form,

$$z_i = n_i e^{-i\sigma(x)}, \quad |x| \to \infty, \qquad \bar{n} \cdot n = 1. \qquad (6.3.17)$$

We can regard this field as a gauge transformation of the "trivial" field configuration $z_i = n_i$, in the same way that in a YM instanton, the gauge connection at infinity (4.3.36) is a gauge transformation of the trivial gauge connection $A_\mu = 0$. In the same way, the behavior (6.3.17) defines a function from the boundary at infinity $\mathbb{S}^1_\infty \subset \mathbb{R}^2$, to the Lie algebra of $U(1)$, which is another circle $\mathbb{S}^1$,

$$\sigma : \mathbb{S}^1_\infty \to \mathbb{S}^1. \qquad (6.3.18)$$

These maps are classified topologically, as we saw in (1.10.14), by a winding number ℓ. As in the case of YM theory, the winding number can be computed as a topological charge,

$$Q = \int q(x) d^2 x, \qquad (6.3.19)$$

where the topological density $q(x)$ is given by

$$q(x) = \frac{1}{2\pi} \epsilon_{\mu\nu} \partial_\mu A_\nu. \qquad (6.3.20)$$

To see this, note that, since

$$\epsilon_{\mu\nu} \partial_\mu A_\nu = i\epsilon_{\mu\nu} \partial_\mu \bar{z} \cdot \partial_\nu z, \qquad (6.3.21)$$

we can write Q as

$$Q = \frac{1}{2\pi i} \int d^2x \, \epsilon_{\mu\nu} \partial_\mu (\bar{z} \cdot \partial_\nu z) . \qquad (6.3.22)$$

By using Stokes' theorem, we can write (6.3.22) as an integral over the boundary at infinity $\mathbb{S}^1_\infty$:

$$Q = \frac{1}{2\pi i} \oint_{\mathbb{S}^1_\infty} dx^\mu \, \bar{z} \cdot \partial_\mu z. \qquad (6.3.23)$$

Plugging in here the boundary behavior (6.3.17), we obtain

$$Q = \frac{1}{2\pi} \oint_{\mathbb{S}^1_\infty} dx^\mu \frac{\partial \sigma}{\partial x^\mu} = \frac{1}{2\pi} \Delta \sigma, \qquad (6.3.24)$$

where $\Delta \sigma$ is the change of σ as we go around $\mathbb{S}^1_\infty$. This is precisely the winding number of the map (6.3.18).

As we saw in the case of YM theory, an important property of instantons is that they minimize the action in their topological sector. To see that this also holds in this model, we note that

$$i\epsilon_{\mu\nu} \overline{D_\mu z} \cdot D_\nu z = i\epsilon_{\mu\nu}(\partial_\mu \bar{z} - iA_\mu \bar{z}) \cdot (\partial_\nu z + iA_\nu z) = \epsilon_{\mu\nu} \partial_\mu A_\nu, \qquad (6.3.25)$$

since

$$- i\epsilon_{\mu\nu} \left(A_\mu \bar{z} \cdot \partial_\nu z - A_\nu \partial_\mu \bar{z} \cdot z \right) \qquad (6.3.26)$$

vanishes due to antisymmetry of $\epsilon_{\mu\nu}$. Therefore, the topological charge can be written as

$$Q = \frac{i}{2\pi} \int d^2x \, \epsilon_{\mu\nu} \overline{D_\mu z} \cdot D_\nu z. \qquad (6.3.27)$$

From the obvious inequality,

$$\left| D_\mu z \mp i\epsilon_{\mu\nu} D_\nu z \right|^2 \geq 0, \qquad (6.3.28)$$

we deduce that

$$\overline{D_\mu z} \cdot D_\mu z + \epsilon_{\mu\rho} \epsilon_{\mu\sigma} \overline{D_\rho z} \cdot D_\sigma z \mp 2i\epsilon_{\mu\nu} \overline{D_\mu z} \cdot D_\nu z \geq 0, \qquad (6.3.29)$$

and since $\epsilon_{\mu\rho} \epsilon_{\mu\sigma} = \delta_{\rho\sigma}$, we find the inequality

$$\overline{D_\mu z} \cdot D_\mu z \geq \pm i\epsilon_{\mu\nu} \overline{D_\mu z} \cdot D_\nu z. \qquad (6.3.30)$$

After integration, we obtain,

$$\int d^2x \, \overline{D_\mu z} \cdot D_\mu z \geq \pm i \int d^2x \, \epsilon_{\mu\nu} \overline{D_\mu z} \cdot D_\nu z, \qquad (6.3.31)$$

i.e.

$$S \geq \frac{2\pi}{g_0^2}|Q|. \tag{6.3.32}$$

This is exactly like the bound (4.4.4) which we found in YM theory. Equality holds only if the bound is saturated, and from here we derive the equation describing instanton configurations in this model:

$$D_\mu z_j \mp i\epsilon_{\mu\nu} D_\nu z_j = 0, \qquad j = 1, \ldots, N. \tag{6.3.33}$$

These are the analogues of the (anti)self-duality conditions (4.4.5), (4.4.6) for instantons in YM theory.

Let us now find the general solution to the equations (6.3.33). To do that, we consider the open subset U_k of $\mathbb{CP}^{N-1}$ where $z_k \neq 0$. On U_k, we can define the inhomogeneous coordinates

$$w_i^{(k)} = \frac{z_i}{z_k}, \qquad i = 1, \ldots, N, \quad i \neq k. \tag{6.3.34}$$

We can invert this relation as

$$z_i = e^{i\Lambda} \frac{w_i^{(k)}}{\|w^{(k)}\|}, \tag{6.3.35}$$

where Λ is the argument of z_k. If we now plug (6.3.34) in (6.3.33), and use the condition (6.3.33) for z_k, we find

$$\partial_\mu w_j^{(k)} \mp i\epsilon_{\mu\nu} \partial_\nu w_j^{(k)} = 0. \tag{6.3.36}$$

It is very easy to interpret these conditions by introducting complex coordinates on the two-dimensional spacetime. Let us define

$$s = x_1 - ix_2, \qquad \bar{s} = x_1 + ix_2. \tag{6.3.37}$$

The conditions (6.3.36) correspond to

$$\partial_{\bar{s}} w_j^{(k)} = 0, \qquad \partial_s w_j^{(k)} = 0, \tag{6.3.38}$$

respectively. These conditions say that $w_j^{(k)}$ are holomorphic (respectively, anti-holomorphic) functions of the complex coordinate s. Using now the conditions at infinity (6.3.17) we conclude that $w_j^{(k)}$ are rational functions, which, after reduction to a common denominator, can be written as

$$w_j^{(k)} = c_j^{(k)} \frac{\prod_{\ell=1}^m (s - a_{\ell,j}^{(k)})}{\prod_{\ell=1}^m (s - a_{\ell,0}^{(k)})}. \tag{6.3.39}$$

The degree of the polynomials m has to be the same for all $j = 1, \ldots, N-1$, due again to the conditions at infinity (6.3.17). Using this result, we can now write down the corresponding field configuration for the z_j, up to a gauge transformation, as

$$z_j = \frac{p_j(s)}{\|p_j(s)\|}, \tag{6.3.40}$$

where

$$p_j(s) = v_j \prod_{\ell=1}^{m} (s - u_j^{(\ell)}), \qquad j = 1, \ldots, N, \tag{6.3.41}$$

are polynomials of degree m with no common roots (if the polynomials have common roots, the configuration is gauge equivalent to another configuration with no common roots). As $|s| \to \infty$, we have

$$z_j \approx \frac{v_j}{\|v_j\|} e^{-im\phi}, \tag{6.3.42}$$

where the angle ϕ is defined by

$$(x_1, x_2) = |x| \left(\cos\phi, \sin\phi \right). \tag{6.3.43}$$

The configuration (6.3.42) has the structure (6.3.17), with $\sigma(x) = m\phi$. Therefore, we conclude from (6.3.24) that it belongs to the topological class with $Q = m > 0$. It is straightforward to verify directly that it solves the condition (6.3.33) with the $-$ sign. A similar argument shows that the conjugate field configuration

$$z_j = \frac{p_j(\bar{s})}{\|p_j(\bar{s})\|}, \tag{6.3.44}$$

has topological charge $Q = -m$. In the case $Q = 1$, it is easy to see that we can write the solution for $z_j(s)$ as

$$z_j(s) = \frac{v_i(s - s_0) + \lambda \mu_i}{\sqrt{\lambda^2 + |s - s_0|^2}}, \tag{6.3.45}$$

where v_i, μ_i are vectors satisfying

$$\|v\| = \|\mu\| = 1, \qquad \bar{v} \cdot \mu = 0, \tag{6.3.46}$$

and λ is a positive, real parameter. We can regard s_0 as the "center" of the instanton in $\mathbb{R}^2$, and λ as its "size."

The $\mathbb{CP}^{N-1}$ model on the two-sphere

It is possible to study the $\mathbb{CP}^{N-1}$ model in more general spacetimes. As in the case of YM theory, eventually we will be interested in spacetimes with a finite volume, where instanton calculus is reliable. A simple example is $\mathbb{S}^2$, the two-dimensional sphere (with radius R), since it can be easily related to $\mathbb{R}^2$ by a stereographic projection. We will label points of $\mathbb{S}^2$ by the coordinates r_a, $a = 1, 2, 3$, where

$$\sum_{a=1}^{3} r_a^2 = R^2. \tag{6.3.47}$$

Then, given a point in $\mathbb{S}^2$ minus the south pole $(0, 0, -R)$, we can map it to $\mathbb{R}^2$ through the stereographic projection (4.5.19),

$$x_\mu = \frac{R r_\mu}{R + r_3}, \qquad \mu = 1, 2, \tag{6.3.48}$$

with inverse

$$r_\mu = \frac{2R^2 x_\mu}{R^2 + x^2}, \qquad \mu = 1, 2, \qquad r_3 = R \frac{R^2 - x^2}{R^2 + x^2}, \tag{6.3.49}$$

where

$$x^2 = x_1^2 + x_2^2. \tag{6.3.50}$$

Therefore, a field $z_i(x)$ on $\mathbb{R}^2$ leads to a field on $\mathbb{S}^2$ defined by

$$\xi_i(r) = z_i(r(x)). \tag{6.3.51}$$

The region at infinity on $\mathbb{R}^2$ is mapped to a neighborhood of the south pole $(0, 0, -R)$ in the two-sphere. An instanton configuration on the plane, with topological charge Q, has a non-zero value for the integral (6.3.24), which is now interpreted as a contour integral around the south pole. Since the integral is non-zero, the integrand should have a pole at $(0, 0, -R)$. Therefore, fields in the plane with non-zero topological charge lead to fields with a singular behavior at the south pole in the sphere.

An efficient way to describe these fields, due to Berg and Lüscher, is the following. Geometrically, one can think about $\mathbb{S}^2$ as the basis of a $U(1) = \mathbb{S}^1$ fibration of the three-sphere $\mathbb{S}^3$. This is the so-called *Hopf fibration* of $\mathbb{S}^3$. Equivalently, since the three-sphere is isomorphic to $SU(2)$ (see (A.1)), we can construct non-trivial sections of the $U(1)$ bundle by considering smooth functions on $\mathbb{S}^3$, satisfying some constraints. This reduces the analysis of non-trivial instantons of the $\mathbb{CP}^{N-1}$ sigma model on $\mathbb{S}^2$, to the study of functions defined on $SU(2)$ and taking values in $\mathbb{CP}^{N-1}$. Some useful ingredients of harmonic analysis on $\mathbb{S}^3$ are recalled in Appendix A. To write the resulting functions in detail, we will take the $\mathbb{S}^1$ in $SU(2)$ to be the circle generated by σ_3, where σ_a, $a = 1, 2, 3$ denote as usual the Pauli matrices. We now decompose an element of $SU(2)$ as in (A.4),

$$g = U(x)e^{i\tau\sigma_3/2}, \qquad -2\pi \leq \tau \leq 2\pi, \tag{6.3.52}$$

where $U(x)$ can be written as

$$U(x) = \frac{1}{\sqrt{R^2 + x^2}} (R + ix_2\sigma_1 - ix_1\sigma_2), \tag{6.3.53}$$

where x_μ, $\mu = 1, 2$ are the coordinates of $\mathbb{R}^2$ given in (6.3.48). In terms of the Euler angles t_i, $i = 1, 2, 3$ appearing in (A.4), we have

$$\begin{aligned} x_1 &= R\tan(t_1/2)\sin(t_2), \\ x_2 &= R\tan(t_1/2)\cos(t_2), \\ \tau &= t_2 + t_3. \end{aligned} \tag{6.3.54}$$

Now, instead of considering fields $\xi_i(r)$, we consider smooth fields

$$\xi : SU(2) \to \mathbb{S}^{2N-1} \tag{6.3.55}$$

i.e. we consider fields whose argument is an element $g \in SU(2)$ and satisfying

$$\|\xi_j(g)\| = 1, \tag{6.3.56}$$

since as we saw after (6.3.5), in order to describe the $\mathbb{CP}^{N-1}$ model, we can consider fields taking values in the sphere $\mathbb{S}^{2N-1}$. However, the general configuration of this type does not reduce to an instanton on $\mathbb{S}^2$. We have to make sure that the circle generated by $\mathbb{S}^3$ is somehow modded out. It turns out that the appropriate constraint on $\xi_i(g)$ is the homogeneity condition

$$\xi_i\left(ge^{i\omega\sigma_3}\right) = e^{-ik\omega}\xi_i(g), \tag{6.3.57}$$

where $k \in \mathbb{Z}$. From this field, we recover the original field $z_i(x)$ as

$$z_i(x) = \xi_i(U(x)), \tag{6.3.58}$$

where $U(x)$ is given in (6.3.53). The homogeneity property (6.3.57) guarantees that the resulting field $z_i(x)$ has topological charge k. Indeed, let us write x as in (6.3.43). Then, as $|x| \to \infty$, we have

$$U(x) \approx -i\sigma_2 e^{i\phi\sigma_3}, \tag{6.3.59}$$

therefore, by (6.3.57),

$$\xi_i(g) \approx \xi\left(-i\sigma_2 e^{i\phi\sigma_3}\right) = e^{-ik\phi}n_i, \qquad n_i = \xi_i(-i\sigma_2), \tag{6.3.60}$$

and

$$z_i \approx n_i e^{-ik\phi}, \qquad |x| \to \infty, \tag{6.3.61}$$

which is a field configuration of charge k, as can easily be seen by comparing it with (6.3.42).

We should now formulate the action principle for the $\mathbb{CP}^{N-1}$ model in terms of these fields defined on $SU(2)$. To do that, we consider the differential operators on $SU(2)$ defined in (A.17) and (A.20). Using the change of coordinates (6.3.54), as well as the complex coordinates (6.3.37), we find

$$L_+ = -\frac{e^{i\tau}}{R}\left\{(R^2 + |s|^2)\partial_s - i\bar{s}\partial_\tau\right\},$$

$$L_- = \frac{e^{-i\tau}}{R}\left\{(R^2 + |s|^2)\partial_{\bar{s}} + is\partial_\tau\right\}, \qquad (6.3.62)$$

$$L_3 = -i\partial_\tau,$$

where

$$L_\pm = L_1 \pm iL_2. \qquad (6.3.63)$$

Notice that, for fields which satisfy the condition (6.3.57), one has

$$L_3\xi_i(g) = -i\partial_\tau \xi_i \left(U(x)e^{i\tau\sigma_3/2}\right) = -i\partial_\tau \left(e^{-ik\tau/2}\xi_i \left(U(x)\right)\right)$$

$$= -\frac{k}{2}\xi_i(g). \qquad (6.3.64)$$

Under complex conjugation, these operators transform as

$$\bar{L}_a = -L_a, \qquad a = 1, 2, 3. \qquad (6.3.65)$$

They are self-adjoint with respect to the natural Hermitian inner product of functions on $SU(2)$. Let us now define the operators

$$J_a\xi_i = L_a\xi_i - \left(\bar{\xi} \cdot L_a\xi\right)\xi_i, \qquad a = 1, 2, 3, \qquad (6.3.66)$$

and

$$J_\pm = J_1 \pm iJ_2. \qquad (6.3.67)$$

Note that, on the fields satisfying (6.3.57), we have, by using (6.3.64), that

$$J_3\xi_i = 0. \qquad (6.3.68)$$

Since $\bar{\xi} \cdot \xi = 1$, we find, using (6.3.65), that

$$\bar{\xi} \cdot L_a\xi = \overline{L_a\xi} \cdot \xi. \qquad (6.3.69)$$

Using this property it is easy to verify that the operators J_a are also self-adjoint. In terms of the original fields z_i, the operators $J_\pm$ read

$$J_\pm\xi_i = \mp\frac{e^{\pm i\tau}}{R}\left(R^2 + |s|^2\right)D_\pm z_i, \qquad (6.3.70)$$

where

$$D_+z_i = D_s z_i = \partial_s z_i - (\bar{z} \cdot \partial_s z)\, z_i,$$
$$D_-z_i = D_{\bar{s}} z_i = \partial_{\bar{s}} z_i - (\bar{z} \cdot \partial_{\bar{s}} z)\, z_i. \tag{6.3.71}$$

We can now write the action functional of the $\mathbb{CP}^{N-1}$ model in terms of the fields on $\mathbb{S}^3$. We have that

$$\int_{SU(2)} \|J_\pm\xi\|^2 dg = \sum_{a=1}^{2} \int_{SU(2)} \|J_a\xi\|^2 dg \mp \int_{SU(2)} \bar{\xi} \cdot L_3\xi\, dg, \tag{6.3.72}$$

where we used the property of self-adjointness for the L_a and the J_a, as well as the commutator

$$[J_1, J_2] = iJ_3 + (L_2\bar{\xi}) \cdot (L_1\xi) - (L_1\bar{\xi}) \cdot (L_2\xi) \tag{6.3.73}$$

acting on the field ξ. In (6.3.72), the measure on $SU(2)$ is the Haar measure, and it is normalized as

$$\int_{SU(2)} dg = 1. \tag{6.3.74}$$

It follows that

$$\sum_{a=1}^{2} \int_{SU(2)} dg\, \|J_a\xi\|^2 = \frac{1}{2} \int_{SU(2)} dg\, \left(\|J_+\xi\|^2 + \|J_-\xi\|^2\right)$$

$$= \int_{SU(2)} d^3t \sqrt{G} \frac{\left(|R|^2 + |s|^2\right)^2}{4R^2} \overline{D_\mu z} \cdot D_\mu z$$

$$= \frac{1}{4\pi} \int_{\mathbb{R}^2} dx_1 dx_2\, \overline{D_\mu z} \cdot D_\mu z, \tag{6.3.75}$$

where we changed variables from $x_{1,2}, \tau$ to $t_i = 1, 2, 3$, we used (A.10) (with $r = 1$) and (6.3.70), and we integrated over τ in the last line. We conclude that the action for the $\mathbb{CP}^{N-1}$ model, when written in terms of $SU(2)$ fields, is given by

$$S = \frac{4\pi}{g_0^2} \sum_{a=1}^{2} \int_{SU(2)} dg\, \|J_a\xi\|^2. \tag{6.3.76}$$

The equations for (anti)instantons read, in this new language,

$$J_\mp \xi_i = 0, \tag{6.3.77}$$

as follows immediately from (6.3.70).

The effective action at large N

We will now come back to the theory on $\mathbb{R}^2$ and we study quantum aspects of the $\mathbb{CP}^{N-1}$ model at large N. As in the case of the non-linear sigma model, we introduce a Lagrange multiplier α to impose the constraint (6.3.6), as well as the analogue of a theta term. The resulting action is

$$S = \int d^2x \left[\frac{1}{g_0^2} \overline{D_\mu z} \cdot D_\mu z - \frac{i\alpha}{g_0^2} (z_i \bar{z}_i - 1) + \frac{i\theta}{2\pi} \epsilon_{\mu\nu} \partial_\mu A_\nu \right]. \tag{6.3.78}$$

We define the 't Hooft parameter as in (6.2.4). We will study the theory in the limit in which N is large and g_0^2 is small, so that t_0 is fixed. We will compute the effective action at large N, and we will set $\theta = 0$. In (6.3.78) we treat A_μ and α as auxiliary fields. When we perform the functional integral over the field α, we impose the constraint (6.3.6) on the fields z_i. The functional integral over the field A_μ can also be easily performed, since it is an auxiliary field appearing quadratically in the action. Up to an overall normalization factor in the partition function, the effect of performing the Gaussian integral over A_μ is to enforce its EOM, and therefore it sets A_μ to be given by (6.3.8).

As in the $O(N)$ sigma model, we can obtain an effective action for the fields α, A_μ by first doing the path integral over the N complex scalar fields z_i. Since this is a Gaussian path integral, we obtain

$$\frac{1}{[\det \Delta]^N} \tag{6.3.79}$$

where

$$\Delta = -\frac{N}{t_0} D_\mu D^\mu - \frac{Ni\alpha}{t}. \tag{6.3.80}$$

After writing the determinant as the exponential of a trace of a log, we find the effective action

$$S_{\text{eff}} = N \text{Tr} \log \left(-(\partial_\mu + iA_\mu)^2 - i\alpha \right) + \frac{iN}{t_0} \int \alpha(x) d^2x, \tag{6.3.81}$$

which depends on the fields A_μ and α. In this effective action, N plays the role of $1/\hbar$. For large N we can then expand this action around a saddle point, as in the $O(N)$ sigma model. We will look for vacua of the form,

$$A_\mu = 0, \qquad \alpha = \text{constant}, \tag{6.3.82}$$

as one expects from rotation and translation invariance. The EOM for α is obtained from

$$\frac{\delta}{\delta\alpha(x)} \left[\frac{iN}{t_0} \int \alpha(x) d^2x + N \text{Tr} \log \left(-(\partial_\mu + iA_\mu)^2 - i\alpha \right) \right] = 0. \tag{6.3.83}$$

Since we are assuming that $A_\mu = 0$ in the vacuum, this EOM is identical to the equation (6.2.12) in the $O(N)$ sigma model. Its solution is also identical, i.e.

$$\alpha = im^2, \qquad m^2 > 0. \tag{6.3.84}$$

As in the $O(N)$ model, m^2 is given by (6.2.28), in terms of a dynamically generated scale defined by (6.2.27). We conclude that, at large N, the $\mathbb{CP}^{N-1}$ model is also asymptotically free, and the expectation value for α in (6.3.78) again gives a mass for the z_i fields.

Let us now analyze the expansion of the effective action around the saddle point (6.3.82). We will set

$$A_\mu \to \frac{1}{\sqrt{N}} A_\mu, \qquad \alpha \to im^2 + \frac{\alpha}{\sqrt{N}}, \tag{6.3.85}$$

where the N-dependent normalizations of the fluctuations are chosen for convenience. As for the case of the $O(N)$ model, the calculation of the quadratic terms is better done in terms of Feynman diagrams. The functional determinant (6.3.79) is obtained by integrating out the z_i fields at one-loop. Therefore, the quadratic terms we are looking for can be computed in the theory with action (6.3.78), by considering one-loop diagrams with two external legs corresponding to the A_μ, α fields. The Feynman rules are shown in Fig. 6.3. The quadratic term in α can be computed using the same diagram that we had in the $O(N)$ sigma model, with the only difference that we now have an extra factor of 2 (this is due to the fact that there are now $2N$ real fields around the loop). The mixing term in α, A_μ turns out to vanish. The quadratic term in A_μ, A_ν in the effective action can be written, in momentum space, as

$$\frac{1}{2} \int \frac{d^2 p}{(2\pi)^2} \tilde{A}^\mu(p) \tilde{\Gamma}^A_{\mu\nu}(p) \tilde{A}^\nu(-p), \tag{6.3.86}$$

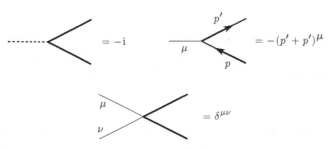

Figure 6.3 The Feynman rules for the $\mathbb{CP}^{N-1}$ model. The dashed line corresponds to the α field, the thick lines correspond to the z_i field, and the thin lines correspond to the A_μ field.

Figure 6.4 The two Feynman diagrams contributing to $\tilde{\Gamma}^A_{\mu\nu}(p)$.

where $\tilde{\Gamma}^A_{\mu\nu}(p)$ is given by the sum of Feynman diagrams in Fig. 6.4. It reads

$$\tilde{\Gamma}^A_{\mu\nu}(p) = 2\delta_{\mu\nu} \int \frac{d^2q}{(2\pi)^2} \frac{1}{(q^2 + m^2)}$$
$$- \int \frac{d^2q}{(2\pi)^2} \frac{(p_\mu + 2q_\mu)(p_\nu + 2q_\nu)}{(q^2 + m^2)((p+q)^2 + m^2)}. \qquad (6.3.87)$$

The quantity $\tilde{\Gamma}^A_{\mu\nu}(p)$ has a very easy interpretation: it is the photon vacuum polar-
ization for scalar QED in two dimensions. The computation of (6.3.87) can also
be found in Appendix C. Although the two diagrams diverge separately, their sum
turns out to be finite, after using dimensional regularization. One finds,

$$\tilde{\Gamma}^A_{\mu\nu}(p) = \left(\delta_{\mu\nu} - \frac{p_\mu p_\nu}{p^2}\right)\left\{(p^2 + 4m^2)f(p) - \frac{1}{\pi}\right\}, \qquad (6.3.88)$$

where $f(p)$ is given in (6.2.40). Since

$$f(p) = \frac{1}{4\pi m^2} - \frac{p^2}{24\pi m^4} + \mathcal{O}(p^4) \qquad (6.3.89)$$

near $p^2 = 0$, the quadratic term in $\tilde{A}$ is of the form

$$(\delta_{\mu\nu}p^2 - p_\mu p_\nu)(c + \mathcal{O}(p^2)), \qquad (6.3.90)$$

where

$$c = \frac{1}{12\pi m^2}. \qquad (6.3.91)$$

This structure is a consequence of gauge invariance, and leads to the standard gauge
field kinetic energy

$$(\partial_\mu A_\nu - \partial_\nu A_\mu)^2 \qquad (6.3.92)$$

written in momentum space. In other words, the quantum corrections have generated a kinetic term for A_μ! The excitations associated to the fields z_i and $\bar{z}_i$ can
be regarded as (scalar) quarks and antiquarks of the model, and they will interact
through the gauge field A_μ. A $U(1)$ gauge field in two dimensions leads to a linear
potential between charges. Therefore, an extra consequence of the emergence of a
dynamical gauge field in this model is *confinement* of the charges.

Topological susceptibility at large N

Another truly non-perturbative effect that can be seen at large N in the $\mathbb{CP}^{N-1}$ model is a non-zero value for the topological susceptibility. The two-dimensional version of (4.3.34) is

$$U(p) = \int d^2x \, e^{ipx} \langle q(x)q(0) \rangle = \int \frac{d^2p'}{(2\pi)^2} \langle \tilde{q}(p)\tilde{q}(-p') \rangle, \qquad (6.3.93)$$

where $q(x)$ is the topological density defined in (6.3.25). The topological suscep-tibility is given by (4.3.33), where $U(k)$ is now given by (6.3.93). Let us now evaluate χ_t at leading order in the $1/N$ expansion. The Fourier transform of $q(p)$ is given by

$$\tilde{q}(p) = -\frac{i}{2\pi\sqrt{N}} \epsilon_{\mu\nu} p_\mu \tilde{A}_\nu. \qquad (6.3.94)$$

Therefore,

$$\langle \tilde{q}(p)\tilde{q}(-p') \rangle = \frac{1}{4\pi^2 N} \epsilon_{\mu\nu} \epsilon_{\rho\sigma} p_\mu p'_\rho \langle \tilde{A}_\nu(p)\tilde{A}_\sigma(-p') \rangle. \qquad (6.3.95)$$

To compute the two-point function appearing here, we have to fix the gauge. We choose for convenience the Landau gauge, and we immediately deduce from (6.3.88) that

$$\langle \tilde{A}_\nu(p)\tilde{A}_\sigma(-p') \rangle = (2\pi)^2 \delta(p-p') \left(\delta_{\mu\nu} - \frac{p_\mu p_\nu}{p^2} \right) D_A(p), \qquad (6.3.96)$$

where

$$D_A(p) = \left\{ (p^2 + 4m^2)f(p) - \frac{1}{\pi} \right\}^{-1}. \qquad (6.3.97)$$

The $(2\pi)^2$ factor in (6.3.96) comes from the kinetic term (6.3.86) in momentum space. Since

$$\epsilon_{\mu\nu}\epsilon_{\rho\sigma} p_\mu p_\rho \left(\delta_{\nu\sigma} - \frac{p_\nu p_\sigma}{p^2} \right) = (\delta_{\mu\rho}\delta_{\nu\sigma} - \delta_{\mu\sigma}\delta_{\nu\rho}) p_\mu p_\rho \left(\delta_{\nu\sigma} - \frac{p_\nu p_\sigma}{p^2} \right)$$

$$= p^2, \qquad (6.3.98)$$

we find

$$\langle \tilde{q}(p)\tilde{q}(-p') \rangle = \frac{p^2}{4\pi^2 N} (2\pi)^2 D_A(p) \delta(p-p'). \qquad (6.3.99)$$

Therefore,

$$\int \frac{d^2p'}{(2\pi)^2} \langle \tilde{q}(p)\tilde{q}(-p') \rangle = \frac{p^2}{N} D_A(p) = \frac{3m^2}{\pi N} + \mathcal{O}(p^2), \qquad (6.3.100)$$

and the topological susceptibility reads, at leading order in the $1/N$ expansion,

$$\chi_t^{\text{large } N} = \frac{3m^2}{\pi N}. \tag{6.3.101}$$

This is a rather remarkable result, since this quantity vanishes order by order in perturbation theory, as we saw in the context of YM theory in (4.3.33). However, in the $1/N$ expansion, it has a non-zero value. This is yet another example of the ability of the large N limit to produce non-perturbative results. The analytic result (6.3.101) has been verified in lattice calculations. As we will show in the next chapter, the fact that the large N expansion leads to a non-vanishing topological susceptibility will play an important role in QCD.

6.4 Bibliographical notes

The book [37] is an excellent collection of original articles on the $1/N$ expansion. The $O(N)$ model and the $\mathbb{CP}^{N-1}$ models are discussed in many books, for example [50, 167]. A useful discussion of renormalization at large N in these models can be found in [145].

Classical aspects of the $\mathbb{CP}^{N-1}$ model are discussed for example in [153, 157]. The $1/N$ expansion of the model was worked out in [60, 61, 187], which also discussed the inclusion of quark fields. Multi-instanton configurations on the two-sphere are described in [33], where the one-loop partition function in a general multi-instanton sector is also computed in detail. The calculation of the topological susceptibility of the $\mathbb{CP}^{N-1}$ model on the lattice is reviewed in [182].

7

The $1/N$ expansion in Quantum Chromodynamics

7.1 Introduction

In the previous chapter we studied the $1/N$ expansion in two-dimensional models in which the basic field has N components. As in those models, the key idea in the $1/N$ expansion of gauge theories is to realize that, in addition to the gauge coupling constant g, there is a hidden variable N, the rank of the gauge group. In YM theory with gauge group $SU(N)$, the gauge connection is a *matrix* field and it has $N^2 - 1$ components, while in the two-dimensional models we studied before, the fields are *vector* fields and they have N components. As a consequence, the $1/N$ expansion in gauge theories is very different from the expansion in vector models, and it was first worked out by 't Hooft. This expansion can be understood in diagrammatic terms, and it can be regarded as a reorganization (and resummation) of the standard perturbative series, by using the double-line diagrams (also called fatgraphs) introduced by 't Hooft.

The $1/N$ expansion of YM theory and QCD has led to many insights. A particularly spectacular prediction is the Witten–Veneziano formula, relating the mass of the η' to the topological susceptibility of pure YM theory, which we will explain in some detail in this chapter. In contrast to the two-dimensional models of the last chapter, the exact solution of planar YM theory or planar QCD in four dimensions has been elusive. However, in recent years, it has been found that the planar solution of some supersymmetric gauge theories in four and three dimensions is encoded in a string theory, and this has led to an explosion in the development and study of large N gauge theories. The next chapters might hopefully be used as an elementary introduction to this active and fascinating area of research.

7.2 Fatgraphs

The first thing to do when considering a $1/N$ expansion is to find the right scaling of the couplings of the theory. In the two-dimensional models of the previous chapter, we found that the coupling which stays fixed as N grows large is the 't Hooft parameter (6.2.4). Similarly, if g_0 is the bare coupling constant of QCD introduced in (5.3.1), we will define the (bare) *'t Hooft parameter* of QCD as

$$t_0 = g_0^2 N. \tag{7.2.1}$$

The 't Hooft large N limit is then defined as

$$N \rightarrow \infty, \quad g_0^2 \rightarrow 0, \quad t_0 \text{ fixed}. \tag{7.2.2}$$

As we will see, the theory is non-trivial in this limit. A first indication of this is the scaling behavior of the one-loop beta function of QCD, (4.2.40)–(4.2.41). If the quarks are in the fundamental representation, we have

$$\mu \frac{dg}{d\mu} = -\left(\frac{11}{3} N - \frac{2}{3} N_f\right) \frac{g^3}{16\pi^2}, \tag{7.2.3}$$

where g is the renormalized coupling constant. If we denote by $t = g^2 N$ the renormalized 't Hooft parameter, we find,

$$\mu \frac{dt}{d\mu} = -\left(\frac{11}{3} - \frac{2}{3} \frac{N_f}{N}\right) \frac{t^2}{8\pi^2}. \tag{7.2.4}$$

In the 't Hooft large N limit, the first term in the beta function survives, while the second term (due to the quark loops) is suppressed as $1/N$, if N_f is kept finite as N grows. We expect then to have a non-trivial theory in the 't Hooft limit, and we learn already that quarks give a subleading contribution as compared to gluons. We will see in a moment, by using diagrammatic techniques, that after introducing the 't Hooft parameter, all interesting quantities in QCD have an expansion in powers of $1/N$, and the large N limit (7.2.2) extracts the leading term in this expansion. We note for future use that the rescalings (4.2.30), (5.3.6) read, in terms of the 't Hooft parameter,

$$A_\mu \rightarrow \sqrt{\frac{t_0}{N}} A_\mu, \quad \psi_f \rightarrow \sqrt{\frac{t_0}{N}} \psi_f. \tag{7.2.5}$$

To understand the N dependence at the diagrammatic level, let us note that it comes from the *group factors* associated to Feynman diagrams. In general, a single Feynman diagram gives rise to a polynomial in N involving different powers of N. Therefore, the standard Feynman diagrams, which are good for keeping track of powers of the coupling constants, are not good for keeping track of powers of N. If we want to keep track of the N dependence we have to "split" each diagram into different pieces which correspond to a definite power of N. To do that, one has to write the Feynman diagrams of the theory as "fatgraphs" or double-line graphs, as first found by 't Hooft. To see how this works, we will re-analyze the Feynman

$$i \longrightarrow j \qquad \delta^i_j$$

Figure 7.1 The quark propagator.

rules by paying particular attention to their group structure. Since $U(N)$ is slightly simpler to analyze from this point of view than $SU(N)$, we will consider both groups in what follows.

The simplest ingredient is the quark propagator, which is of the form

$$\langle \psi^i(x)\overline{\psi}_j(y)\rangle = \frac{g_0^2}{2}\delta^i_j S(x-y).\tag{7.2.6}$$

Here, $S(x-y)$ is the spacetime dependent part of the propagator, and $i, j = 1,\ldots,N$ are color indices. From the point of view of the color structure, this propagator only couples quarks with the same color index, so it is proportional to δ^i_j, as shown in (7.2.6). We will represent this diagrammatically by a single line, where the color at the beginning of the line is the same as at the end of the line due to the δ^i_j factor, see Fig. 7.1.

Let us now analyze the gluon propagator. In the double-line notation, the most natural normalization of the generators of the Lie algebra (4.2.4) is $\alpha = 1$ (the normalization (4.2.5) can always be recovered by an appropriate rescaling). With this normalization, the propagator for the gluon field is of the form,

$$\langle A^a_\mu(x) A^b_\nu(y)\rangle = \frac{g_0^2}{2}\delta^{ab} D_{\mu\nu}(x-y), \qquad a,b=1,\ldots,d(G).\tag{7.2.7}$$

Here, $d(G)$ is the dimension of the gauge group, which is N^2-1 for $SU(N)$ and N^2 for $U(N)$, and $D_{\mu\nu}(x-y)$ encodes the spacetime dependence. Instead of treating the gauge connection as a field with a single index $a = 1,\ldots,d(G)$, it is preferable to treat it as an $N \times N$ matrix with two indices in the N and $\overline{N}$ representations, i.e.

$$\left(A_\mu\right)^i_j = A^a_\mu (T_a)^i_j,\tag{7.2.8}$$

where we write explictily the matrix index structure of the generators of the Lie algebra. These generators satisfy, in the case of $U(N)$,

$$\sum_{a=1}^{d(G)} (T_a)^i_j (T_a)^k_l = \delta^i_l \delta^k_j.\tag{7.2.9}$$

For $SU(N)$, since the generators are traceless, we find instead

$$\sum_{a=1}^{d(G)} (T_a)^i_j (T_a)^k_l = \delta^i_l \delta^k_j - \frac{1}{N}\delta^i_j \delta^k_l.\tag{7.2.10}$$

The above identities follow from the decomposition of the tensor product of the fundamental and the antifundamental representations, which in the $U(N)$ case gives the adjoint representation, while in the $SU(N)$ case it gives the adjoint representation, plus the trivial representation, represented in (7.2.10) by the last term on the right hand side. If we regard the gluon as an $N \times N$ Hermitian matrix, its propagator reads, in the $U(N)$ case,

$$\langle (A_\mu)^i_j (x) (A_\nu)^k_l (y) \rangle = \frac{g_0^2}{2} D_{\mu\nu} (x - y) \, \delta^i_l \, \delta^k_j. \tag{7.2.11}$$

This can be obtained directly from the quadratic term of the YM action. Indeed, from the point of view of the matrix structure, this term involves the trace of the product of two gauge connections, and therefore it leads directly to (7.2.11). For $SU(N)$, we find instead

$$\langle (A_\mu)^i_j (x) (A_\nu)^k_l (y) \rangle = \frac{g_0^2}{2} D_{\mu\nu} (x - y) \left(\delta^i_l \, \delta^k_j - \frac{1}{N} \delta^i_j \, \delta^k_l \right). \tag{7.2.12}$$

The group structure of the gluon propagator of $U(N)$ can be represented by a *double line*, as in Fig. 7.2. In the case of $SU(N)$, we have to add the second term on the right hand side of (7.2.10). This leads to the graphic representation shown in Fig. 7.3.

We can also write the interaction vertices in the double-line notation. The three-gluon vertex involves the structure constants f_{abc} of the Lie algebra, which are defined by

$$[T_a, T_b] = \mathrm{i} f_{abc} T_c. \tag{7.2.13}$$

By multiplying by T_d and taking a trace, we find the relation

$$\mathrm{i} f_{abc} = \mathrm{Tr} \, (T_a T_b T_c) - \mathrm{Tr} \, (T_b T_a T_c). \tag{7.2.14}$$

The first term, which is the trace of three generators of the Lie algebra, can be interpreted as a cubic vertex. Indeed, it comes from

$$\mathrm{Tr}(A_\mu A_\nu A_\rho) = A^a_\mu A^b_\nu A^c_\rho \, \mathrm{Tr}(T_a T_b T_c), \tag{7.2.15}$$

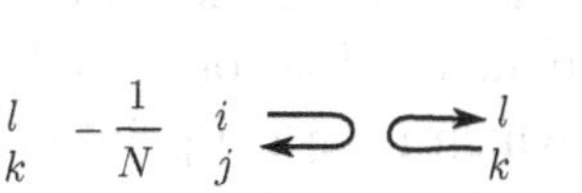

Figure 7.2 The gluon propagator in the double-line notation for $U(N)$.

Figure 7.3 The gluon propagator in the double-line notation for $SU(N)$.

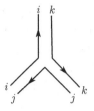

Figure 7.4 The cubic vertex (7.2.16) in the double-line notation.

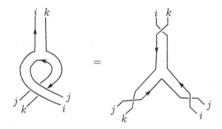

Figure 7.5 The twisted vertex (7.2.17) in the double-line notation.

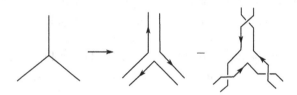

Figure 7.6 The standard cubic vertex of QCD becomes a sum of two fatgraphs.

which in the double-line notation leads to the index structure

$$\sum_{i,j,k}(A_\mu)^i_{\ j}\,(A_\nu)^j_{\ k}\,(A_\rho)^k_{\ i},\qquad(7.2.16)$$

and it can be depicted as in Fig. 7.4. The second trace on the right hand side of (7.2.14) gives an additional term

$$-\sum_{i,j,k}(A_\nu)^i_{\ j}\,(A_\mu)^j_{\ k}\,(A_\rho)^k_{\ i},\qquad(7.2.17)$$

which can also be represented as a double-line vertex. Notice however that this one is "twisted," in comparison to the previous one, see Fig. 7.5. We conclude that the cubic vertex of YM theory, which from the point of view of the group structure corresponds to the structure constant of the Lie algebra, leads to two vertices in the double-line notation: a direct vertex and a twisted vertex, as shown in Fig. 7.6.

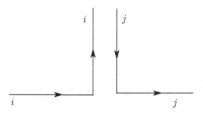

Figure 7.7 The interaction vertex between a quark (horizontal line) and a gluon (vertical double-line), in the double-line notation.

$$i \longrightarrow l \quad -\frac{1}{N} \quad i \longrightarrow l \quad = \left(N - \frac{1}{N}\right) \quad i \longrightarrow l$$

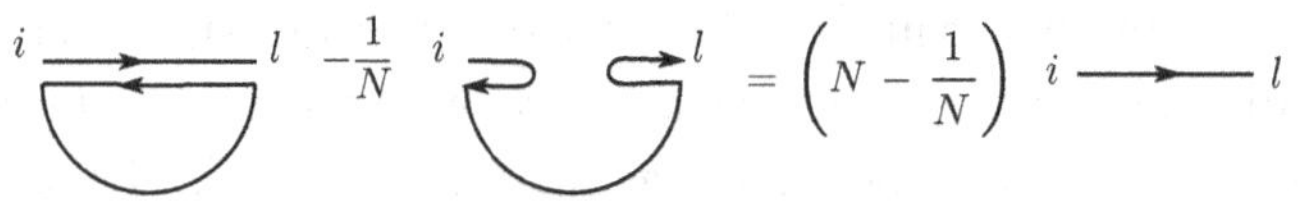

Figure 7.8 The quadratic Casimir in the fundamental representation can be computed by joining the lines k and j in Fig. 7.3.

Our final rule concerns the quark–gluon vertex describing the interaction between a quark bilinear and a gluon. The group structure of this vertex is

$$\overline{\psi}_j(x) \left(A_\mu\right)^j_{\ i}(x)\psi^i(x), \tag{7.2.18}$$

therefore it can be represented, in the double-line notation, by the graph of Fig. 7.7.

The double-line notation re-expresses the Feynman diagrams of QCD as a sum of diagrams with double lines or "fatgraphs." This notation is also very useful for calculations of group factors. As an example of this technique, let us calculate the quadratic Casimir of the fundamental and the adjoint representation. If we look at the definition of C_2(fund) in (4.2.11), we see that it can be obtained from the gluon propagator by contracting two indices and summing over then. This creates a closed loop in the first diagram of Fig. 7.3, which gives a factor of N, times the quark propagator. In the second diagram we get a factor of $1/N$, times the quark propagator, see Fig. 7.8. After including the normalization factor α, we find

$$C_2(\text{fund})\delta^i_l = \alpha \left(N - \frac{1}{N}\right)\delta^i_l, \tag{7.2.19}$$

which is the expected result. To analyze the Casimir in the adjoint representation of $SU(N)$, we notice that we can compute

$$- f_{abc}f_{bdc} \tag{7.2.20}$$

by joining two cubic vertices, which after the resolution in Fig. 7.6 leads to four pairings: one pairing between the direct vertices, one pairing between the twisted vertices, and two pairings between the twisted and the direct vertices. The first two

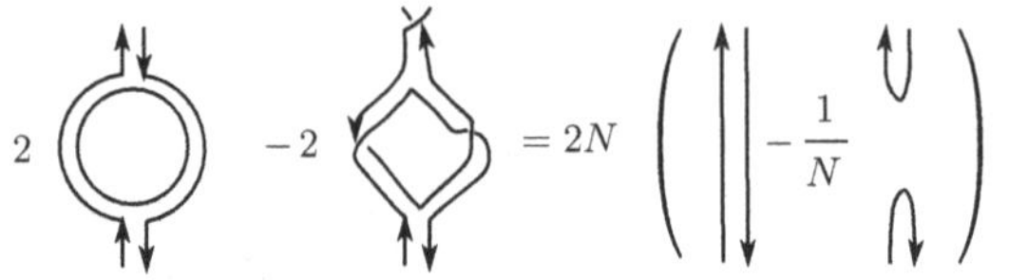

Figure 7.9 The quadratic Casimir in the adjoint representation can be computed by joining two cubic vertices.

pairings give the same result, which is the first contribution in the left hand side of Fig. 7.9, while the two pairings of the twisted and the direct vertices give the second contribution. All these pairings are done with the $SU(N)$ propagators, but it is easy to see that the contribution of the second term in (7.2.10) cancels among the pairings. The right hand side of Fig. 7.9 is proportional to the gluon propagator for the remaining indices, which is δ_{cd}. After taking into account the factors of α we find,

$$f_{abc} f_{dbc} = 2N\alpha\delta_{cd}, \qquad (7.2.21)$$

which is again the expected result.

We will now classify fatgraphs with no external lines according to their topological properties. This will make clear how to reorganize the perturbative expansion in powers of $1/N$, in the 't Hooft limit. A fatgraph is characterized by the number of propagators or edges E, the number of vertices V, and the number of closed loops h. By (7.2.11), each propagator gives a factor of g_0^2, while each interaction vertex gives a power of g_0^{-2}. Finally, each closed loop involves a sum over a color index and gives a factor of N. Therefore, each fatgraph comes with a factor of

$$N^h g_0^{2(E-V)}. \qquad (7.2.22)$$

In terms of the 't Hooft parameter this is

$$N^{V-E+h} t_0^{E-V}. \qquad (7.2.23)$$

Interestingly, this has an interpretation in terms of the topology of two-dimensional surfaces. Orientable two-dimensional surfaces (also called Riemann surfaces) can be classified topologically by the number of handles, or genus g, and the number of boundaries b. We can think about every fatgraph as an orientable Riemann surface obtained by gluing polygons: each edge represents a side of a polygon, and a closed loop is the perimeter of a polygon, therefore it is in one-to-one correspondence with a face. A double-line is then interpreted as an instruction to glue polygons: we identify one edge of a polygon with one edge of another polygon if they both lie on the same double line. Finally, each closed quark loop (which is a single line) can be interpreted as a boundary for the surface, while fatgraphs made only out of double-lines are interpreted as Riemann surfaces with no boundary. The orientation of the Riemann surface is inherited from the orientation of the arrows in the double-line diagrams. Euler's relation says that the combination $V - E + h$ (the number of vertices, minus the number of edges, plus the number of faces) is a topological invariant of the surface, called the Euler characteristic χ. The Euler characteristic only depends on the genus g and the number of boundaries b of the surface:

$$\chi = V - E + h = 2 - 2g - b. \tag{7.2.24}$$

Therefore the power of N in (7.2.23) is

$$N^{2-2g-b} = N^\chi. \tag{7.2.25}$$

The fatgraphs with $g = 0$, which are interpreted as spheres with holes, are called *planar* graphs, while the fatgraphs with $g > 0$ are called *non-planar*. It is easy to see that each conventional Feynman diagram with no external lines gives rise to many different fatgraphs with different genera.

Another way to interpret the fatgraphs in terms of Riemann surfaces is by associating to each vertex a triangle (mathematically, this means that we look at the dual discretization of the surface), as in Fig. 7.10. In this construction, the contraction of two double-line propagators is interpreted as an instruction to glue the edges of the triangle, and each fatgraph with no external lines is interpreted as a triangulation of a Riemman surface. The number of vertices in the Feynman diagram is the number of triangles in the triangulation, the edges of the diagram are in one-to-one correspondence with the edges of the triangulation, and the number of closed loops corresponds to the vertices. In this interpretation we obtain of course

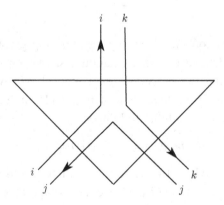

Figure 7.10 The trivalent vertex can be associated to a triangle. Fatgraphs with no external lines are then interpreted as triangulated Riemann surfaces.

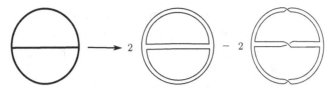

Figure 7.11 The two fatgraphs associated to the theta diagram. The first one has $g = 0$, while the second one has $g = 1$.

the same counting (7.2.25). Note that the twisted vertex Fig. 7.5 can also be interpreted as a triangle, but with different instructions to glue the edges. We will see concrete examples of this in a moment.

We can now use the double-line notation to compute the group factor $W(\Gamma)$ of any diagram Γ in QCD. Notice that, from the point of view of the group theory structure, the quartic vertex of YM can be reduced to two cubic vertices joined by an extra edge, therefore any diagram can be re-written in terms of trivalent diagrams. Given a trivalent diagram Γ, with V vertices, we use Fig. 7.6 to get a sum over the 2^V possible "resolutions" of the vertices, and we contract them with the double-line propagators for $U(N)$ or $SU(N)$, depicted in Fig. 7.2 and Fig. 7.3, respectively. Once this is done, we obtain, from the diagrammatic point of view, a formal linear combination of fatgraphs $\Gamma_{g,h}$, classified topologically by their genus g and the number of closed loops h:

$$\Gamma \to \sum_{g,h} p_{g,h}(\Gamma)\Gamma_{g,h}. \tag{7.2.26}$$

Here, the numbers $p_{g,h}(\Gamma)$ are integers counting the multiplicity of each fatgraph. The group factor of Γ is then given by

$$W(\Gamma) = \sum_{g,h} p_{g,h}(\Gamma)N^h. \tag{7.2.27}$$

Example 7.1 Let us consider the simplest two-loop graph made out of cubic vertices, the so-called theta or sunset diagram, shown on the left hand side of Fig. 7.11. By following the procedure explained above, we find two different graphs, as shown on the right hand side of Fig. 7.11: an "untwisted" graph $\Gamma_{0,3}$, and a "twisted" graph $\Gamma_{1,1}$, with multiplicities

$$p_{0,3}(\Gamma) = -p_{1,1}(\Gamma) = 2. \tag{7.2.28}$$

The graph $\Gamma_{0,3}$ has three edges, two vertices, and three closed lines, i.e. $E = 3$, $V = 2$, $h = 3$. Using Euler's formula (7.2.24) we indeed find that it is a planar fatgraph with $g = 0$ and $h = 3$. We can also look at this fatgraph from the dual point of view, in which each vertex is interpreted as a triangle. The fact that we are contracting

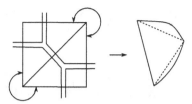

Figure 7.12 The fatgraph made out of the contraction of two direct or two twisted vertices can be interpreted as a sphere, obtained by gluing two triangles.

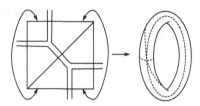

Figure 7.13 If we contract a direct vertex with a twisted vertex, the gluing instructions change accordingly and we obtain a torus.

two direct or twisted vertices means that the sides of the triangles have to be glued as shown in Fig. 7.12. We then obtain a sphere made out of two triangles. The fatgraph $\Gamma_{1,1}$ has instead $E = 3$, $V = 2$, $h = 1$, therefore by using (7.2.24) we find that it has indeed $g = 1$. When constructing the corresponding Riemann surface, we again have to glue two triangles. However, since one of the graphs is twisted, the gluing instructions are now as shown in Fig. 7.13. The resulting surface is a torus, made out of two triangles. By using (7.2.27), the group factor of the sunset diagram can then be computed as (in the conventions where $\alpha = 1$).

$$W(\Gamma) = f_{abc} f_{abc} = 2N^3 - 2N = 2N(N^2 - 1). \tag{7.2.29}$$

Notice that the weight of $\Gamma_{0,3}$ in the $1/N$ expansion is (discarding numerical factors)

$$N^3 g_0^2 = N^2 t_0, \tag{7.2.30}$$

while the weight of $\Gamma_{1,1}$ is

$$N g_0^2 = t_0. \tag{7.2.31}$$

The difference between the factors N^3 and N in the two diagrams can be traced back to the different index structure in the corresponding contractions. For $\Gamma_{0,3}$, we have the index structure

$$\sum_{i,j,k,m,n,p} \langle \left(A_{\mu_1}\right)^i_j \left(A_{\mu_2}\right)^m_n \rangle \langle \left(A_{\mu_3}\right)^j_k \left(A_{\mu_4}\right)^p_m \rangle \langle \left(A_{\mu_5}\right)^k_i \left(A_{\mu_6}\right)^n_p \rangle \propto N^3, \qquad (7.2.32)$$

while for $\Gamma_{1,1}$ we have

$$\sum_{i,j,k,m,n,p} \left(A_{\mu_1}\right)^i_j \left(A_{\mu_2}\right)^m_n \rangle \langle \left(A_{\mu_3}\right)^j_k \left(A_{\mu_4}\right)^n_p \rangle \langle \left(A_{\mu_5}\right)^k_i \left(A_{\mu_6}\right)^p_m \rangle \propto N. \qquad (7.2.33)$$

□

Example 7.2 In Fig. 7.14 we show a diagram with a quark line and a cubic gluon vertex. The corresponding group factor is

$$f_{abc}\,\text{Tr}(T^a T^b T^c) = N(N^2 - 1), \qquad (7.2.34)$$

again in the conventions where $\alpha = 1$. This can be verified by noting that

$$f_{abc}\,\text{Tr}(T^a T^b T^c) = \frac{1}{2} f_{abc}\,\text{Tr}([T^a, T^b]T^c) = \frac{1}{2} f_{abc} f_{abc}, \qquad (7.2.35)$$

and using (7.2.29). To compute this group factor in terms of fatgraphs, we resolve the cubic vertex according to Fig. 7.6. The subleading pieces of the $SU(N)$ propagators cancel, and we are left with two diagrams, with multiplicity one, as shown on the right hand side of Fig. 7.14. The first graph has $h = 3$, $V = 4$ and $E = 6$, and it corresponds to a Riemann surface with $\chi = 1$, i.e. to a sphere ($g = 0$) with one boundary ($b = 1$) due to the quark line. It gives a contribution of N^3 to the group factor, and its weight in the $1/N$ expansion is

$$N^3 g_0^4 = N t_0^2. \qquad (7.2.36)$$

The second graph has $h = 1$, $V = 4$ and $E = 6$, and it corresponds to a Riemann surface with $\chi = -1$, i.e. to a torus ($g = 1$) with one boundary ($b = 1$). It gives a contribution of $-N$ to the group factor, and its weight in the $1/N$ expansion is (up to the overall sign)

$$N g_0^4 = \frac{t_0^2}{N}. \qquad (7.2.37)$$

Figure 7.14 A QCD diagram with one closed quark line (left) and the two fatgraphs it generates (right).

By adding the contributions of the two fatgraphs to the group factor, we recover the result (7.2.34). □

It is possible to generalize the large N counting developed in this section to the other classical gauge groups $SO(N)$ and $Sp(N)$. For these gauge groups, the double-line notation leads to the appearance of non-orientable surfaces.

7.3 Quantum Chromodynamics at large N

We can now use the diagrammatic representation in terms of fatgraphs to analyze the large N dependence of various quantities of interest in $SU(N)$ gauge theories. We have seen that, when we reorganize the perturbative expansion in terms of fatgraphs, the Feynman diagrams become two-dimensional surfaces labelled by two topological quantities: the genus g and the number of boundaries b, with a weight (7.2.25). The largest values of χ are 2 in the case of closed surfaces, corresponding to $g = 0$, $b = 0$, and 1 for the surfaces with boundaries, corresponding to $g = 0$ and $b = 1$. Immediately, we have the following.

1. The leading connected vacuum-to-vacuum graphs are of order N^2. They are planar graphs made out of gluons.
2. The leading connected vacuum-to-vacuum graphs with quark lines are of order N. They are planar graphs with only one quark loop forming the boundary of the graph.

To see in detail how this works, let us focus on the free energy of $SU(N)$ or $U(N)$ YM theory, i.e. the logarithm of the partition function of the theory:

$$F(N, g_0) = \log Z(N, g_0). \tag{7.3.1}$$

In principle, this free energy can be computed on different spacetime geometries. If we consider the Euclidean theory on $\mathbb{R}^{d-1} \times \mathbb{S}^1_\beta$, where d is the total dimension and $\mathbb{S}^1_\beta$ is a circle of radius β, the free energy is in fact the thermal free energy on $\mathbb{R}^{d-1}$ and with inverse temperature β. Let us consider the contribution of the quantum fluctuations to the free energy, around the trivial configuration where $A_\mu = 0$, $\psi = 0$ (since we are only interested in the group structure of the expansion, we will use bare quantities and assume that some UV regularization is used to make sure that the diagrams are finite). Diagrammatically, the free energy is given by the sum over all connected diagrams with no external lines. It follows from the large N counting that the free energy has a series expansion of the form

$$F(N, g_0) = \sum_{g=0}^{\infty} F_g(t_0) N^{2-2g}, \tag{7.3.2}$$

where

$$F_g(t_0) = \sum_{h \geq 0} a_{g,h} t_0^{2g-2+h} \tag{7.3.3}$$

is a sum over all fatgraphs of genus g. The quantity $F_g(t_0)$ is sometimes called the genus g free energy. In the large N limit (7.2.2), the free energy is dominated by the planar piece, $F_0(t_0)$, which is given by the sum of all fatgraphs with the topology of the sphere.

Note that the expansion around $t_0 = 0$ in (7.3.3) can be obtained from the standard loop expansion. To see this, let us denote by $\mathcal{A}_n^{(c)}$ the set of connected Feynman diagrams appearing in the standard perturbative expansion of the free energy, and

$$n = E - V = L - 1, \tag{7.3.4}$$

where E is the number of edges (propagators), V is the number of vertices, and L is the number of loops. We then have the perturbative expansion,

$$F(N, g_0) = \sum_{n=1}^{\infty} \sum_{\Gamma \in \mathcal{A}_n^{(c)}} \mathcal{I}_\Gamma W(\Gamma) g_0^{2n}, \tag{7.3.5}$$

where $\mathcal{I}_\Gamma$ is the Feynman integral associated to the diagram Γ, and $W(\Gamma)$ is the group factor, computed for $SU(N)$ or $U(N)$. Notice that, in this expansion, we have only written the terms where the number of loops is greater than or equal to two. The tree level contribution (when present) and the one-loop contribution have a very different dependence on g_0, and in general they are not analytic at $g_0 = 0$: the tree level contribution is often of the form A/g_0^2, while the one-loop contribution leads to terms of the form $\log g_0$ in the free energy. Let us now see how to translate (7.3.5) in terms of the $1/N$ expansion. After expressing each standard Feynman diagram as a formal sum of fatgraphs $\Gamma_{g,h}$, as in (7.2.26), we find the following expression for the free energy:

$$F(N, g_0) = \sum_{g=0}^{\infty} \sum_{\Gamma_{g,h}} \mathcal{I}_\Gamma p_{g,h}(\Gamma) N^{h(\Gamma_{g,h})} g_0^{2(E(\Gamma)-V(\Gamma))}, \tag{7.3.6}$$

where $E(\Gamma)$, $V(\Gamma)$ are the numbers of edges and vertices in Γ (which is the single-line diagram underlying the fatgraphs $\Gamma_{g,h}$). If we now use Euler's relation (7.2.24), we find that the series (7.3.6) can be put in the form (7.3.2), and the coefficients $a_{g,h}$ in (7.3.3) are given by

$$a_{g,h} = \sum_{\Gamma_{g,h}} \mathcal{I}_\Gamma p_{g,h}(\Gamma), \tag{7.3.7}$$

where we sum over all fatgraphs $\Gamma_{g,h}$ with given g, h. Therefore, from each standard diagram appearing in perturbation theory, we generate fatgraphs of different genera by following the rules explained in the previous section, and $F_g(t_0)$ gathers the contribution of all fatgraphs of genus g. In particular, it is clear that each $F_g(t_0)$ is obtained by a partial resummation of the perturbative series of YM theory.

Let us now study the large N rules for correlation functions. Let G_i be a gauge invariant operator made out of gluons only. We will assume that G_i cannot be split into pieces which are separately gauge invariant. An example of such an operator is $\text{Tr}\, F_{\mu\nu} F^{\mu\nu}$. We consider now the modified Lagrangian,

$$\mathcal{L}_{\text{QCD}} + N \sum_i J_i G_i, \tag{7.3.8}$$

where J_i are sources. Due to the overall factor of N, the counting rules for powers of N in the new Lagrangian are the same as before. On the other hand, we know that the sum of connected vacuum-to-vacuum graphs with these sources, which we denote by $F(J)$, is the generating functional of *connected* correlation functions. We then have that

$$\langle G_1(x_1) \cdots G_r(x_r) \rangle^{(c)} = \frac{1}{(\mathrm{i}N)^r} \frac{\delta^r F(J)}{\delta J_1(x_1) \cdots \delta J_r(x_r)} \bigg|_{J=0}. \tag{7.3.9}$$

Since the leading contribution to this generating functional is again of order N^2, we conclude that

$$\langle G_1 \cdots G_r \rangle^{(c)} \sim N^{2-r} \tag{7.3.10}$$

at leading order in N. By using the full $1/N$ expansion of free energies, we can also deduce the full $1/N$ expansion of the correlation function in pure YM theory,

$$W^{(r)}(N, t_0) = \langle G_1 \cdots G_r \rangle^{(c)} = \sum_{g=0}^{\infty} W_g^{(r)}(t_0) N^{2-2g-r} \tag{7.3.11}$$

where

$$W_g^{(r)}(t_0) = \sum_{h \geq 0} W_{h,g}^{(r)} t_0^h \tag{7.3.12}$$

is the sum over all fatgraphs contributing to the connected correlation function, and with a fixed topology (for simplicity, we have not written down the spacetime dependence of the correlator). As in the case of the free energy, it is easy to see that the above expansion is a reorganization of the standard perturbative expansion of the correlation function.

Similarly, we can consider gauge invariant operators M_i involving quark bilinears, for example $\bar{\psi}(x)\psi(x)$. We assume again that M_i cannot be split into separate gauge invariant pieces. We consider the perturbed Lagrangian,

$$\mathcal{L}_{\text{QCD}} + N \sum_i J_i M_i, \tag{7.3.13}$$

where J_i are sources, and we deduce, as before, that

$$\langle M_1(x_1) \cdots M_r(x_r) \rangle^{(c)} = \frac{1}{N^r} \frac{\delta^r F(J)}{\delta J_1(x_1) \cdots \delta J_r(x_r)}\bigg|_{J=0}. \tag{7.3.14}$$

The leading contribution to this generating functional is of order N, and it involves a quark loop at the boundary, where we insert the bilinears. We conclude that, at leading order in N,

$$\langle M_1 \cdots M_r \rangle^{(c)} \sim N^{1-r}. \tag{7.3.15}$$

We now use these results to derive counting rules for scattering amplitudes. Gluon operators G_i can create "glueball" states, i.e. bound states made out of gluons only. Quark bilinears M_i can create meson states. To have an appropriate normalization for the states, we would like their inner product to be of order one at large N. In the case of glueball states, we can set

$$G_i |0\rangle \sim |G_i\rangle, \tag{7.3.16}$$

since in this case,

$$\langle G_1 | G_2 \rangle \sim \langle G_1 G_2 \rangle^{(c)} \sim \mathcal{O}(N^0). \tag{7.3.17}$$

Therefore G_i creates glueball states with unit amplitude. However, since

$$\langle M_1 M_2 \rangle^{(c)} \sim \mathcal{O}(1/N), \tag{7.3.18}$$

the appropriately normalized meson state is

$$|M_i\rangle \sim \sqrt{N} M_i |0\rangle. \tag{7.3.19}$$

We can now see that meson and glueball interactions are suppressed by factors of N. An r-glueball vertex is suppressed by N^{2-r}, and each additional glueball adds a $1/N$ suppression. Similarly, a normalized r meson vertex will be suppressed as

$$\langle \sqrt{N} M_1 \cdots \sqrt{N} M_r \rangle^{(c)} \sim N^{1-r/2} \tag{7.3.20}$$

and each additional meson adds a $1/\sqrt{N}$ suppression. Finally, mixed glueball–meson correlators will be suppressed as

$$\langle G_1 \cdots G_s \sqrt{N} M_1 \cdots \sqrt{N} M_r \rangle^{(c)} \sim N^{1-s-r/2}. \tag{7.3.21}$$

Equation (7.3.10) shows that the interaction between glueballs is of order $1/N$. Equations (7.3.20) and (7.3.21) show that the interaction between mesons and between mesons and glueballs are both of order $1/\sqrt{N}$. In other words, if we regard $1/N$ as a coupling constant, we have reorganized QCD into a theory of weakly interacting glueballs and mesons.

Example 7.3 Consider for example the correlators

$$\langle 0|\mathrm{Tr}(FF)|M\rangle, \qquad \langle 0|\mathrm{Tr}(FF)|G\rangle. \qquad (7.3.22)$$

Using the rules above we find

$$\langle 0|\mathrm{Tr}(FF)|M\rangle \sim \frac{1}{\sqrt{N}}, \qquad \langle 0|\mathrm{Tr}(FF)|G\rangle \sim \mathcal{O}(1). \qquad (7.3.23)$$

We can obtain counting rules for the rescaled fields of the Lagrangian by using (7.2.5). In terms of these rescaled fields, we have $\mathrm{Tr}(\hat{F}\hat{F}) \sim N\mathrm{Tr}(FF)$, therefore

$$\langle 0|\mathrm{Tr}(\hat{F}\hat{F})|M\rangle \sim \sqrt{N}, \qquad \langle 0|\mathrm{Tr}(\hat{F}\hat{F})|G\rangle \sim N. \qquad (7.3.24)$$

We can also deduce the large N scaling of F_π. The pion decay constant is defined by (5.3.75)–(5.3.76). The left hand side of (5.3.75) has the structure

$$\langle 0|M_1|M_2\rangle \sim 1/\sqrt{N}. \qquad (7.3.25)$$

However, the current appearing on the left hand side of (5.3.75) is written in terms of rescaled fields, and therefore it involves an extra factor of N. We finally obtain

$$F_\pi \sim \sqrt{N}. \qquad (7.3.26)$$

$\square$

From the above simple scaling arguments, one can extract qualitative lessons for the spectrum of QCD at large N. A first result is that, at large N, both mesons and glueballs are free, stable and non-interacting. Their masses have a smooth large N limit, and their number is infinite. To understand why this result is true for mesons, let us consider the two-point function of a current J made of quark bilinears (and that can therefore create a meson, like in (5.3.75)). As for any other two-point function, its spectral representation expresses it as a sum over poles, plus a more complicated part coming from multi-particle states. However, at large N, only the sum over poles contributes, i.e.

$$\langle J(k)J(-k)\rangle = \sum_n \frac{a_n^2}{k^2 - m_n^2}. \qquad (7.3.27)$$

Here the sum is over one-particle meson states $|n\rangle$ with masses m_n, and

$$a_n = \langle 0|J|n\rangle, \qquad (7.3.28)$$

up to a kinematic factor. This can be established by noticing that the Feynman diagrams that contribute to this correlator at large N are diagrams with one single-quark loop at the boundary. Therefore, when we cut this diagram to detect intermediate states, we find exactly one $\bar{\psi}\,\psi$ pair. This state is a single meson

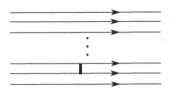

Figure 7.15 A gluon exchange between two quarks inside a baryon.

(assuming that quarks are confined). From (7.3.27) we can also deduce that the spectrum of mesons contains an infinite number of states whose masses are of order $\mathcal{O}(1)$ at large N. This is because in (7.3.27) all the ingredients have a finite limit at large N (if we normalize the currents as in (7.3.20), the right hand side is independent of N, and the same thing is true for (7.3.28)). The number of states must be infinite, since at large k^2 we know from asymptotic freedom that the two-point function is logarithmic in k^2. The logarithmic behavior can only be obtained at large k^2 from the right hand side if the number of terms in the sum is infinite, otherwise we would find a k^{-2} behavior (we will see an explicit example of this infinite tower of mesons at large N in two-dimensional QCD, in Chapter 9).

So far we have only discussed mesons and glueballs in QCD. Baryons can also be analyzed in the large N limit, although their structure is quite different. Indeed, in $SU(N)$ theory, a baryon is an N-quark state, and it can be schematically represented as

$$\epsilon_{i_1 \cdots i_N} \psi^{i_1} \dots \psi^{i_N}, \tag{7.3.29}$$

where $i_1, \ldots, i_N$ are color indices and we have used the $SU(N)$ invariant ϵ-symbol. Therefore, a baryon contains N quarks, one of each color, since all the indices on the ϵ-symbol must be different for it to be non-zero. The basic interaction between the quarks inside the baryon is due to the exchange of a gluon between two quarks, which can be represented by the diagram in Fig. 7.15. It then involves two vertices of the form (7.7). Since each interaction is proportional to g_0, a gluon exchange is of order $g_0^2 \sim 1/N$. However, there are $\sim N^2$ pairs of quarks which can interact, therefore the contribution of such diagrams is of order N. In addition, there might be interactions involving n quarks. At leading order in $1/N$, the corresponding fatgraph can always be obtained by cutting a fatgraph with a single quark loop at n different points of the loop. Such a fatgraph has a factor of N^{1-n}, since a fatgraph with a single quark loop is of order N, and by cutting it n times we have eliminated n index sums over colors. Since there are $\sim N^n$ ways of getting n quarks out of N, the n-body interaction also gives a contribution of order N. Notice in particular that the masses of baryons will scale like N as N becomes large, since all n-body interactions scale as N. How can we study such a complicated interacting system? One possible approach, proposed by Witten in 1979, is to look at a baryon as a many-body system with N particles, in which we have n-body interactions of strength N^{1-n}. Such a many-body system can be analyzed using the Hartree–Fock approximation. Moreover, this approximation is expected to become exact when $N \to \infty$. We will see an explicit example of this situation in our analysis of two-dimensional QCD. Besides the Hartree–Fock approach, it is also possible to study baryons in the $1/N$ expansion using other approaches (like the existence of a spin-flavor symmetry), and the results turn out to be in good agreement with what is observed in particle physics.

Although the large N picture of QCD is qualitatively good, it is hard to make it precise. According to this picture, the main features of QCD should be captured by considering the planar sector of the theory, which is the dominating one when $N \to \infty$. For example, the leading contribution to the free energy at large N should be given by $F_0(t_0)$. However, this function is given by a sum over all planar diagrams of the theory. The behavior of this function at small 't Hooft parameter t_0 can be easily obtained by working out the first orders of perturbation theory. In order to have a full understanding of the planar sector beyond perturbation theory, we need to be able to resum all planar diagrams in the calculation of correlation functions. This is well beyond our current abilities. There are indications from lattice calculations, however, that the $1/N$ picture of the QCD spectrum is a very reasonable one. Lattice results are obtained by calculating the relevant physical quantities for different values of N and extrapolating the results to $N \to \infty$. There is good evidence from these calculations that observables scale with N as expected from the 't Hooft counting. For example, the masses of mesons have a finite limit as N becomes large. Moreover, these calculations indicate that the large N results are quite close to the results obtained for $SU(3)$. In other words, the large N limit of QCD seems to be a good approximation to the physically realized theory with gauge group $SU(3)$.

One interesting aspect of the large N limit of YM theory and of QCD is that it is essentially semiclassical. We have seen a similar phenomenon in the study of large N sigma models. In the effective action describing these models (like for example (6.2.9)), N played the role of $\hbar$ and the large N limit was described by the saddle point of this effective action. To see that something similar happens in YM theory, let us consider for example a gluon operator G (the argument can be extended to QCD in a straightforward way). The strength of quantum fluctuations can be measured by the ratio of the quadratic dispersion of this operator to its average, i.e. by

$$\frac{\Delta G}{\langle G \rangle},$$

(7.3.30)

where

$$(\Delta G)^2 = \langle G^2 \rangle - \langle G \rangle^2 = \langle G^2 \rangle^{(c)}. \tag{7.3.31}$$

Due to the large N rule for correlation functions in (7.3.10), we have that

$$\frac{\Delta G}{\langle G \rangle} \sim \mathcal{O}\left(1/N^2\right). \tag{7.3.32}$$

Therefore, at large N, quantum fluctuations are suppressed and the theory is classical. As pointed out by Witten, this means that there should be a classical gauge configuration A_μ^{cl} (up to gauge transformations) which encodes the large N limit of the theory, in the sense that any correlation function can be computed by simply evaluating it on the classical configuration A_μ^{cl}. For example, one should have

$$\langle G \rangle \approx G(A_\mu^{\text{cl}}), \tag{7.3.33}$$

at large N. This classical configuration is sometimes called the *master field*. Finding an explicit algorithm for the master field is tantamount to solving the large N limit of the theory, and by expanding around this configuration one should be able to generate the $1/N$ corrections. In the example of large N sigma models, the master field is easy to find: it is just the saddle-point configuration of the effective action (in the case of the sigma model, this is just the constant field configuration (6.2.19)). In some very simple examples involving matrix-valued fields, like matrix models and matrix QM, the master field can be found in an explicit way, as we will show in the next chapter. For YM in two dimensions, the master field can also be described explicitly. For four-dimensional YM theory and conventional QCD, the master field is not known. One of the biggest theoretical surprises in the study of the large N expansion has been that, in some supersymmetric versions of YM theory, the master field can be described in detail, but in terms of a theory involving gravity and strings. 't Hooft had already speculated that the Riemann surfaces appearing in the $1/N$ expansion are related to string theory, and this idea has been made precise through the AdS/CFT correspondence of Maldacena. We refer the reader to the references at the end of this chapter for further reading on this fascinating topic.

7.4 θ-dependence at large N

It is also interesting to understand the interplay between the dependence on the theta angle and the large N dependence. To do this, let us consider the YM Lagrangian together with the topological density:

$$\mathcal{L}_\theta = -\frac{N}{4t_0} F_{\mu\nu}^a F^{\mu\nu a} - \theta q(x). \tag{7.4.1}$$

If we want to have a non-trivial dependence on the θ angle, both terms in the above Lagrangian should scale similarly as N grows, since $q(x)$ has the same color structure as the standard YM Lagrangian. We conclude that the natural scaling variable at large N is not θ, but

$$\bar{\theta} = \frac{\theta}{N}. \tag{7.4.2}$$

Let us now consider the vacuum energy density (4.3.11). Since this is obtained from vacuum-to-vacuum diagrams, it scales at large N like N^2, i.e.

$$E(\theta) - E(0) \approx N^2 e(\bar{\theta}), \tag{7.4.3}$$

and by comparing this with the expansion (4.3.15), (4.3.16), we conclude that

$$e(\bar{\theta}) = \frac{1}{2} \chi_t^\infty \bar{\theta}^2 \left(1 + \sum_{n=1}^{\infty} \bar{b}_{2n} \bar{\theta}^{2n} \right), \tag{7.4.4}$$

where χ_t^∞ and $\bar{b}_{2n}$ are defined by the $1/N$ expansion,

$$\chi_t = \chi_t^\infty \left(1 + \mathcal{O}\left(\frac{1}{N^2}\right) \right), \qquad b_{2n} = \frac{\bar{b}_{2n}}{N^{2n}} \left(1 + \mathcal{O}\left(\frac{1}{N^2}\right) \right). \tag{7.4.5}$$

In particular, the topological susceptibility has a finite limit at large N, which we have denoted by χ_t^∞.

Note that a similar scaling holds for the $\mathbb{CP}^{N-1}$ model considered in the previous chapter: by looking at the action (6.3.78), we see that the natural scaling variable for the N dependence is also $\bar{\theta}$. However, since the vacuum energy is of order N, instead of order N^2, we have instead

$$E(\theta) - E(0) \approx N e(\bar{\theta}), \tag{7.4.6}$$

where $e(\bar{\theta})$ has the same form as in (7.4.4), and

$$\chi_t = \frac{\chi_t^\infty}{N} \left(1 + \mathcal{O}\left(\frac{1}{N}\right) \right). \tag{7.4.7}$$

This is precisely what we found in the explicit calculation (6.3.101), with

$$\chi_t^\infty = \frac{3m^2}{\pi}. \tag{7.4.8}$$

There is however one puzzle to resolve concerning the interplay of the $1/N$ expansion and the θ angle: the large N scaling behavior of the theta dependence seems to be incompatible with the periodicity of $E(\theta)$, since a smooth function of θ/N which is invariant under $\theta \to \theta + 2\pi$ must be a constant. A resolution of

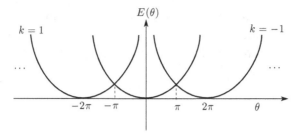

Figure 7.16 The function (7.4.11) and its different branches, which are parabolae
with vertices at $\theta = -2\pi k, k \in \mathbb{Z}$. The vacuum energy $E(\theta)$ is obtained by taking
the minimum of the different branches, for a given value of θ.

this puzzle was proposed by Witten in 1980: the function $e(\overline{\theta})$ should be a multi-branched function, and each branch should describe a vacuum state. All these vacua become stable at large N, and the energy of each vacuum is of the form

$$N^2 e \left(\frac{\theta + 2\pi k}{N} \right), \qquad k \in \mathbb{Z}. \tag{7.4.9}$$

The true vacuum energy density is obtained by picking the value of k which minimizes the energy for a given value of θ, i.e.

$$E(\theta) - E(0) \approx N^2 \min_k e \left(\frac{\theta + 2\pi k}{N} \right). \tag{7.4.10}$$

This function is periodic under $\theta \to \theta + 2\pi$, but is not smooth, and it jumps at the values of θ where we change from one branch to the other. Since $\theta = 0$ is a minimum of the energy, we have that $e(\overline{\theta}) \approx \overline{\theta}^2$ near $\overline{\theta} = 0$. We conclude that, at fixed θ, and at leading order in the $1/N$ expansion, we must have

$$E(\theta) - E(0) \approx \frac{\chi_t^\infty}{2} \min_k (\theta + 2\pi k)^2. \tag{7.4.11}$$

For $0 \leq \theta \leq \pi$, the minimum is obtained by considering the branch with $k = 0$. However, jumps occur at $\theta = \pi$, where the minimum involves the branch with $k = -1$, see Fig. 7.16.

7.5 The $U(1)$ problem at large N: the Witten–Veneziano formula

As we explained in Section 5.4, the $U(1)_A$ axial symmetry is broken by anomalies, and the anomaly is proportional to the topological density $q(x)$. This means that this symmetry is not really spontaneously broken, and there is no Goldstone boson associated to it. However, there is a pseudo-scalar meson, the η', which has the right quantum numbers to be such a Goldstone boson. Of course, this particle is not massless, and its mass is much larger than the masses of the octet of mesons. As we mentioned in Section 5.4, one could say that this large mass is due to the anomaly, but this statement has to be made precise. We will show now that, in the context of the large N expansion, we can think about the η' as a pseudo-Goldstone boson, where the symmetry breaking parameter is $1/N$ (in addition to the quark masses). In that sense, we would expect that, when the quark masses are zero, $m_{\eta'}^2 \sim \mathcal{O}(1/N)$, in the same way that the squared masses of the mesons are proportional to the masses of the quarks. It was shown by Witten (and subsequently by Veneziano) that this insight can be made very precise, and one can obtain a remarkable expression for the mass of the η' at leading order in $1/N$ known as the Witten–Veneziano formula. This formula has been spectacularly confirmed by lattice gauge theory calculations. We will now give a derivation of this formula, based on the effective Lagrangian including the η' discussed in Section 5.4.

The starting point for the Witten–Veneziano formula is the following: when quarks are massless, the anomalous contribution to the divergence of the $U(1)$ current *vanishes* in the large N limit (7.2.2). Indeed, when we express all quantities in terms of normalized fields (7.2.5), one has from (5.4.49) that

$$\partial_\mu J^\mu = -\frac{2N_f t_0}{N} q(x). \tag{7.5.1}$$

Therefore, at large N the η' is a true Goldstone boson, and we can regard $1/N$ as a symmetry breaking parameter. We conclude that, at leading order in the $1/N$ expansion, the chiral symmetry breaking pattern is indeed given by (5.4.1). Since there is now a full symmetry between the nine mesons, we can study their low-energy dynamics by using a chiral Lagrangian for the field (5.4.2) with $F_{\eta_0} = F_\pi$. However, we want to incorporate the $U(1)_A$ anomaly in the Lagrangian in order to calculate the corresponding correction in $1/N$. To do this, we introduce a current for the topological density in the QCD Lagrangian:

$$- \theta(x)q(x). \tag{7.5.2}$$

We know that, under an axial transformation (5.4.24), the source $\theta(x)$ changes as dictated by (5.4.51),

$$\theta(x) \to \theta(x) + 2N_f \alpha(x). \tag{7.5.3}$$

On the other hand, we have that

$$\text{Tr} \log U \to \text{Tr} \log U - 2iN_f \alpha(x). \tag{7.5.4}$$

Therefore, the field

$$X \equiv \text{Tr} \log U + i\theta(x), \tag{7.5.5}$$

as well as any function of X, is invariant under the full symmetry group. According to the logic of effective field theory, we have to write down the most general Lagrangian which is invariant under the symmetries. This is

$$
\mathcal{L}_{\text{eff}} = - W_0(X) + W_1(X)\text{Tr}(D_\mu U D^\mu U^\dagger)
$$
$$
+ W_2(X)\text{Tr}(\chi U^\dagger + U\chi^\dagger) + \cdots , \tag{7.5.6}
$$

where the dots denote terms which do not contribute to the masses of the particles. By parity invariance, all functions of X appearing here must be even (since they encode the dependence on the theta angle), and we must have, in order to recover (5.3.59),

$$
W_1(0) = W_2(0) = \frac{F_\pi^2}{4}. \tag{7.5.7}
$$

Let us now compute the mass matrix, by expanding the Lagrangian (7.5.6). The term involving $W_2(X)$ incorporates the contribution due to the quark masses, and is just given by (5.4.11) with $\xi = 1$. But we have a new mass term coming from $-W_0(X)$, which gives, after expanding to second order,

$$
-\frac{1}{2} W_0''(0) \left(\frac{i\sqrt{2N_f}}{F_\pi} \eta_0 \right)^2 , \tag{7.5.8}
$$

i.e. it gives a contribution to the last entry of the mass matrix (5.4.11) of the form

$$
\epsilon = -\frac{2N_f}{F_\pi^2} W_0''(0). \tag{7.5.9}
$$

Interestingly, $W_0''(0)$ can be related to a quantity computed in *pure* YM theory, i.e. in the theory without quarks. Indeed, in the theory with no quarks but with a source for the topological charge of the form (7.5.2), the effective Lagrangian (7.5.6) reduces to $-W_0(X)$. In particular, if we take a constant θ angle, we find, at finite but large volume,

$$
Z(\theta) \approx e^{-iW_0(i\theta)V_4}, \tag{7.5.10}
$$

where V_4 is the volume of spacetime. Since we are working now in Minkowski signature, we find, by comparing with (4.3.11), and in the limit of infinite volume, that

$$
E(\theta) = W_0(i\theta). \tag{7.5.11}
$$

Therefore, $-W_0''(0)$ is nothing but the second derivative of the vacuum energy density with respect to the theta angle, i.e. it is the topological susceptibility of pure YM theory in the large N limit. We conclude that

$$
\epsilon = \frac{2N_f}{F_\pi^2} \chi_t^\infty. \tag{7.5.12}
$$

We can now use this information to obtain the Witten–Veneziano formula. Let us set $N_f = 3$ and come back to the mass matrix (5.4.11). We will assume for simplicity that $m_u = m_d$, so that all mixings in the mass matrix are eliminated except for those of the η_8 and the η_0. These two states mix, i.e. the eigenvectors of the mass matrix will be certain linear combinations of the η_8 and the η_0, and the corresponding particles will be called the η and the η', with masses m_η^2 and $m_{\eta'}^2$. This is the mixing that we implemented in Table 5.1 through the mixing angle θ. The mass matrix for the $\eta - \eta'$ is then

$$
\mathcal{M}_{\eta-\eta'} = \begin{pmatrix} \frac{4}{3}m_K^2 - \frac{1}{3}m_\pi^2 & -\frac{2\sqrt{2}}{3}(m_K^2 - m_\pi^2) \\ -\frac{2\sqrt{2}}{3}(m_K^2 - m_\pi^2) & \frac{2}{3}m_K^2 + \frac{1}{3}m_\pi^2 + \epsilon \end{pmatrix}. \tag{7.5.13}
$$

The eigenvalues of (7.5.13) are m_η^2, $m_{\eta'}^2$, and we have, by computing $\mathrm{Tr}\, \mathcal{M}_{\eta-\eta'}$,

$$
m_\eta^2 + m_{\eta'}^2 = 2m_K^2 + \frac{6}{F_\pi^2}\chi_t^\infty. \tag{7.5.14}
$$

This is the Witten–Veneziano formula, which relates the mass of the η' to the topological susceptibility of pure YM theory. The formula above should be understood as giving the first two terms in the $1/N$ expansion of the left hand side. On the right hand side, both m_K^2 (which is of order $\mathcal{O}(N^0)$) and F_π^2 (which is of order $\mathcal{O}(N)$ by (7.3.26)) are computed in the large N limit. As compared to our analysis in Section 5.4, we see that the right hand side of (7.5.14) now involves a $1/N$ correction which is proportional to the topological susceptibility at large N. This expresses the fact that $1/N$ is regarded here as a symmetry breaking parameter. Indeed, as the masses of the quarks go to zero, the mass of the η' is proportional to $1/N$. The equation (7.5.14) encodes in a precise, quantitative way, the relation between the axial anomaly and the mass of the η' particle. Note that this resolution of the $U(1)$ problem requires a non-vanishing value for the topological susceptibility of pure YM theory, and in particular a non-zero value for its large N limit. Although χ_t^{YM} vanishes order by order in perturbation theory, we have already seen in the analysis of the $\mathbb{CP}^{N-1}$ sigma model that it is possible to find a non-vanishing value at large N. The picture of the η' developed by Witten and Veneziano requires that this is also the case in YM theory.

It turns out that the Witten–Veneziano formula, albeit obtained in the large N limit, is approximately true in the real world, where $N = 3$. In order to test the formula, one should calculate the topological susceptibility of pure YM theory. At present, this can only be done in the framework of lattice gauge theory, where one finds, for $N = 3$,

$$
\chi_t^{\mathrm{YM}} \approx (191 \pm 5\,\mathrm{MeV})^4. \tag{7.5.15}
$$

On the other hand, by plugging the experimental values of the meson masses into (7.5.14) we get

$$\frac{F_\pi^2}{6} \left(m_{\eta'}^2 + m_\eta^2 - 2m_K^2 \right) \approx (180\,\text{MeV})^4, \tag{7.5.16}$$

which is quite close to (7.5.15). This remarkable agreement indicates that the axial anomaly, when implemented quantitatively through the Witten–Veneziano formula, solves the $U(1)$ problem. Note that the Witten–Veneziano formula should have $1/N$ corrections, but these are relatively small, since as we mentioned before, the large N limit of QCD seems to be a reasonable approximation to the $SU(3)$ theory of the real world. Let us finally mention that the mixing angle θ between η_8 and η_0 appearing in the last two entries of Table 5.1 can be computed from the matrix (7.5.13), using the meson masses, and it leads to the value $\theta \approx -17°$ quoted in Section 5.3.

7.6 Bibliographical notes

The $1/N$ expansion in QCD was introduced by 't Hooft in [173]. Classical reviews of this topic are [50, 189], and an early, lucid invitation to the subject can be found in [191]. More modern reviews include [24, 134]. The books [89, 156, 167] all have chapters on the $1/N$ expansion in QCD.

The calculation of group factors inspired by the formalism of fatgraphs was introduced in [57] and developed in [20, 58]. The double-line notation for $SO(N)$ and $Sp(N)$ groups is developed for example in [48]. Lattice studies of the $1/N$ expansion are reviewed in [126]. The theta dependence at large N is reviewed in [182]. The multi-branch structure of the vacuum energy as a function of θ was first proposed in [192] and is further discussed in [195], where it was tested explicitly in some examples by using the AdS/CFT correspondence. Baryons at large N were first discussed in detail in [189], and a review of more recent developments can be found in [134]. The idea of a "master field" for large N gauge theories was proposed by Witten in [190]. A review of the algebraic approach to the master field, with applications to QCD in two dimensions, can be found in [99]. The connection between large N supersymmetric gauge theories and string theories mentioned in the text has triggered an enormous literature in recent years. A review of this connection can be found in [6].

The Witten–Veneziano formula was originally proposed in [180, 188]. The derivation in [188] involves some subtleties which are addressed in [96, 165]. Modern implementations of the Witten–Veneziano formula in the lattice can be found in [67, 96, 131]. The calculation of the topological susceptibility of $SU(3)$ YM theory cited in this chapter is due to [67]. The derivation of the Witten–Veneziano formula presented here, based on the effective Lagrangian, goes back to [72], but we have used the slightly updated framework and notation of [107]. There are two-dimensional versions of the Witten–Veneziano formula which clarify some of its aspects, see for example [71, 96, 165].

8

Matrix models and matrix Quantum Mechanics at large N

8.1 Introduction

As we have seen in the previous chapter, the $1/N$ expansion leads to a new point of view in the study of YM theory and QCD, which captures some important aspects of their dynamics. However, quantitative results are difficult to obtain since we are unable to calculate the planar limit exactly, i.e. we are unable to resum all planar diagrams. In this and the next chapter we will consider toy models in lower dimensions where we can in fact perform such a resummation. We will begin by a drastic simplification: we will go from four to zero dimensions, and consider theories made out of constant fields in the adjoint representation of $U(N)$, i.e. theories based on Hermitian matrices, also known as matrix models. In these models, the path integral reduces to a conventional multi-dimensional integral. One might think that they would be too simple to be relevant in any way to actual gauge theories. However, the problem of resumming the diagrams with a fixed topology in matrix models is still a non-trivial one, and has many applications in many branches of physics and mathematics. Moreover, it has been shown that some quantities in *bona fide* interacting QFTs (albeit with some special properties, like supersymmetry) reduce to matrix models. Therefore, the problem of resumming the fatgraphs in matrix models has unexpected applications in the study of these highly symmetric QFTs, and the large N techniques we will introduce in this chapter are relevant in the study of these gauge theories.

A closely related class of toy models for large N QCD is matrix QM, in which one considers simple quantum mechanical systems (typically in one spatial dimension) and promotes the dynamical variables to Hermitian matrices. This system also has some interesting applications and provides a framework where one can use exact results in QM to give insights in the $1/N$ expansion.

8.2 Hermitian matrix models

Matrix models are in a sense the simplest examples of quantum gauge theories, namely, they are quantum gauge theories in zero dimensions in which the spacetime dependence has been removed. This means that the basic fields have just the group structure of the gauge connection, i.e. they are matrices in the adjoint representation of the gauge group. If the gauge group is $U(N)$, these are $N \times N$ Hermitian matrices. For simplicity we will consider in this chapter one-matrix models, i.e. models where there is a single matrix M. Our gauge group will be $U(N)$, so M will be an $N \times N$ Hermitian matrix. In order to define a theory for M, we should specify an action incorporating gauge invariance in a suitable way. The simplest action we can consider is of the form,

$$\frac{1}{g_s} \text{Tr} V(M), \tag{8.2.1}$$

where

$$V(M) = \frac{1}{2} M^2 + \sum_{p \geq 3} \frac{g_p}{p} M^p, \tag{8.2.2}$$

and g_s and g_p are coupling constants. Equation (8.2.2) is often called the potential of the matrix model, and the action (8.2.1) is clearly invariant under the symmetry

$$M \to UMU^\dagger, \tag{8.2.3}$$

where U is any $U(N)$ matrix. This can be regarded as the zero-dimensional version of the gauge transformation (4.2.23).

The quantum theory based on the action (8.2.1) can be defined by using a path integral. However, since the field M is spacetime independent, the path integral reduces to an ordinary integral. The partition function of the matrix model theory with action (8.2.1) is then given by

$$Z = \frac{1}{\text{vol}(U(N))} \int dM \, e^{-\frac{1}{g_s} \text{Tr} V(M)}. \tag{8.2.4}$$

Here, $\text{vol}(U(N))$ is the volume of the gauge group. The measure in the "path integral" is the Haar measure

$$dM = 2^{\frac{N(N-1)}{2}} \prod_{i=1}^{N} dM_{ii} \prod_{1 \leq i < j \leq N} d\text{Re} \, M_{ij} d\text{Im} \, M_{ij}. \tag{8.2.5}$$

The numerical factor in (8.2.5) is introduced to obtain a convenient normalization. The natural observables in the matrix model are gauge invariant operators, i.e. functions of the matrix M, $f(M)$, satisfying

$$f\left(UMU^\dagger\right) = f(M), \qquad U \in U(N). \tag{8.2.6}$$

The (normalized) vev of such an observable is defined as

$$\langle f(M) \rangle = \frac{\int dM \, f(M) \, e^{-V(M)/g_s}}{\int dM \, e^{-V(M)/g_s}}. \tag{8.2.7}$$

A particularly simple example of the one-matrix model is the *Gaussian matrix model*, defined by the potential (8.2.2) with $g_p = 0$ for $p \geq 3$. Its partition function will be denoted by

$$Z_G = \frac{1}{\text{vol}(U(N))} \int dM \, e^{-\frac{1}{2g_s} \text{Tr} M^2}, \tag{8.2.8}$$

and we will also denote by $\langle f(M) \rangle_G$ the normalized vev of a gauge invariant function $f(M)$. The main interest of the Gaussian matrix model is that, as in any other QFT, it can be taken as the starting point to perform a perturbation expansion. For example, if one wants to evaluate the partition function (8.2.4) of a general matrix model with action (8.2.2), one expands the exponential of $\sum_{p \geq 3}(g_p/g_s) \text{Tr} M^p/p$ in (8.2.4), and computes Z as a power series in the coupling constants g_p. The evaluation of each term of the series involves the computation of vevs in the Gaussian matrix model. Of course, this computation can be interpreted in terms of Feynman diagrams, and as usual the perturbative expansion of the free energy $F = \log Z$ will only involve connected vacuum bubbles.

Since we are dealing with the quantum theory of a matrix-valued field, we can follow the same strategy that we developed in QCD and re-express the perturbative expansion of the free energy F in terms of fatgraphs. Since there is no spacetime dependence, the $1/N$ counting captures all the information about the diagrammatics. For example, the propagator of the theory is a simplified version of (7.2.11),

$$\langle M_{ij} M_{kl} \rangle_G = g_s \delta_{il} \delta_{jk} \tag{8.2.9}$$

and can also be represented as in Fig. 7.2. The vertices of the theory are also very easy to write down. For example, the cubic vertex

$$\frac{g_3}{g_s} \text{Tr} M^3 = \frac{g_3}{g_s} \sum_{i,j,k} M_{ij} M_{jk} M_{ki} \tag{8.2.10}$$

can be represented in the double-line notation as in Fig. 7.4. A vertex of order p can be represented in a similar way by drawing p double lines joined together. Note that, in contrast to the YM cubic vertex, there is no "twisted" graph contribution.

As in QCD, a single graph in perturbation theory leads to many fatgraphs. Let us consider for example the contraction of two cubic vertices, which is needed in the evaluation of

$$\langle (\text{Tr} M^3)^2 \rangle_G. \tag{8.2.11}$$

As we already know from the analysis in the previous chapter, in the double-line notation the contraction can be done in two different ways. The first way leads to the first fatgraph in the right hand side of Fig. 7.11, with $g = 0$, and gives a factor of $g_s^3 N^3$. The second way leads to the second fatgraph in the right hand side of Fig. 7.11, with $g = 1$, and gives a factor of $g_s^3 N$. However, the weight of each contraction in the calculation of (8.2.11) is different from the calculation in (7.2.29). As is easily seen, there are 12 contractions corresponding to the planar diagram, but only three possible contractions for the fatgraph with $g = 1$. We conclude that

$$\langle (\mathrm{Tr}\, M^3)^2 \rangle_{\mathrm{G}} = g_s^3 (12N^3 + 3N),$$
(8.2.12)

which gives the first term in the perturbative expansion of the free energy for the cubic matrix model with $g_p = 0$, $p \geq 4$,

$$F - F_{\mathrm{G}} = \frac{2}{3} g_s g_3^2 N^3 + \frac{1}{6} g_s g_3^2 N + \cdots.$$
(8.2.13)

Here, $F_{\mathrm{G}} = \log Z_{\mathrm{G}}$ is the free energy of the Gaussian matrix model.

A fatgraph appearing in the calculation of the free energy of the matrix model with potential (8.2.2) is characterized topologically by the number of propagators or edges E, the number of vertices with p legs V_p, and the number of closed loops h. The total number of vertices is $V = \sum_p V_p$. Each propagator gives a power of g_s, while each interaction vertex with p legs gives a power of g_p/g_s. The factor associated to the fatgraph is then,

$$g_s^{E-V} N^h \prod_p g_p^{V_p}.$$
(8.2.14)

As in the previous chapter, we can regard the fatgraph as a Riemann surface. By using again Euler's relation (7.2.24) we can write (7.2.22) as

$$g_s^{2g-2+h} N^h \prod_p g_p^{V_p} = g_s^{2g-2} t^h \prod_p g_p^{V_p},$$
(8.2.15)

where we have introduced the 't Hooft parameter

$$t = N g_s.$$
(8.2.16)

As in (7.3.2), the free energy of the matrix model, after subtracting the Gaussian free energy, can be written as

$$F - F_{\mathrm{G}} = \sum_{g=0}^{\infty} F_g(t) g_s^{2g-2},$$
(8.2.17)

where

$$F_g(t) = \sum_{h=1}^{\infty} a_{g,h} t^h, \tag{8.2.18}$$

and the coefficients $a_{g,h}$ (which depend on the coupling constants of the model g_p) take into account the multiplicities of the different fatgraphs. Note that the subtraction of the Gaussian matrix model free energy guarantees that we only have contributions starting at two loops. In (8.2.17), in contrast to (7.3.2), we have done the expansion in g_s rather than $1/N$, and g_s plays the role of g_0^2 in gauge theories. Of course, both expansions are equivalent when t is fixed. The leading contribution to the free energy, at fixed 't Hooft parameter, comes again from planar diagrams, with $g = 0$. As in the analysis of QCD, the free energy at genus g, $F_g(t)$, is given by an infinite series where we sum over all possible numbers of holes of the fatgraphs, h, weighted by t^h.

There is an alternative way of writing the matrix model partition function which is very useful. The original matrix model variable has N^2 real parameters, and therefore in the partition function (8.2.4) we have to integrate over N^2 real variables. However, many of these parameters can be removed by using the gauge symmetry (8.2.3). We can for example take advantage of our gauge freedom to diagonalize the matrix M

$$M \rightarrow UMU^\dagger = D, \tag{8.2.19}$$

with $D = \mathrm{diag}(\lambda_1, \ldots, \lambda_N)$. This leaves only N parameters, the eigenvalues of M. We can regard (8.2.19) as a gauge fixing, and use standard Faddeev–Popov techniques in order to compute the gauge-fixed integral. The gauge fixing (8.2.19) leads to the delta function constraint

$$\delta(^U M) = \prod_{i<j} \delta^{(2)} \left[(^U M)_{ij} \right], \tag{8.2.20}$$

where $^U M = UMU^\dagger$. We then introduce the gauge invariant function $\mathcal{J}(M)$ through the equation

$$(\mathcal{J}(M))^{-1} = \int dU \, \delta(^U M). \tag{8.2.21}$$

It then follows that the integral of any gauge invariant function $f(M)$ can be written as

$$\int dM \, f(M) = \int dM \, f(M) \mathcal{J}(M) \int dU \, \delta\left(^U M\right)$$

$$= \Omega_N \int \prod_{i=1}^{N} d\lambda_i \, \mathcal{J}(\lambda) f(\lambda), \tag{8.2.22}$$

where we have used the gauge invariance of $\mathcal{J}(M)$, and

$$\Omega_N = \int dU \qquad (8.2.23)$$

is proportional to the volume of the gauge group $U(N)$, as we will see shortly. We have to evaluate the factor $\mathcal{J}(\lambda)$, which can be obtained from (8.2.21) by choosing M to be diagonal. If the gauge-fixing condition is given by a constraint of the form $F(M) = 0$, the standard Faddeev–Popov formula gives

$$\mathcal{J}(M) = \det\left(\frac{\delta F\left(^U M\right)}{\delta A}\right)_{F=0} \qquad (8.2.24)$$

where we write $U = e^A$, and A is an anti-Hermitian matrix. In our case, the constraint imposes that the off-diagonal terms of $^U M$ vanish, and we evaluate $\mathcal{J}(M)$ on a diagonal matrix D. Since

$$\left(^U D\right)_{ij} = (U D U^\dagger)_{ij} = A_{ij}(\lambda_j - \lambda_i) + \cdots, \qquad i \neq j, \qquad (8.2.25)$$

(8.2.24) leads immediately to

$$\mathcal{J}(\lambda) = \Delta^2(\lambda), \qquad (8.2.26)$$

where

$$\Delta(\lambda) = \prod_{i<j}(\lambda_i - \lambda_j) \qquad (8.2.27)$$

is the Vandermonde determinant of the eigenvalues. Finally, we fix the factor Ω_N as follows. The Gaussian matrix integral can be computed explicitly by using the Haar measure (8.2.5), and is simply

$$\int dM\, e^{-\frac{1}{2g_s}\operatorname{Tr} M^2} = (2\pi g_s)^{N^2/2}. \qquad (8.2.28)$$

On the other hand, by (8.2.22) this should equal

$$\Omega_N \int \prod_{i=1}^{N} d\lambda_i\, \Delta^2(\lambda) e^{-\frac{1}{2g_s}\sum_{i=1}^{N}\lambda_i^2}. \qquad (8.2.29)$$

The integral over eigenvalues can be evaluated in various ways (for example using orthogonal polynomials or the Selberg integral), and one finds

$$\int \prod_{i=1}^{N} d\lambda_i\, \Delta^2(\lambda) e^{-\frac{1}{2g_s}\sum_{i=1}^{N}\lambda_i^2} = g_s^{N^2/2}(2\pi)^{N/2} G_2(N+2), \qquad (8.2.30)$$

where $G_2(z)$ is the Barnes function, defined by

$$G_2(z+1) = \Gamma(z)G_2(z), \quad G_2(1) = 1. \qquad (8.2.31)$$

Comparing these results, we find that

$$\Omega_N = \frac{(2\pi)^{\frac{N(N-1)}{2}}}{G_2(N+2)}. \tag{8.2.32}$$

Using now the known result for the volume of the $U(N)$ group,

$$\mathrm{vol}(U(N)) = \frac{(2\pi)^{\frac{1}{2}N(N+1)}}{G_2(N+1)}, \tag{8.2.33}$$

we see that

$$\frac{1}{\mathrm{vol}(U(N))} \int \mathrm{d}M \, f(M) = \frac{1}{N!} \frac{1}{(2\pi)^N} \int \prod_{i=1}^{N} \mathrm{d}\lambda_i \, \Delta^2(\lambda) f(\lambda). \tag{8.2.34}$$

The factor $1/N!$ on the right hand side of (8.2.34) has an obvious interpretation: after fixing the gauge symmetry of the matrix integral by imposing the diagonal gauge, there is still a residual symmetry given by the Weyl group of $U(N)$, which is the symmetric group S_N acting by permutations of the eigenvalues. The "volume" of this discrete gauge group is just its order, $|S_N| = N!$. Since we are dividing by the volume of the gauge group, we find in front of the resulting integral the inverse of the volume of this residual gauge symmetry. As a particular case of the above formula, it follows that one can write the partition function (8.2.4) as

$$Z = \frac{1}{N!} \frac{1}{(2\pi)^N} \int \prod_{i=1}^{N} \mathrm{d}\lambda_i \, \Delta^2(\lambda) e^{-\frac{1}{g_s} \sum_{i=1}^{N} V(\lambda_i)}. \tag{8.2.35}$$

The partition function of the Gaussian matrix model (8.2.8) is given essentially by the inverse of the volume factor, and its exact value (for all N) is given by

$$Z_G = g_s^{N^2/2} (2\pi)^{-N/2} G_2(N+1). \tag{8.2.36}$$

Its free energy to all orders can be computed by using the asymptotic expansion of the Barnes function

$$\log G_2(N+1) = \frac{N^2}{2} \log N - \frac{1}{12} \log N - \frac{3}{4} N^2 + \frac{1}{2} N \log 2\pi + \zeta'(-1)$$

$$+ \sum_{g=2}^{\infty} \frac{B_{2g}}{2g(2g-2)} N^{2-2g}, \tag{8.2.37}$$

where B_{2g} are the Bernoulli numbers. Therefore, we find the following expression for the total free energy of the Gaussian matrix model:

$$F_G = \frac{N^2}{2}\left(\log(Ng_s) - \frac{3}{2}\right) - \frac{1}{12}\log N + \zeta'(-1)$$

$$+ \sum_{g=2}^{\infty} \frac{B_{2g}}{2g(2g-2)} N^{2-2g}. \tag{8.2.38}$$

If we re-express this free energy in terms of the 't Hooft parameter (8.2.16), we see that, if we discard the terms $\log(g_s)/12 + \zeta'(-1)$, it has the structure (8.2.17). The genus g free energies $F_g(t)$ are given by:

$$F_0(t) = \frac{1}{2}t^2\left(\log t - \frac{3}{2}\right),$$

$$F_1(t) = -\frac{1}{12}\log t, \tag{8.2.39}$$

$$F_g(t) = \frac{B_{2g}}{2g(2g-2)}t^{2-2g}, \quad g > 1.$$

Notice however that these free energies do not come from summing diagrams, but rather from a one-loop calculation, and therefore the $F_g(t)$ do not have the structure (8.2.18). In particular, they are singular at $t = 0$.

The computation of the functions $F_g(t)$ in closed form for a generic matrix model is of course much simpler than in a *bona fide* QFT, but it is still a difficult task, since each genus g free energy is a sum of an infinite number of fatgraphs. However, Brézin, Itzykson, Parisi and Zuber showed in 1980 that $F_0(t)$ can be obtained in closed form for any polynomial matrix model. We will now explain their result.

Let us consider a general matrix model with action $V(M)$, and let us write the partition function after reduction to eigenvalues (8.2.35) as follows:

$$Z = \frac{1}{N!}\int \prod_{i=1}^{N} \frac{d\lambda_i}{2\pi} e^{N^2 S_{\text{eff}}(\lambda)}, \tag{8.2.40}$$

where the effective action is given by

$$S_{\text{eff}}(\lambda) = -\frac{1}{tN}\sum_{i=1}^{N} V(\lambda_i) + \frac{2}{N^2}\sum_{i<j}\log|\lambda_i - \lambda_j|. \tag{8.2.41}$$

Notice that, since a sum over N eigenvalues is roughly of order N, the effective action is of order $\mathcal{O}(1)$. As $N \to \infty$ with t fixed, we expect the integral (8.2.40) to be dominated by a saddle-point configuration that extremizes the effective action. This is yet another indication of the semiclassical nature of the large N limit, which we have already encountered in the study of sigma models and of QCD. Varying $S_{\text{eff}}(\lambda)$ with respect to the eigenvalue λ_i, we obtain the EOM

$$\frac{1}{2t} V'(\lambda_i) = \frac{1}{N} \sum_{j \neq i} \frac{1}{\lambda_i - \lambda_j}, \quad i = 1, \ldots, N. \tag{8.2.42}$$

This equation can be given a simple interpretation: we can regard the eigenvalues as coordinates of a system of N classical particles moving on the real line. Equation (8.2.42) says that these particles are subject to an effective potential

$$V_{\text{eff}}(\lambda_i) = V(\lambda_i) - \frac{2t}{N} \sum_{j \neq i} \log |\lambda_i - \lambda_j|, \tag{8.2.43}$$

which involves a logarithmic Coulomb repulsion between eigenvalues coming from the Vandermonde interaction. When the 't Hooft parameter is small, the potential term dominates over the Coulomb repulsion, and the particles tend to be at a critical point x_* of the potential: $V'(x_*) = 0$. As t grows, the Coulomb repulsion will force the eigenvalues to spread out along the real axis until they reach an equilibrium position, in which the attraction of the potential exactly cancels the repulsion. The equilibrium condition is precisely (8.2.42).

To encode the information about the equilibrium distribution of the particles, it is convenient to define a *density of eigenvalues* (for finite N) as

$$\rho(\lambda) = \left\langle \frac{1}{N} \operatorname{Tr} \delta(\lambda - M) \right\rangle = \frac{1}{N} \sum_{i=1}^{N} \langle \delta(\lambda - \lambda_i) \rangle. \tag{8.2.44}$$

In the large N limit, it is reasonable to expect that this density of eigenvalues becomes a continuous function $\rho_0(\lambda)$. If the potential grows sufficiently fast at large distances, it can compensate for the Coulomb repulsion and prevent the eigenvalues from spreading out to infinity. In this situation, we also expect the density of eigenvalues to have a compact support. The simplest solution occurs when $\rho_0(\lambda)$ vanishes outside an interval $\mathcal{C} = [b, a]$. This is the so-called *one-cut solution*. Based on the considerations above, the interval $[b, a]$ should be centered around an extremum x_* of the potential. In particular, as $t \to 0$, the interval $[b, a]$ should collapse to the point x_*.

We can now write the saddle-point equation in terms of continuum quantities, by using the rule

$$\frac{1}{N} \sum_{i=1}^{N} f(\lambda_i) \rightarrow \int_{\mathcal{C}} f(\lambda) \rho_0(\lambda) d\lambda. \tag{8.2.45}$$

Note that the density of eigenvalues $\rho_0(\lambda)$ satisfies the normalization condition

$$\int_{\mathcal{C}} \rho_0(\lambda) d\lambda = 1. \tag{8.2.46}$$

Equation (8.2.42) becomes, at large N,

$$\frac{1}{2t}V'(\lambda) = P \int_C \frac{\rho_0(\lambda')d\lambda'}{\lambda - \lambda'}, \tag{8.2.47}$$

where P denotes the principal value of the integral. This is an integral equation that allows one in principle to compute $\rho_0(\lambda)$, given the potential $V(\lambda)$, as a function of the 't Hooft parameter t and the coupling constants. Once $\rho_0(\lambda)$ is known, one can easily compute $F_0(t)$: in the saddle-point approximation, the free energy is given by

$$\frac{1}{N^2}F = S_{\text{eff}}(\rho_0) + \mathcal{O}(N^{-2}), \tag{8.2.48}$$

where the effective action in the continuum limit is a functional of ρ_0:

$$S_{\text{eff}}(\rho_0) = -\frac{1}{t}\int_C \rho_0(\lambda)V(\lambda)d\lambda + \int_{C\times C} \rho_0(\lambda)\rho_0(\lambda') \log|\lambda - \lambda'|d\lambda\,d\lambda'. \tag{8.2.49}$$

Therefore, the planar free energy is given by

$$F_0(t) = t^2 S_{\text{eff}}(\rho_0). \tag{8.2.50}$$

We can obtain (8.2.42) directly in the continuum formulation by computing the extremum of the functional

$$S(\rho_0, \xi) = S_{\text{eff}}(\rho_0) + \xi\left(\int_C \rho_0(\lambda)d\lambda - 1\right) \tag{8.2.51}$$

with respect to ρ_0. Here, ξ is a Lagrange multiplier that imposes the normalization condition of the density of eigenvalues. It is in general a function of the 't Hooft parameter. This leads to

$$\frac{1}{t}V(\lambda) = 2\int \rho_0(\lambda') \log|\lambda - \lambda'|d\lambda' + \xi(t), \tag{8.2.52}$$

which can also be obtained by integrating (8.2.47) with respect to λ. The Lagrange multiplier ξ appears in this way as an integration constant that only depends on the coupling constants. It can be computed by evaluating (8.2.52) at a convenient value of λ (say, $\lambda = 0$ if $V(\lambda)$ is a polynomial). Since the effective action is evaluated for the density of eigenvalues that solves (8.2.47), one can simplify the expression to

$$F_0(t) = -\frac{t}{2}\int_C \rho_0(\lambda)V(\lambda)d\lambda - \frac{1}{2}t^2\xi(t). \tag{8.2.53}$$

It is convenient to introduce the effective potential on an eigenvalue as,

$$V_{\text{eff}}(\lambda) = V(\lambda) - 2t\int \rho_0(\lambda') \log|\lambda - \lambda'|d\lambda'. \tag{8.2.54}$$

This is of course the continuum counterpart of (8.2.43). In terms of this object, the saddle-point equation (8.2.52) says that the effective potential is *constant* on the interval $\mathcal{C}$:

$$V_{\text{eff}}(\lambda) = t\xi(t), \qquad \lambda \in \mathcal{C}. \tag{8.2.55}$$

Once $\rho_0(\lambda)$ is known, we can compute some useful correlation functions in the planar limit. For example,

$$\frac{1}{N}\langle \text{Tr } M^\ell \rangle = \int_\mathcal{C} \lambda^\ell \rho_0(\lambda) d\lambda. \tag{8.2.56}$$

We then see that the planar limit is characterized by a *classical* density $\rho_0(\lambda)$. In this limit, the quantum averages can be computed as moments of this density. This implements in a very concrete way the idea of the master field that we discussed in the context of QCD: in the case of matrix models, the master field is encoded in the function $\rho_0(\lambda)$.

To obtain an explicit expression for the density of eigenvalues we have to solve the saddle-point equation (8.2.47). This equation is a singular integral equation which has been studied in detail in other contexts of physics. A very elegant way to solve it is to introduce an auxiliary function called the *resolvent*. The resolvent is defined as a correlator in the matrix model:

$$\omega(p) = \frac{1}{N}\left\langle \text{Tr}\frac{1}{p - M} \right\rangle, \tag{8.2.57}$$

which is in fact a generating functional of the correlation functions (8.2.56):

$$\omega(p) = \frac{1}{N}\sum_{k=0}^{\infty}\langle \text{Tr}M^k \rangle p^{-k-1}. \tag{8.2.58}$$

Since it is a generating functional of connected correlators, it admits an expansion of the form (7.3.11), with $r = 1$. After taking into account the overall factor of $1/N$ in (8.2.57), we deduce that the resolvent has the genus expansion

$$\omega(p) = \sum_{g=0}^{\infty} g_s^{2g}\omega_g(p), \tag{8.2.59}$$

and the genus zero piece can be written in terms of the density of eigenvalues as

$$\omega_0(p) = \int \frac{\rho_0(\lambda)}{p - \lambda} d\lambda. \tag{8.2.60}$$

The genus zero resolvent (8.2.60) has three important properties. First of all, due to the normalization property of the density of eigenvalues (8.2.46), it has the asymptotic behavior

$$\omega_0(p) \sim \frac{1}{p}, \qquad p \to \infty. \tag{8.2.61}$$

Second, as a function of p, it is analytic on the whole complex plane, but it has a branch cut at the interval C. The discontinuity in crossing the branch cut can be computed by standard contour deformation arguments. We have

$$\omega_0(p + i\epsilon) = \int_{\mathbb{R}} \frac{\rho_0(\lambda)}{p + i\epsilon - \lambda} d\lambda = \int_{\mathbb{R}-i\epsilon} \frac{\rho_0(\lambda)}{p - \lambda} d\lambda$$

$$= P \int \frac{\rho_0(\lambda)}{p - \lambda} d\lambda + \int_{C_\epsilon} \frac{\rho_0(\lambda)}{p - \lambda} d\lambda, \tag{8.2.62}$$

where $0 < \epsilon \ll 1$, and C_ϵ is a semicircle contour around $\lambda = p$ in the lower half-plane, oriented counterclockwise. This can be evaluated as a residue, and we finally obtain,

$$\omega_0(p + i\epsilon) = P \int \frac{\rho_0(\lambda)}{p - \lambda} d\lambda - \pi i \rho_0(p). \tag{8.2.63}$$

Similarly,

$$\omega_0(p - i\epsilon) = \int_{\mathbb{R}+i\epsilon} \frac{\rho_0(\lambda)}{p - \lambda} d\lambda = P \int \frac{\rho_0(\lambda)}{p - \lambda} d\lambda + \pi i \rho_0(p). \tag{8.2.64}$$

One then finds the key equation

$$\rho_0(\lambda) = -\frac{1}{2\pi i} \left(\omega_0(\lambda + i\epsilon) - \omega_0(\lambda - i\epsilon) \right). \tag{8.2.65}$$

Another way to derive this equation is to start from the definition of $\rho(\lambda)$ and use the basic identity

$$\delta(x) = \lim_{\epsilon \to 0} \frac{1}{\pi} \frac{\epsilon}{x^2 + \epsilon^2} = \lim_{\epsilon \to 0} \text{Im} \frac{1}{\pi} \frac{1}{x - i\epsilon}. \tag{8.2.66}$$

This leads to

$$\rho(\lambda) = \frac{1}{\pi} \text{Im} \frac{1}{N} \left\langle \text{Tr} \frac{1}{\lambda - M - i\epsilon} \right\rangle, \tag{8.2.67}$$

and from the definition of the resolvent we just find,

$$\rho(\lambda) = -\frac{1}{2\pi i} \left(\omega(\lambda + i\epsilon) - \omega(\lambda - i\epsilon) \right) = -\frac{1}{\pi} \text{Im} \, \omega(\lambda), \tag{8.2.68}$$

which is an equation to all orders in N. This shows that the density of eigenvalues also has a $1/N$ expansion, whose large N limit is the function $\rho_0(\lambda)$.

From these equations we deduce that, if the resolvent at genus zero is known, the planar density of eigenvalues follows from (8.2.65), and one can compute the planar free energy. On the other hand, by using again (8.2.62) and (8.2.64) we can compute

$$\omega_0(p+i\epsilon)+\omega_0(p-i\epsilon) = 2P\int \frac{\rho(\lambda)}{p-\lambda}d\lambda, \qquad (8.2.69)$$

and we then find the equation

$$\omega_0(p+i\epsilon)+\omega_0(p-i\epsilon) = \frac{1}{t}V'(p), \qquad (8.2.70)$$

which determines the resolvent in terms of the potential. In this way we have reduced the original problem of computing $F_0(t)$ to the problem of computing $\omega_0(\lambda)$. There is in fact a closed form expression for the planar resolvent which is very useful and valid for a very general class of potentials, not only polynomial,

$$\omega_0(p) = \frac{1}{2t}\oint_C \frac{dz}{2\pi i}\frac{V'(z)}{p-z}\left(\frac{(p-a)(p-b)}{(z-a)(z-b)}\right)^{1/2}, \qquad (8.2.71)$$

where C denotes a closed contour around the interval $[b,a]$. This equation is easily proved by converting (8.2.70) into a discontinuity equation:

$$\widehat{\omega}_0(p+i\epsilon)-\widehat{\omega}_0(p-i\epsilon) = \frac{1}{t}\frac{V'(p)}{\sqrt{(p-a)(p-b)}}, \qquad (8.2.72)$$

where $\widehat{\omega}_0(p) = \omega_0(p)/\sqrt{(p-a)(p-b)}$. The discontinuity equation determines $\omega_0(p)$ to be given by (8.2.71) up to an analytic function of p, but because of the asymptotics (8.2.61), this function must vanish. The asymptotics of $\omega_0(p)$ also gives two more conditions. By taking $p\to\infty$, one finds that the right hand side of (8.2.71) behaves like $c+d/p+\mathcal{O}(1/p^2)$. Requiring the asymptotic behavior (8.2.61) imposes $c=0$ and $d=1$, and this leads to

$$\oint_C \frac{dz}{2\pi i}\frac{V'(z)}{\sqrt{(z-a)(z-b)}} = 0,$$
$$\oint_C \frac{dz}{2\pi i}\frac{zV'(z)}{\sqrt{(z-a)(z-b)}} = 2t. \qquad (8.2.73)$$

These equations are enough to determine the endpoints of the cuts, a and b, as functions of the 't Hooft parameter t and the coupling constants of the model.

When $V(z)$ is a polynomial, one can find a very convenient expression for the resolvent: if we deform the contour in (8.2.71) to infinity, we pick up a pole at $z=p$, and another one at infinity, and we get

$$\omega_0(p) = \frac{1}{2t}\left(V'(p)-M(p)\sqrt{(p-a)(p-b)}\right), \qquad (8.2.74)$$

where the *moment function*

$$M(p) = \oint_\infty \frac{dz}{2\pi i} \frac{V'(z)}{z-p} \frac{1}{\sqrt{(z-a)(z-b)}}, \tag{8.2.75}$$

can be written as a contour integral around $z = 0$

$$M(p) = \oint_0 \frac{dz}{2\pi i} \frac{V'(1/z)}{1-pz} \frac{1}{\sqrt{(1-az)(1-bz)}}. \tag{8.2.76}$$

These formulae, together with the expressions (8.2.73) for the endpoints of the cut, completely solve the one-matrix model with one cut in the planar limit, for polynomial potentials. The function

$$y(p) = M(p)\sqrt{(p-a)(p-b)} \tag{8.2.77}$$

appearing in (8.2.74) is sometimes called the *spectral curve* of the matrix model, since it has the form of an equation for an algebraic curve.

Example 8.1 *The Gaussian matrix model.* Let us now apply this technology to the simplest case, the Gaussian model with $V(M) = M^2/2$. Let us first look for the position of the endpoints from (8.2.73). Deforming the contour to infinity and changing $z \rightarrow 1/z$, we find that the first equation in (8.2.73) becomes

$$\oint_0 \frac{dz}{2\pi i} \frac{1}{z^2} \frac{1}{\sqrt{(1-az)(1-bz)}} = 0, \tag{8.2.78}$$

where the contour is now around $z = 0$. Therefore $a + b = 0$, in accord with the symmetry of the potential. Taking this into account, the second equation becomes:

$$\oint_0 \frac{dz}{2\pi i} \frac{1}{z^3} \frac{1}{\sqrt{1-a^2z^2}} = 2t, \tag{8.2.79}$$

and gives

$$a = 2\sqrt{t}. \tag{8.2.80}$$

We see that the support of the density of eigenvalues $[-a, a] = [-2\sqrt{t}, 2\sqrt{t}]$ opens as the 't Hooft parameter grows, and as $t \rightarrow 0$ it collapses to the minimum of the potential at the origin, as expected. We immediately find from (8.2.74)

$$\omega_0(p) = \frac{1}{2t}\left(p - \sqrt{p^2 - 4t}\right), \tag{8.2.81}$$

and from the discontinuity equation we derive the density of eigenvalues

$$\rho_0(\lambda) = \frac{1}{2\pi t}\sqrt{4t - \lambda^2}. \tag{8.2.82}$$

The graph of this function is a semicircle of radius $2\sqrt{t}$, and the above eigenvalue density gives the famous *Wigner–Dyson semicircle law*. The spectral curve (8.2.77) is in this case

$$y^2 = p^2 - 4t. \tag{8.2.83}$$

Once we know $\rho_0(\lambda)$, we can compute $\xi(t)$. Evaluating (8.2.55) at $\lambda = 0$, we find

$$\xi(t) = -2 \int d\lambda \rho(\lambda) \log |\lambda| = \frac{1 - \log t}{2} \tag{8.2.84}$$

and

$$-\frac{t}{4} \int d\lambda \rho(\lambda)\lambda^2 = -\frac{t^2}{4}. \tag{8.2.85}$$

Therefore,

$$F_0(t) = \frac{1}{2}t^2 \log t - \frac{3}{4}t^2. \tag{8.2.86}$$

This is in precise agreement with the first line of (8.2.39). $\qquad\square$

Example 8.2 We now consider the so-called quartic matrix model, with potential

$$V(z) = \frac{1}{2}z^2 + gz^4. \tag{8.2.87}$$

In this case, it is customary to write the support of the density of eigenvalues as $[-2a, 2a]$. The resolvent can be easily computed from (8.2.71) and reads,

$$\omega_0(z) = \frac{1}{2t} \left(z + 4gz^3 - \left(1 + 8ga^2 + 4gz^2\right)\sqrt{z^2 - 4a^2}\right). \tag{8.2.88}$$

The density of eigenvalues is given by the discontinuity of this function,

$$\rho_0(\lambda) = \frac{1}{2\pi t}\left(1 + 8ga^2 + 4g\lambda^2\right)\sqrt{4a^2 - \lambda^2}. \tag{8.2.89}$$

In order to determine the position of the endpoints as a function of g, we notice that

$$\oint_C \frac{dz}{2\pi i} \frac{zV'(z)}{\sqrt{z^2 - 4a^2}} = 2(a^2 + 12a^4g), \tag{8.2.90}$$

and equating this to $2t$ we find,

$$a^2 = \frac{1}{24g}\left(-1 + \sqrt{1 + 48gt}\right). \tag{8.2.91}$$

As expected, $a \to 0$ as $t \to 0$. The spectral curve is given by

$$y(z) = \left(1 + 8ga^2 + 4gz^2\right)\sqrt{z^2 - 4a^2}. \tag{8.2.92}$$

The free energy at genus zero can be computed by using (8.2.53). First of all, we have

$$-\frac{t}{2}\int_{\mathcal{C}} d\lambda \, \rho(\lambda)V(\lambda) = -\frac{1}{4}a^4(1 + 20a^2 g + 72a^4 g^2). \tag{8.2.93}$$

In order to determine the integration constant ξ, we evaluate (8.2.52) at $\lambda = 0$ to obtain

$$-t^2\frac{\xi}{2} = t^2 \int_{\mathcal{C}} d\lambda \, \rho(\lambda) \log|\lambda| = \frac{ta^2}{2}\left(-1 - 6a^2 g + \left(2 + 24a^2 g\right)\log(a)\right). \tag{8.2.94}$$

Adding both contributions and using (8.2.91) one finds,

$$F_0(g) = -\frac{1}{24}(9t^2 + 10ta^2 - a^4) + \frac{t^2}{2}\log a^2. \tag{8.2.95}$$

It is useful however to subtract the part of the free energy that corresponds to the Gaussian model. This is obtained by evaluating the above quantity at $g = 0$, i.e. at $a^2 = t$, which agrees exactly with the Gaussian limit (8.2.39), and one finds:

$$\mathcal{F}_0(g) = F_0(g) - F_0(g = 0) = -\frac{1}{24}(a^2 - t)(9t - a^2) + \frac{t^2}{2}\log\frac{a^2}{t}. \tag{8.2.96}$$

This function has the following expansion around $t = 0$,

$$\mathcal{F}_0(g) = -t^2 f_0(gt), \tag{8.2.97}$$

where

$$f_0(z) = \sum_{k=1}^{\infty} a_k z^k, \qquad a_k = (-12)^k \frac{(2k-1)!}{k!(k+2)!}, \tag{8.2.98}$$

and the first few terms in the expansion of $\mathcal{F}_0(g)$ are

$$\mathcal{F}_0(g) = -t^2 \left(2\,(gt) - 18\,(gt)^2 + 288\,(gt)^3 + \mathcal{O}(g^4)\right). \tag{8.2.99}$$

The result for the coefficient a_k can be derived from (8.2.91) and (8.2.96) by using for example the Lagrange inversion formula, and it can be directly checked against perturbative calculations. For example, it is easy to see, by doing perturbation theory, that

$$F - F_G = -\frac{g}{g_s}\langle \mathrm{Tr}\, M^4\rangle_G + \frac{g^2}{2g_s^2}\langle(\mathrm{Tr}\, M^4)^2\rangle_G^{(c)} + \mathcal{O}(g^3)$$
$$= -gg_s(2N^3 + N) + g^2 g_s^2 \left(18N^4 + 30N^2\right) + \mathcal{O}(g^3). \tag{8.2.100}$$

The term proportional to gN^3 comes from the planar diagram in the first line of Fig. 8.1. There are two possible contractions which preserve planarity, and this leads to the coefficient -2 in the first term of the right hand side of (8.2.99). The term proportional to $g^2 N^4$ comes from two different planar diagrams, shown in the

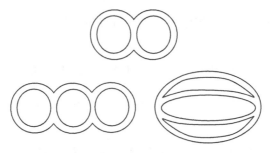

Figure 8.1 The planar diagrams contributing to the free energy of the quartic matrix model, up to order g^2.

second line of Fig. 8.1. For the first one there are 32 possible contractions, while for the second one there four possible contractions, and this leads to the coefficient $18 = (32 + 4)/2$ in the second term of the series in (8.2.99). ☐

There are many possible generalizations of the simple one-matrix model with one cut that we have considered in this section. One can consider, for example, situations in which the support of the density of eigenvalues is a disjoint union of intervals in the real axis (or, even more generally, of segments in the complex plane). This is the so-called multi-cut case, and it has many important applications. One can also consider multi-matrix models, involving more than one matrix. Finally, we have given an explicit solution just for the planar free energy $F_0(t)$, but in fact there are systematic procedures to compute the full $1/N$ expansion of the free energy and of a large class of correlation functions. The interested reader can find a guide to these developments in the bibliographical notes at the end of this chapter.

8.3 Unitary matrix models

In the previous section we have considered matrix models based on Hermitian matrices. It is also useful to consider models based on unitary matrices. These arise naturally, for example, when considering lattice gauge theories, in which the basic variables are unitary matrices. In fact, in two dimensions, one can reduce YM theory on the lattice to a simple unitary matrix model.

Let us then consider matrix models based on a single unitary matrix $U \in U(N)$, and defined by the partition function

$$Z = \int dU \exp(S(U)), \tag{8.3.1}$$

where the action $S(U)$ is of the form

$$S(U) = N \sum_{l \geq 1} \left(g_l \text{tr}\, U^l + \overline{g}_l \text{tr}\, U^{\dagger l} \right). \tag{8.3.2}$$

We will set

$$g_l = \frac{1}{2l}(\beta_l - i\gamma_l). \tag{8.3.3}$$

As in the case of Hermitian matrix models, we can use the gauge symmetry of the model,

$$U \rightarrow VUV^\dagger \tag{8.3.4}$$

to write it in terms of the eigenvalues $e^{i\phi_i}$ of U, where $\phi_i \in [-\pi, \pi], i = 1, \ldots, N$. A computation similar to the one we did in the Hermitian case leads to

$$Z = \int \prod_{i=1}^N d\phi_i \, e^{NS(\phi_i)} \prod_{i<j} 4\sin^2\left(\frac{\phi_i - \phi_j}{2}\right), \tag{8.3.5}$$

up to an overall constant. In this equation,

$$S(\phi) = \sum_{l \geq 1}\left(\frac{\beta_l}{l}\cos l\phi + \frac{\gamma_l}{l}\sin l\phi\right). \tag{8.3.6}$$

We can also write, as in (8.2.40),

$$Z = \int \prod_i d\phi_i e^{N \sum_{i=1}^N S_{\text{eff}}(\phi_i)}, \tag{8.3.7}$$

where

$$S_{\text{eff}}(\phi_i) = S(\phi_i) + \frac{1}{2N}\sum_{j \neq i}\log\left[4\sin^2\left(\frac{\phi_i - \phi_j}{2}\right)\right]. \tag{8.3.8}$$

We can now analyze the planar limit of this model by mimicking what we did in the Hermitian case. At large N, the matrix model is described by a density of eigenvalues $\rho_0(\phi)$ which verifies the normalization condition

$$\int_{-\pi}^{\pi}\rho_0(\phi)d\phi = 1. \tag{8.3.9}$$

In terms of this density, the planar free energy, defined by

$$F_0 = \lim_{N \to \infty}\frac{1}{N^2}\log Z, \tag{8.3.10}$$

is given by

$$F_0 = \frac{1}{2}\int_{-\pi}^{\pi}d\phi\int_{-\pi}^{\pi}d\psi \rho_0(\phi)\rho_0(\psi)\log\left[4\sin^2\left(\frac{\phi - \psi}{2}\right)\right]$$
$$+ \int_{-\pi}^{\pi}d\phi\rho(\phi)S(\phi) + \xi\left(\int_{-\pi}^{\pi}\rho_0(\phi)d\phi - 1\right), \tag{8.3.11}$$

where ξ is a Lagrange multiplier which imposes the constraint (8.3.9). Taking a functional variation with respect to ρ_0 gives the equation

$$S(\phi) + \int_{-\pi}^{\pi} d\psi \, \rho_0(\psi) \log\left[4\sin^2\left(\frac{\phi - \psi}{2}\right)\right] + \xi = 0, \qquad (8.3.12)$$

and by acting with a derivative with respect to ϕ we obtain the analogue of (8.2.47)

$$S'(\phi) + \int_{-\pi}^{\pi} d\psi \rho_0(\psi) \cot\left(\frac{\phi - \psi}{2}\right) = 0. \qquad (8.3.13)$$

We can also define an effective potential,

$$V_{\text{eff}}(\phi) = -S(\phi) - \int d\psi \, \rho_0(\psi) \log\left[4\sin^2\left(\frac{\phi - \psi}{2}\right)\right], \qquad (8.3.14)$$

which due to (8.3.12) is constant on the support of ρ_0. Once we have found the equilibrium density, the planar free energy is obtained by computing (8.3.11) for this density. Using (8.3.12) we can simplify the computation to

$$F_0 = \frac{1}{2} \int_{-\pi}^{\pi} d\phi \rho_0(\phi) S(\phi) - \frac{\xi}{2}. \qquad (8.3.15)$$

To find an explicit expression for the density of eigenvalues, we introduce a planar resolvent, as in the Hermitian case

$$\omega_0(z) = \frac{1}{2} \int d\psi \, \rho_0(\psi) \cot\left(\frac{z - \psi}{2}\right). \qquad (8.3.16)$$

This resolvent has the following properties: first of all, it is a periodic function

$$\omega_0(z + 2\pi) = \omega_0(z). \qquad (8.3.17)$$

Second, as in the Hermitian case, it has a discontinuity along the support of $\rho_0(z)$,

$$\omega_0(z \pm i\epsilon) = -\frac{1}{2} S'(z) \mp \pi i \rho_0(z). \qquad (8.3.18)$$

Finally, for any z with $\text{Im } z \neq 0$, one has

$$\omega_0(z) \to \mp \frac{i}{2}, \qquad |\text{Im } z| \to \infty. \qquad (8.3.19)$$

The sign depends on whether $\text{Im } z$ is larger or smaller than zero, respectively. This last property follows from the fact that

$$\cot\left(\frac{z - \psi}{2}\right) = -i\frac{e^{-i\phi_\psi} + e^{-\text{Im } z + i\phi_\psi}}{e^{-i\phi_\psi} - e^{-\text{Im } z + i\phi_\psi}}, \qquad (8.3.20)$$

where

$$\phi_\psi = \frac{1}{2}(\text{Re } z - \psi). \qquad (8.3.21)$$

As in the Hermitian case, the planar resolvent gives information about vevs of single trace operators $\text{Tr}\, U^n$. To see this, we write

$$\lambda = e^{iz}, \qquad \mu = e^{i\phi}, \tag{8.3.22}$$

so that

$$\cot\left(\frac{z-\phi}{2}\right) = i + \frac{2i\mu}{\lambda - \mu}, \tag{8.3.23}$$

and

$$\omega_0(\lambda) = \frac{i}{2} + i\int d\phi \frac{\rho_0(\mu)\mu}{\lambda - \mu} = \frac{i}{2} + i\sum_{n\geq 1}\lambda^{-n}\int d\phi\, \rho_0(\mu)\mu^n, \tag{8.3.24}$$

therefore $\omega_0(\lambda)$ is, up to a factor i, the generating functional of the vevs of traces of powers of U,

$$\omega_0(\lambda) = \frac{i}{2} + i\sum_{n\geq 1}\lambda^{-n}\left\langle \frac{1}{N}\text{Tr}\, U^n\right\rangle. \tag{8.3.25}$$

Let us now try to solve the planar limit of this theory. If we use the Fourier expansion,

$$\cot\left(\frac{\phi-\psi}{2}\right) = \sum_{n\geq 1}(\sin(n\phi)\cos(n\psi) - \cos(n\phi)\sin(n\psi)), \tag{8.3.26}$$

we immediately find that

$$\rho_0(\phi) = \frac{1}{2\pi} + \frac{1}{2\pi}\sum_{l\geq 1}(\beta_l \cos l\phi + \gamma_l \sin l\phi) \tag{8.3.27}$$

solves (8.3.13). The corresponding planar free energy can be computed by using (8.3.15). The constant ξ can be found by evaluating (8.3.12) at $\phi = 0$. This gives

$$\xi = -S(0) - \int_0^{2\pi} d\psi \rho_0(\psi) \log\left[4\sin^2\frac{\psi}{2}\right]. \tag{8.3.28}$$

If we use the Fourier series

$$\log\left[4\sin^2\frac{\psi}{2}\right] = -2\sum_{n\geq 1}\frac{1}{n}\cos(n\psi), \tag{8.3.29}$$

we find that $\xi = 0$, therefore

$$F_0 = \sum_{l\geq 1}\frac{1}{4l}(\beta_l^2 + \gamma_l^2). \tag{8.3.30}$$

It can be shown that there are no further corrections to this result in the $1/N$ expansion. If this were the whole story, the unitary matrix model would be exceedingly simple. However, it was noticed by Gross and Witten, and by Wadia, that the density (8.3.27) cannot be the solution for all values of the coupling. This is simply due to the fact that, for general values of the couplings β_l, γ_l, (8.3.27) is not positive, which is a necessary property of densities of eigenvalues. This means that, for a certain range of the couplings, the density must have a different functional form. In other words, there must be different *phases* for the general unitary matrix model, described by different functional forms of the density of eigenvalues (and, correspondingly, with different forms for the free energies).

In order to understand this issue in more detail, let us focus on a simple model, first studied by Gross and Witten, and Wadia, which we will call the Gross–Witten–Wadia (GWW) model. This is the unitary matrix model with action

$$S(\phi) = \beta \cos \phi. \tag{8.3.31}$$

Notice that β, which we will take by convention to be a positive parameter, can be regarded as the inverse of the 't Hooft parameter. The density of eigenvalues (8.3.27) reduces to

$$\rho_0(\phi) = \frac{1}{2\pi}(1 + \beta \cos \phi). \tag{8.3.32}$$

This function is positive as long as $\beta < 1$, and therefore, in that range, it describes appropriately the planar limit of the model. The planar free energy in this region of parameter space is given by the specialization of (8.3.30), i.e.

$$F_0^s(\beta) = \frac{\beta^2}{4}, \qquad \beta < 1. \tag{8.3.33}$$

The superscript s indicates that, in terms of the 't Hooft parameter, $\beta < 1$ is the strong coupling region. For $\beta = 1$, (8.3.32) vanishes at $\phi = \pi$, and for $\beta > 1$ it becomes negative around $\phi = \pi$. This is clearly not acceptable for a density, and as we just explained, it indicates that there must be a phase transition at $\beta = 1$. In order to compute the resolvent for $\beta \geq 1$, we look for a function $\omega_0(z)$ satisfying the properties listed above. It is easy to see that, in the case of the action (8.3.31),

$$\omega_0(z) = \frac{\beta}{2}\left(\sin z - 2\cos\frac{z}{2}\sqrt{\sin^2\frac{z}{2} - \sin^2\frac{\alpha_c}{2}}\right) \tag{8.3.34}$$

satisfies all the required properties. To determine α_c, we write

$$\sin z = \frac{1}{2i}(u - u^{-1}), \tag{8.3.35}$$

and consider the limit $u \to +\infty$, which corresponds to $z \to -i\infty$. In this limit we have

$$\omega_0(z) \approx \frac{i}{2}\beta \sin^2\frac{\alpha_c}{2} + \mathcal{O}(u^{-1}) \tag{8.3.36}$$

and the asympotic behavior (8.3.19) imposes

$$\sin^2 \frac{\alpha_c}{2} = \frac{1}{\beta}. \tag{8.3.37}$$

The resulting density of eigenvalues is

$$\rho_0(\phi) = \frac{\beta}{\pi} \cos\left(\frac{\phi}{2}\right) \sqrt{\frac{1}{\beta} - \sin^2 \frac{\phi}{2}}, \qquad \beta \geq 1. \tag{8.3.38}$$

The planar free energy can be obtained from (8.3.15). The integrals appearing in this calculation can be evaluated by performing the change of variable $x = \sin(\phi/2)$ and by using that

$$\int_0^c (1 - 2x^2)\sqrt{c^2 - x^2}\, dx = \frac{\pi}{8} c^2 (2 - c^2),$$

$$\int_0^c \log(2x)\sqrt{c^2 - x^2}\, dx = \frac{\pi}{8} c^2 \left(\log(c^2) - 1\right). \tag{8.3.39}$$

Evaluating (8.3.12) for $\phi = 0$, we find

$$\xi = 1 + \log \beta - \beta, \tag{8.3.40}$$

and the planar free energy is given by

$$F_0^{\mathrm{w}}(\beta) = \beta - \frac{1}{2} \log \beta - \frac{3}{4}, \qquad \beta > 1. \tag{8.3.41}$$

The superscript w means that this corresponds to the weak coupling region of the 't Hooft parameter.

We conclude that, at large N, the unitary matrix model undergoes a phase transition as a function of the inverse 't Hooft parameter β. This transition is often called the Gross–Witten–Wadia (GWW) phase transition. In the strong coupling phase $\beta < 1$, the density of eigenvalues has the form (8.3.32) and it fills the full circle. For this reason, this is called the *ungapped phase*. In the weak coupling phase $\beta > 1$, the support of the density of eigenvalues is strictly smaller than the circle, since $\alpha_c < \pi$, and a "gap" appears. For this reason, this is sometimes called the *gapped phase*. The behavior of the density of eigenvalues in the different phases is shown schematically in Fig. 8.2. It is easy to see that the planar free energy, as well as its first and second derivatives, are continuous at $\beta = 1$, but the third derivative is discontinuous. The phase transition at $\beta = 1$ is therefore of third order. As we will see in Chapter 10, this phase transition can be thought of as triggered by large N instantons of the unitary matrix model.

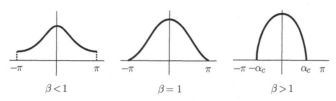

Figure 8.2 The density of eigenvalues $\rho_0(\theta)$ as we go through the GWW phase transition.

8.4 Matrix Quantum Mechanics

In the previous sections we have studied matrix models, which can be regarded as toy models for YM theory in zero dimensions. In these toy models, only the matrix structure of the gauge connection is kept. This represents a vast simplification of the problem, and we have shown that the planar vacuum diagrams of the theory can be resummed into a single function, and the master field can be explicitly constructed in terms of the density of eigenvalues.

We will now consider a different toy model for the $1/N$ expansion, this time in one dimension. This is a quantum mechanical model where the degrees of freedom are the time dependent entries of a Hermitian $N \times N$ matrix $M(t)$, and it is sometimes called *matrix Quantum Mechanics* (MQM). On top of its pedagogical value as a simplified version of gauge theories at large N, different versions of MQM appear in various other contexts in theoretical physics. They provide for example a discretized version of string theories in one dimension, and they appear as dimensional reductions of more complicated gauge theories.

MQM is described by the Euclidean Lagrangian

$$L_{\mathrm{E}} = \mathrm{Tr}\left[\frac{1}{2}\dot{M}^2 + V(M)\right], \tag{8.4.1}$$

where $V(M)$ is a polynomial in M. This Lagrangian can be regarded as a generalization of the standard QM of a one-dimensional particle in a polynomial potential, or as a one-dimensional field theory for a quantum field $M(t)$, taking values in the adjoint representation of $U(N)$. It also has the gauge symmetry (8.2.3), where U is a constant unitary matrix. We will assume that the potential $V(M)$ is of the form

$$V(M) = \frac{1}{2}M^2 + V_{\mathrm{int}}(M), \tag{8.4.2}$$

where $V_{\mathrm{int}}(M)$ is an interaction term of the form

$$V_{\mathrm{int}}(M) = \sum_{p \geq 3}\frac{g_p}{pN^{p/2-1}}M^p. \tag{8.4.3}$$

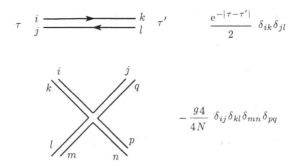

$$\frac{e^{-|\tau-\tau'|}}{2}\,\delta_{ik}\delta_{jl}$$

$$-\frac{g_4}{4N}\,\delta_{ij}\delta_{kl}\delta_{mn}\delta_{pq}$$

Figure 8.3 The Feynman rules for MQM with a quartic interaction.

In this expression, g_p are coupling constants, and the factors of N are introduced in such a way that we have a well-defined large N limit, as we will see in a moment. We can now study this one-dimensional field theory by doing perturbation theory in the coupling constants g_p. This is very similar to what we did in Section 1.2. The Feynman rules are as in ordinary one-dimensional QM, with the only difference that we now have "group factors," due to the fact that M is matrix valued. The propagator of MQM is

$$\frac{e^{-|\tau|}}{2}\,\delta_{ik}\delta_{jl}. \tag{8.4.4}$$

For a theory with a quartic interaction

$$V_{\text{int}}(M) = \frac{g_4}{4N}M^4, \tag{8.4.5}$$

the Feynman rules are illustrated in Fig. 8.3. For the interaction due to the term M^p in (8.4.3), the corresponding vertex has p legs. One can use these rules to compute the perturbation series of the ground state energy of MQM, exactly as we did in Section 1.2. Each Feynman diagram now has a group factor which depends on N. As in the case of QCD, this means that each conventional Feynman diagram gives rise to various fatgraphs. For example, in the case of a quartic interaction, the relevant Feynman diagrams, up to order g_4^3, were already shown in Fig. 1.3, and the planar fatgraphs corresponding to the diagrams 1, 2a and 2b are those depicted in Fig. 8.1.

Let $\Gamma_{g,h}$ be a vacuum-to-vacuum connected fatgraph appearing in the perturbative series for the ground state energy, and let V_p denote the number of vertices with p legs in this fatgraph. This fatgraph leads to a factor

$$N^{h+\sum_{p\geq3}(1-p/2)V_p}\prod_p g_p^{V_p} = N^{2-2g}\prod_p g_p^{V_p}, \tag{8.4.6}$$

Table 8.1 *Multiplicities of the planar quartic diagrams shown in Fig. 1.3*

Diagram	1	2a	2b	3a	3b	3c	3d
Multiplicities	2	16	2	256/3	32/3	64	128

where we have used Euler's relation (7.2.24) as well as

$$\sum_{p\geq3} V_p = V, \qquad \sum_{p\geq3} pV_p = 2E. \tag{8.4.7}$$

We see that the factors of N in (8.4.3) have been chosen in such a way that the fatgraphs have the standard 't Hooft weight (7.2.25) (with $b = 0$, since in this case there are no fields in the fundamental representation), and we conclude that the ground state energy has the $1/N$ expansion

$$E(g_p, N) = \sum_{g=0}^{\infty} N^{2-2g} \mathcal{E}_g(g_p). \tag{8.4.8}$$

We will call $\mathcal{E}_g(g_p)$ the genus g ground state energy. It is given by a sum over connected fatgraphs with a fixed genus g. For example, in the case of the quartic interaction (8.4.5) we have

$$\mathcal{E}_g(g_4) = \sum_{h=0}^{\infty} a_{g,h} \left(\frac{g_4}{4}\right)^{h+2g-2}, \tag{8.4.9}$$

and the coefficient $a_{g,h}$ is

$$a_{g,h} = \sum_{\Gamma_{g,h}} s_{\Gamma_{g,h}} \mathcal{I}_\Gamma. \tag{8.4.10}$$

In this expression, the sum is over connected fatgraphs $\Gamma_{g,h}$ with genus g and h boundaries, $s_{\Gamma_{g,h}}$ is the corresponding multiplicity, and $\mathcal{I}_\Gamma$ is the Feynman integral associated to the underlying, standard Feynman diagram, which is the same one appearing in (1.2.19).

Let us illustrate these considerations with an explicit calculation, again in the theory with a quartic interaction (8.4.5). The first few planar diagrams contributing to the genus zero ground state energy $\mathcal{E}_0(g_4)$ are obtained by thickening the graphs shown in Fig. 1.3, and their multiplicities are shown in Table 8.1 (some of them were discussed when considering the quartic matrix model in the previous section). Using these multiplicities, we can now compute the first corrections to the planar ground state energy. The Feynman integrals were already computed in (1.2.24) for conventional QM. Putting them together with the multiplicities, we obtain

$$\mathcal{E}_0(g_4) = \frac{1}{2} + \frac{1}{8}g_4 - \frac{17}{256}g_4^2 + \frac{75}{1024}g_4^3 + \cdots .$$

(8.4.11)

As in the case of matrix models, we should ask whether we can resum in closed form all the planar diagrams of MQM contributing to the ground state energy. Remarkably, this problem was also solved by Brézin, Itzykson, Parisi and Zuber by using a free fermion formulation. This goes as follows. After quantization of the system we obtain a Hamiltonian operator

$$H = \mathrm{Tr}\left[-\frac{1}{2}\frac{\partial^2}{\partial M^2} + V(M)\right],$$

(8.4.12)

where

$$\mathrm{Tr}\frac{\partial^2}{\partial M^2} = \sum_{a,b=1}^{N}\frac{\partial^2}{\partial M_{ab}\partial M_{ba}}.$$

(8.4.13)

In order to study the spectrum of this Hamiltonian, it is useful to exploit the underlying $U(N)$ symmetry and write M as a unitary transformation of a diagonal matrix,

$$M = U^\dagger \Lambda U,$$

(8.4.14)

where

$$\Lambda = \mathrm{diag}(\lambda_1, \lambda_2, \dots, \lambda_N).$$

(8.4.15)

The eigenvalue variables λ_i, $i = 1, \dots, N$, are invariant under the $U(N)$ symmetry, up to the action of the Weyl group, which acts as the permutation group of the N eigenvalues. We can then regard them as "radial variables," while the matrix U encodes the "angular" variables. We expect the ground state of the system to be invariant under the global $U(N)$ symmetry. The wavefunction of a $U(N)$ invariant state should only depend on the eigenvalues λ_i and in addition it should be invariant under the Weyl group. Therefore, it is represented by a *symmetric* function of the N eigenvalues,

$$\Psi(\lambda_i).$$

(8.4.16)

Such states are called "singlet" states. Let us now show that, when acting on singlet states, the differential operator (8.4.13) has the form,

$$-\frac{1}{2}\mathrm{Tr}\frac{\partial^2}{\partial M^2} = -\frac{1}{2}\frac{1}{\Delta(\lambda)}\sum_{a=1}^{N}\left(\frac{\partial}{\partial\lambda_a}\right)^2\Delta(\lambda),$$

(8.4.17)

where $\Delta(\lambda)$ is the Vandermonde determinant (8.2.27). To see this, we first note that we can write the matrix U in (8.4.14) in terms of the normalized eigenfunctions of M. If

$$M_{jk}u_k^a = \lambda_a u_j^a, \tag{8.4.18}$$

with

$$\left(u_i^a\right)^* u_i^b = \delta^{ab}, \tag{8.4.19}$$

then

$$U_{ij} = \left(u_j^i\right)^*. \tag{8.4.20}$$

Here, repeated indices are summed over. We now want to express the derivative with respect to M in terms of a derivative with respect to λ_a, i.e.

$$\frac{\partial}{\partial M_{ij}} = \sum_{a=1}^N \frac{\partial \lambda_a}{\partial M_{ij}} \frac{\partial}{\partial \lambda_a}. \tag{8.4.21}$$

To calculate the derivative appearing here, we use standard quantum mechanical perturbation theory at first order: under a small change in M, δM, we will have

$$\delta \lambda_a = \left(u_i^a\right)^* (\delta M)_{ij} u_j^a = U_{ai} (\delta M)_{ij} U_{ja}^\dagger, \tag{8.4.22}$$

and

$$\delta u_i^a = -\sum_{b \neq a} \frac{u_i^b \left(u_j^b\right)^* (\delta M)_{jk} u_k^a}{\lambda_b - \lambda_a}, \tag{8.4.23}$$

which leads to

$$\delta U_{ia}^\dagger = -\sum_{b \neq a} \frac{U_{ib}^\dagger U_{bj} (\delta M)_{jk} U_{ka}^\dagger}{\lambda_b - \lambda_a}. \tag{8.4.24}$$

We conclude from these equations that

$$\frac{\partial \lambda_a}{\partial M_{ij}} = U_{ai} U_{ja}^\dagger,$$

$$\frac{\partial U_{ia}^\dagger}{M_{jk}} = \sum_{b \neq a} \frac{U_{ib}^\dagger U_{bj} U_{ka}^\dagger}{\lambda_a - \lambda_b}, \tag{8.4.25}$$

$$\frac{\partial U_{ia}}{M_{jk}} = \sum_{b \neq a} \frac{U_{ij} U_{kb}^\dagger U_{ba}}{\lambda_i - \lambda_b}.$$

Therefore, going back to (8.4.22), we find that, when acting on singlet states,

$$\frac{\partial}{\partial M_{ij}} = \sum_{a=1}^N U_{ai} U_{ja}^\dagger \frac{\partial}{\partial \lambda_a}. \tag{8.4.26}$$

We can now calculate

$$\text{Tr}\frac{\partial^2}{\partial M^2} = \sum_{i,j=1}^{N} \frac{\partial^2}{\partial M_{ji}\partial M_{ij}}$$

$$= \sum_{i,j=1}^{N} \left\{ \left[\sum_{a=1}^{N} U_{aj}U_{ia}^{\dagger}\frac{\partial}{\partial\lambda_a} \right] \left[\sum_{b=1}^{N} U_{bi}U_{jb}^{\dagger}\frac{\partial}{\partial\lambda_b} \right] \right.$$

$$\left. + \sum_{b=1}^{N} \frac{\partial}{\partial M_{ji}}\left(U_{bi}U_{jb}^{\dagger}\right)\frac{\partial}{\partial\lambda_b} \right\}. \tag{8.4.27}$$

Since

$$\frac{\partial}{\partial M_{ji}}\left(U_{ai}U_{jb}^{\dagger}\right) = 2\delta_{ab}\sum_{c\neq a}\frac{1}{\lambda_a - \lambda_c}, \tag{8.4.28}$$

we finally obtain (8.4.17).

We now introduce a completely *antisymmetric* wavefunction

$$\Phi(\lambda) = \Delta(\lambda)\Psi(\lambda). \tag{8.4.29}$$

Due to (8.4.17), the Hamiltonian problem for a singlet state becomes

$$\left[-\frac{1}{2}\sum_{i=1}^{N}\frac{\partial^2}{\partial\lambda_i^2} + \sum_{i=1}^{N}V(\lambda_i) \right]\Phi(\lambda) = E\Phi(\lambda). \tag{8.4.30}$$

This is the Schrödinger equation for N non-interacting fermions (since the function (8.4.29) is completely antisymmetric) in the external potential $V(\lambda)$. To have a good large N limit, the potential $V(\lambda)$ should have good large N scaling properties, and as we will see in a moment this means that it can be obtained from a function $v(z)$ as

$$V(\lambda) = Nv\left(\frac{\lambda}{\sqrt{N}}\right), \tag{8.4.31}$$

where $v(\lambda)$ does not contain any N dependence. For the polynomial potentials considered above, this is equivalent to the N scaling of the coefficients in in (8.4.3), and we have

$$v(z) = \frac{1}{2}z^2 + \sum_{p\geq 3}\frac{g_p}{p}z^p. \tag{8.4.32}$$

After rescaling $\lambda \to \sqrt{N}\lambda$, the one-body fermion problem reduces to

$$\left\{ -\frac{1}{2N^2}\frac{d^2}{d\lambda^2} + v(\lambda) \right\}\phi_n(\lambda) = e_n\phi_n(\lambda), \tag{8.4.33}$$

where

$$e_n = \frac{1}{N} E_n, \qquad (8.4.34)$$

and E_n are the energy levels in the original one-body problem. The ground state energy is then given by

$$E(g_p, N) = \sum_{n=1}^{N} E_n = N \sum_{n=1}^{n} e_n = N^2 \mathcal{E}_0 + \cdots, \qquad (8.4.35)$$

where $\mathcal{E}_0$ is independent of N. To calculate $\mathcal{E}_0$ we note that, in this Schrödinger equation, $1/N$ plays the role of $\hbar$, and quantum effects are controlled by $1/N$. Therefore, as in previous examples, the large N limit is equivalent to the semiclassical limit, and we can use the Bohr–Sommerfeld quantization condition to find the energy spectrum at leading order in the $1/N$ expansion. We will write this condition as

$$N J(e_n) = n - \frac{1}{2}, \qquad n \geq 1, \qquad (8.4.36)$$

where

$$J(e) = \frac{1}{\pi} \int_{\lambda_1(e)}^{\lambda_2(e)} d\lambda \sqrt{2(e - V(\lambda))}, \qquad (8.4.37)$$

and $\lambda_{1,2}(e)$ are the turning points of the potential. If we denote

$$\xi = \frac{n - \frac{1}{2}}{N}, \qquad (8.4.38)$$

we see that (8.4.36) defines implicitly a function $e(\xi)$ through

$$J(e(\xi)) = \xi. \qquad (8.4.39)$$

At large N, the spectrum becomes denser and denser, and the variable ξ becomes a continuous variable:

$$\xi \in [0, 1]. \qquad (8.4.40)$$

Therefore, at large N, the sum in (8.4.35) can be approximated by an integral,

$$\sum_{n=1}^{N} \rightarrow N \int_0^1 d\xi, \qquad (8.4.41)$$

and from (8.4.35) we find

$$E(N) \approx N^2 \int_0^1 e(\xi) d\xi, \qquad (8.4.42)$$

which gives the planar ground state energy defined in (8.4.35),

$$\mathcal{E}_0 = \int_0^1 e(\xi) d\xi. \tag{8.4.43}$$

To evaluate this integral, we change variables from ξ to e. The Fermi energy of the system corresponds to the energy of the last fermion, i.e. $n = N$, and at large N it is defined by the condition

$$J(e_F) = 1. \tag{8.4.44}$$

Therefore, at large N, $\xi = 1$ corresponds to $e = e_F$, while $\xi = 0$ corresponds to the minimum value of the potential, $e = \min V(\lambda)$. We then find, by changing variables from ξ to e,

$$\mathcal{E}_0 = \int_{\min V(\lambda)}^{e_F} e J'(e) de, \tag{8.4.45}$$

where

$$J'(e) = \frac{1}{\pi} \int_{\lambda_1(e)}^{\lambda_2(e)} \frac{d\lambda}{\sqrt{2(e - V(\lambda))}}. \tag{8.4.46}$$

An easy calculation gives,

$$\mathcal{E}_0 = e_F - \frac{1}{3\pi} \int_{\lambda_1(e_F)}^{\lambda_2(e_F)} [2(e_F - V(\lambda))]^{3/2} d\lambda. \tag{8.4.47}$$

Example 8.3 As an application of the above formalism, let us consider again the potential with a quartic interaction term (8.4.5). We first compute the Fermi energy, which is defined by (8.4.44). The resulting integral can be computed in terms of elliptic functions. We first write,

$$2e - \lambda^2 - \frac{g_4}{2}\lambda^4 = \frac{g_4}{2}(a^2 - \lambda^2)(b^2 + \lambda^2), \tag{8.4.48}$$

where

$$a^2 = \frac{\sqrt{4eg_4 + 1} - 1}{g_4}, \qquad b^2 = \frac{\sqrt{4eg_4 + 1} + 1}{g_4}. \tag{8.4.49}$$

Note that, for $g_4 > 0$, $\pm a$ give the turning points for a particle of energy e in the quartic potential. We introduce the elliptic modulus

$$k^2 = \frac{a^2}{a^2 + b^2}. \tag{8.4.50}$$

Then, we have that

$$J(e) = \frac{1}{3\pi} (2g_4)^{1/2} (a^2 + b^2)^{1/2} [b^2 K(k) + (a^2 - b^2) E(k)], \tag{8.4.51}$$

where $K(k)$, $E(k)$ are complete elliptic integrals of the first and the second kind, respectively. The implicit function $e_F(g_4)$ is easy to compute as a power series in g_4:

$$e_F(g_4) = 1 + \frac{3g_4}{8} - \frac{17g_4^2}{64} + \frac{375g_4^3}{1024} + \mathcal{O}(g_4^4). \tag{8.4.52}$$

The planar free energy is given by

$$\mathcal{E}_0(g_4) = e_F(g_4) - \frac{1}{3\pi}\left(\frac{g_4}{2}\right)^{3/2} \mathcal{I}(g_4, e_F(g_4)), \tag{8.4.53}$$

and it involves the integral

$$\mathcal{I}(g_4, e) = \int_{-a}^{a} dt \left[(a^2 - t^2)(b^2 + t^2)\right]^{3/2}$$

$$= \frac{2}{35}\sqrt{a^2 + b^2}\left\{2(a^2 - b^2)(a^4 + 6a^2b^2 + b^4)E(k)\right.$$

$$\left. + b^2(2b^4 + 9a^2b^2 - a^4)K(k)\right\}. \tag{8.4.54}$$

The function $\mathcal{E}_0(g_4)$ can be easily expanded in powers of g_4, and one finds

$$\mathcal{E}_0(g_4) = \frac{1}{2} + \frac{g_4}{8} - \frac{17g_4^2}{256} + \frac{75g_4^3}{1024} - \frac{3563g_4^4}{32678} + \mathcal{O}(g_4^5). \tag{8.4.55}$$

The first terms agree with the calculation in planar perturbation theory (8.4.11), and the function (8.4.53) resums all the planar diagrams into an exact function of g_4. □

   In the toy model of MQM we can address quantitatively the question of how good is the planar limit, as an approximation to the system at finite N. To be concrete, we consider the quartic potential analyzed in the previous example, and as a finite N problem we will take the extreme case of $N = 1$. We can then compare the planar solution (8.4.53) to the ground state energy for a single particle in the quartic potential. It turns out that they are quite close, specially for small g_4. For example, for $g_4 = 0.4$, one finds

$$\mathcal{E}_0(g_4) \approx 0.542\ldots \tag{8.4.56}$$

to be compared to the numerical value $0.559\ldots$ for $N = 1$. By looking at the values of the energies over the whole range of values for g_4, one can see that the planar approximation is at most 12% incorrect. This is a relatively good agreement, and indicates that the large N limit might be a good approximation to the finite, small N case. As we mentioned in the last chapter, this seems to be the case also in QCD.

8.5 Bibliographical notes

An overview of the various applications of random matrices can be found in the books [7, 87, 142]. For introductions to matrix models and their large N expansion, one can see [68, 69, 135]. The planar solution to the Hermitian, one-matrix model was found in [40]. Subleading corrections to the $1/N$ expansion were studied in [13, 35, 84, 85]. The Faddeev–Popov approach to the gauge fixing of matrix integrals is explained for example in [35], which also shows how to compute the normalization factor $1/\mathrm{vol}(U(N))$. The calculation of the integral over eigenvalues with orthogonal polynomials can be found in [35], and the approach via Selberg integrals is summarized in [88], where a complete list of references can also be found.

The Gross–Witten–Wadia model and its large N phase transition was studied in [102, 183]. More general unitary matrix models and their phase structure are studied for example in [116].

Matrix QM was studied in [40], where the large N solution was also found. A generalization mimicking QCD, which includes fermions in the fundamental representation of $U(N)$, was studied in [4], which also gives the simple derivation of (8.4.17) that we have presented. A unitary version of matrix QM was studied in [184]. Applications of MQM to lower dimensional string theory are reviewed for example in [94].

9

Large N Quantum Chromodynamics in two dimensions

9.1 Introduction

In the toy models of the preceding chapter, we have studied theories in zero dimensions where the only degrees of freedom were matrix valued, i.e. gluon-like. However, some of the important insights of the large N expansion in QCD concern the quark sector, for example the properties of mesons. Is there a toy model where we can see quantitatively how the large N expansion works in this sector? Such a model was proposed by 't Hooft shortly after his pioneering paper on the $1/N$ expansion, and it is QCD in two dimensions. In this model, the pure gluon sector is now trivial, but the interaction between the quark sector and the gluon sector is very rich. 't Hooft noted that the planar diagrams for the quark–gluon interaction could be resummed, and he derived an integral equation which determines the large N limit of the spectrum of meson masses.

't Hooft's solution of planar QCD in two dimensions is not only a pedagogical illustration of the $1/N$ expansion: it is an analytic triumph, providing an exact solution of an interacting planar theory, and it is also a beautiful toy model for the strong interactions. Many of the ideas put forward by 't Hooft in this solution continue to find applications in the analysis of gauge theories, for example in recent investigations of interacting conformal field theories in three dimensions. This chapter will be devoted to a detailed presentation of some aspects of 't Hooft's solution to two-dimensional QCD in the large N limit.

9.2 The fermion propagator

We consider the standard QCD Lagrangian (5.3.1) in two dimensions. In this chapter we will use the rescaled fields (4.2.30), (5.3.6), but for obvious reasons of economy we will omit the hats. We will also omit the flavor indices, and we will assume that all the quark flavors have the same mass m. Note that, in two dimensions, the coupling constant g_0^2 has the dimensions of a mass. Our conventions for the gamma matrices, in Minkowski space, are

$$\gamma^0 = \sigma_3, \quad \gamma^1 = i\sigma_2, \quad \gamma_5 = \gamma^0\gamma^1 = \sigma_1. \tag{9.2.1}$$

Following the work of Bars and Green, we will do the analysis in the gauge

$$A_1^a = 0, \tag{9.2.2}$$

which is sometimes called the axial or Coulomb gauge. This is not the original gauge used by 't Hooft, but it has some conceptual advantages. In this gauge, the only non-zero component of the gauge field is $F_{01}^a = -\partial_1 A_0^a$, and the Lagrangian reads

$$\mathcal{L} = \frac{1}{2}\left(\partial_1 A_0^a\right)^2 + \rho^a A_0^a + \overline{\psi}\left(i\slashed{\partial} - m\right)\psi, \tag{9.2.3}$$

where

$$\rho^a = g_0 \overline{\psi}_i \gamma^0 (T_a)_j^i \psi^j, \tag{9.2.4}$$

and $i, j = 1, \ldots, N$ are color indices. The main simplification in this model is that the gluon self-interaction has disappeared. The EOM for A_0^a is simply

$$\partial_1^2 A_0^a = \rho^a. \tag{9.2.5}$$

Let us denote $x^0 = t$, $x^1 = x$. Equation (9.2.5) can be solved explicitly as

$$A_0^a(x, t) = \frac{1}{2}\int_{-\infty}^{\infty} dy |x - y| \rho^a(y, t) - x F^a(t) + B^a(t), \tag{9.2.6}$$

where $F^a(t)$, $B^a(t)$ are arbitrary functions of t. The above solution can be checked directly by using that

$$\partial_x |x - y| = \theta(x - y) - \theta(y - x) \equiv \epsilon(x - y), \tag{9.2.7}$$

where

$$\epsilon(x - y) = \begin{cases} 1, & \text{if } x > y, \\ -1, & \text{if } x < y. \end{cases} \tag{9.2.8}$$

The electric field is then

$$F_{01}^a = -\partial_1 A_0^a = -\frac{1}{2}\int_{-\infty}^{\infty} dy\, \epsilon(x - y)\rho^a(y, t) + F^a(t). \tag{9.2.9}$$

At spatial infinity, the electric field takes the values

$$\lim_{x \to \pm\infty} F_{01}^a(x, t) \equiv F_{01}^a(\pm\infty, t) = F^a(t) \mp \frac{1}{2}Q^a(t), \tag{9.2.10}$$

where we have defined the charge

$$Q^a(t) = \int_{-\infty}^{\infty} dx \rho^a(x, t). \tag{9.2.11}$$

We will denote

$$\Pi^a(t) = F^a(t) - \frac{1}{2} Q^a(t), \tag{9.2.12}$$

so that

$$F_{01}^a(\infty, t) = \Pi^a(t), \qquad F_{01}^a(-\infty, t) = \Pi^a(t) + Q^a(t). \tag{9.2.13}$$

Similarly, we find, for the gauge potential,

$$\lim_{x \to \pm\infty} A_0^a(x, t) = -x F_{01}^a(\pm\infty, t) + B^a(t) \mp \frac{1}{2} Q_1^a, \tag{9.2.14}$$

where

$$Q_1^a(t) = \int_{-\infty}^{\infty} x \rho^a(x, t) dx. \tag{9.2.15}$$

Note that, after fixing the gauge (9.2.2), one can still perform gauge transformations which are x independent but t dependent. They do not change the condition (9.2.2) and are called residual gauge transformations. The field $B^a(t)$ appearing in (9.2.6) can be changed by such a gauge transformation, and we can use this gauge freedom to set

$$B^a(t) = -\frac{1}{2} Q_1^a(t). \tag{9.2.16}$$

The energy-momentum tensor of the theory is given by

$$\Theta_{\mu\nu} = -F_{\mu\lambda}^a F_\nu^{a\,\lambda} + \frac{i}{2} \overline{\psi} \left(D_\mu \gamma_\nu + D_\nu \gamma_\mu \right) \psi - g_{\mu\nu} \mathcal{L}. \tag{9.2.17}$$

The Hamiltonian of the theory can be computed by integrating the Hamiltonian density,

$$\Theta_{00} = \frac{1}{2} \left(F_{01}^a \right)^2 + \overline{\psi} \left(-i \eth_x + m \right) \psi, \tag{9.2.18}$$

where we have denoted

$$\eth_x = \gamma^1 \partial_x. \tag{9.2.19}$$

The integral of the first term in (9.2.18) is given by,

$$\int \left(F_{01}^a \right)^2 dx = -\int A_0^a(x) \rho^a(x) dx + \Pi^a Q_1^a$$

$$= -\frac{1}{2} \int \rho^a(x) |x - y| \rho^a(y) dx\, dy + Q_1^a \left(2\Pi^a + Q^a \right). \tag{9.2.20}$$

In the first line we have integrated by parts, used the boundary conditions (9.2.13), (9.2.14), and dropped an infinite term proportional to $(\Pi^a)^2 + (\Pi^a + Q^a)^2$, which comes from the x-dependent term in (9.2.14). We then find

$$H = \int \overline{\psi}(x) \left(-i\partial_x + m\right) \psi(x)\, dx - \frac{1}{4} \int \rho^a(x)|x - y|\rho^a(y)dx\, dy$$
$$+ Q_1^a \left(\Pi^a + \frac{1}{2}Q^a\right), \tag{9.2.21}$$

which we will take as our definition of the Hamiltonian of the theory. The last term in (9.2.21), due to a background field $F^a = \Pi^a + \frac{1}{2}Q^a$, leads to some complications in the analysis of the quantum theory (for example, this term is not translation invariant). However, Bars and Green showed that it can be put to zero in the quark–antiquark singlet sector. Therefore, we will not consider it anymore in our analysis.

To analyze the quantum theory, we postulate the standard, equal-time, canonical anticommutation relations for the quark fields,

$$\{\psi_\alpha^i(x), \psi_\beta^j(y)\} = \{\psi_{i\alpha}^\dagger(x), \psi_{j\beta}^\dagger(y)\} = 0,$$
$$\{\psi_\alpha^i(x), \psi_{j\beta}^\dagger(y)\} = \delta_j^i \delta_{\alpha\beta}\delta(x - y), \tag{9.2.22}$$

where $\alpha, \beta = 1, 2$ are spinor indices. From (9.2.22) one can deduce that the charges defined in (9.2.11) satisfy the commutation relations,

$$[Q^a, Q^b] = igf^{abc}Q^c, \tag{9.2.23}$$

i.e. they are the generators of gauge transformations on the quark fields. So far, we have not made any use of the large N approximation. If we use (9.2.4) we can write the quartic fermion term in the Lagrangian as

$$\rho^a(x)\rho^a(y) = g_0^2 \overline{\psi}_i(x)\gamma^0\psi^j(x)\overline{\psi}_k(y)\gamma^0\psi^l(y)\, (T_a)_j^i\, (T_a)_l^k$$
$$= \frac{g_0^2}{2} \overline{\psi}_i(x)\gamma^0\psi^j(x)\overline{\psi}_j(y)\gamma^0\psi^i(y)$$
$$- \frac{g_0^2}{2N} \overline{\psi}_i(x)\gamma^0\psi^i(x)\overline{\psi}_j(y)\gamma^0\psi^j(y), \tag{9.2.24}$$

where we used (7.2.10) with the standard normalization (4.2.5). Notice that the second term is subleading with respect to the first one in the large N limit. We now introduce the 't Hooft parameter

$$\gamma = \frac{g_0^2 N}{4\pi}, \tag{9.2.25}$$

which is dimensionful and will be kept fixed as $N \to \infty$, as we explained in Chapter 7. The Hamiltonian that we obtain at large N is then

$$H = \int \overline{\psi}(x) \left(-i\partial_x + m \right) \psi(x) dx$$

$$- \frac{\pi \gamma}{2N} \int \overline{\psi}_i(x) \gamma^0 \psi^j(x) |x - y| \overline{\psi}_j(y) \gamma^0 \psi^i(y) dx dy. \tag{9.2.26}$$

We want to calculate the spectrum of this Hamiltonian, at least at large N, in the singlet sector. We expect to find a tower of bound states describing the mesons of QCD in two dimensions. The diagonalization of the above Hamiltonian is not a simple problem, however, since it is quartic in the fermion fields. This is the kind of situation that one finds in many-body physics, where the Hamiltonian is a sum of a one-body operator, which is quadratic in the fermion fields, and a two-body operator, which is quartic in the fermion fields. Except in some special models, the problem cannot be solved in closed form and one has to use some approximation method, like diagrammatic perturbation theory, or some sort of mean-field approximation, like the Hartree–Fock approximation.

It turns out that, in our problem, the Hartree–Fock approximation becomes *exact* at large N, as first pointed out by Witten. We will later show in detail that this is the case for the above Hamiltonian, but here we will diagonalize (9.2.26) at large N by using a more conventional method from QFT, which is equivalent to a relativistic version of the Hartree–Fock method. The basic idea is to define a vacuum for the quark fields which is annihilated by the large N Hamiltonian. We therefore expand the quark fields in terms of creation and annihilation operators,

$$\psi^i_\alpha(x) = \frac{1}{\sqrt{2\pi}} \int_{-\infty}^{\infty} dk \left[u_\alpha(k) b^i(k) + v_\alpha(-k) d^{\dagger i}(-k) \right] e^{ikx}. \tag{9.2.27}$$

As in the case of the ordinary quantization of the free Dirac field, we require

$$u^\dagger(k) u(k) = v^\dagger(-k) v(-k) = 1,$$
$$u^\dagger(k) v(-k) = 0, \tag{9.2.28}$$
$$u_\alpha(k) u^*_\beta(k) + v_\alpha(-k) v^*_\beta(-k) = \delta_{\alpha\beta}.$$

In the first and second equations, $u(k)$ denotes a matrix with a single row and components $u_\alpha(k)$, and similarly for $v(k)$. Together with the canonical anticommutation relations (9.2.24), the conditions (9.2.28) lead to

$$\{b^i(k), b^{\dagger j}(k')\} = \delta(k - k')\delta^{ij},$$
$$\{d^i(k), d^{\dagger j}(k')\} = \delta(k - k')\delta^{ij}. \tag{9.2.29}$$

The other anticommutators are zero. The vacuum state is defined as

$$b^i(k)|0\rangle = d^i(k)|0\rangle = 0. \tag{9.2.30}$$

We can then define a normal-ordering operation with respect to the vacuum state. Since

$$\langle 0|\psi_\alpha^i(x)\psi_{j\beta}^\dagger(y)|0\rangle = \delta_j^i \int \frac{dk}{2\pi} e^{ik(x-y)} u_\alpha(k)u_\beta^*(k),$$

$$\langle 0|\psi_{i\alpha}^\dagger(x)\psi_\beta^j(y)|0\rangle = \delta_i^j \int \frac{dk}{2\pi} e^{-ik(x-y)} v_\alpha^*(-k)v_\beta(-k),$$

(9.2.31)

it follows that

$$\psi_{i\alpha}^\dagger(x)\psi_\alpha^j(x)\psi_{j\beta(y)}^\dagger\psi_\beta^i(y) = : \psi_{i\alpha}^\dagger(x)\psi_\alpha^j(x)\psi_{j\beta(y)}^\dagger\psi_\beta^i(y) :$$

$$+ \psi_{i\alpha}^\dagger(x)\psi_\beta^i(y)\langle 0|\psi_\alpha^j(x)\psi_{j\beta}^\dagger(y)|0\rangle$$

$$- \psi_{i\alpha}^\dagger(y)\psi_\beta^i(x)\langle 0|\psi_{j\beta}^\dagger(x)\psi_\alpha^j(y)|0\rangle + c\text{-number},$$

(9.2.32)

where the last term involves contraction of all Fermi fields. We can then write the quartic term in the Hamiltonian (9.2.26) as

$$-\frac{\pi\gamma}{2N}\int : \overline{\Psi}_i(x)\gamma^0\psi^j(x)|x-y|\overline{\Psi}_j(y)\gamma^0\psi^i(y) : dx\, dy$$

$$-\frac{\gamma}{4}\int e^{ik(x-y)}|x-y|\psi^\dagger(x)\left[u(k)u^\dagger(k)-v(-k)v^\dagger(-k)\right]\psi(y)\, dx\, dy\, dk.$$

(9.2.33)

Notice that the normal-ordered part is of order $1/N$. Therefore, at large N, we are left with a Hamiltonian which is quadratic in the Fermi fields,

$$H_2 = \int \overline{\Psi}(x)\left(-i\partial_x + m\right)\psi(x)dx$$

$$-\frac{\gamma}{4}\int e^{ik(x-y)}|x-y|\psi^\dagger(x)\left[u(k)u^\dagger(k)-v(-k)v^\dagger(-k)\right]\psi(y)dx\, dy\, dk.$$

(9.2.34)

For the vacuum $|0\rangle$ to be the true vacuum of the theory at large N, we have to require H_2 to be of the form

$$H_2 = \int E(k)\left[b^{\dagger i}(k)b^i(k)+d^{\dagger i}(k)d^i(k)\right]dk,$$

(9.2.35)

for a certain function $E(k)$. Let us see what are the conditions for this to be true. If we plug into (9.2.34) the expansion of the Fermi field (9.2.27) and of its conjugate, we find the distribution defined by the integral

$$\int |x-y|e^{i(k-k_1)x+i(k_2-k)y}dxdy.$$

(9.2.36)

To compute it, we first need the Fourier transform of the Heaviside function,

$$\int e^{ivk}\theta(v)dv = \pi\delta(k) + iP\frac{1}{k},$$

(9.2.37)

where P denotes Cauchy's principal value, as well as the Fourier transform,

$$\int e^{iv(p-k)}|v|dv = -2\,\text{H}\frac{1}{(p-k)^2}.$$ (9.2.38)

Here, H denotes the Hadamard principal value,

$$\text{H}\frac{1}{(p-k)^2} = \frac{d}{dk}\text{P}\frac{1}{p-k},$$ (9.2.39)

and it is calculated as follows,

$$\text{H}\int_a^b \frac{f(t)}{(t-x)^2}dt = \lim_{\epsilon \to 0}\left[\int_a^{x-\epsilon} \frac{f(t)}{(t-x)^2}dt + \int_{x+\epsilon}^b \frac{f(t)}{(t-x)^2}dt - \frac{2f(x)}{\epsilon}\right].$$ (9.2.40)

We will also denote

$$\text{H}\int_a^b \frac{f(t)}{(t-x)^2}dt \equiv \fint_a^b \frac{f(t)}{(t-x)^2}dt.$$ (9.2.41)

We can now calculate (9.2.36). First we change variables

$$u = \frac{1}{2}(x+y), \qquad v = x - y,$$ (9.2.42)

so that

$$\int |x-y|e^{i(k-k_1)x+i(k_2-k)y}dxdy = 2\pi\,\delta(k_1 - k_2)\int e^{i(k-k_1)v}|v|dv$$

$$= -4\pi\,\delta(k_1 - k_2)\text{H}\frac{1}{(k-k_1)^2}.$$ (9.2.43)

Grouping the terms involving two b, we find the equation

$$\left\{p\gamma_5 + m\gamma_0 + \frac{\gamma}{2}\fint \frac{dk}{(p-k)^2}\left[u(k)u^\dagger(k) - v(-k)v^\dagger(-k)\right]\right\}u(p)$$

$$= E(p)u(p).$$ (9.2.44)

The terms involving two d lead to a similar equation, with $v(-p)$ instead of $u(p)$, and $-E(p)$ instead of $E(p)$:

$$\left\{p\gamma_5 + m\gamma_0 + \frac{\gamma}{2}\fint \frac{dk}{(p-k)^2}\left[u(k)u^\dagger(k) - v(-k)v^\dagger(-k)\right]\right\}v(-p)$$

$$= -E(p)v(-p).$$ (9.2.45)

Finally, the crossed b-d terms vanish after using the second equation in (9.2.28).
   In order to solve these equations, we parametrize

$$u(k) = T(k)\begin{pmatrix}1\\0\end{pmatrix}, \qquad v(-k) = T(k)\begin{pmatrix}0\\1\end{pmatrix},$$ (9.2.46)

where $T(k)$ is a unitary matrix. It is easy to see that the eigenvalue equations (9.2.44) and (9.2.45) can be written as a single matrix equation,

$$E(p)T(p)\gamma_0 T^\dagger(p) = p\gamma_5 + m\gamma_0 + \frac{\gamma}{2}\!\!\!\!\fint \frac{dk}{(p-k)^2} T(k)\gamma_0 T^\dagger(k). \qquad (9.2.47)$$

We will also parametrize

$$T(k) = \exp\left(-\frac{\theta(k)}{2}\gamma^1\right), \qquad (9.2.48)$$

so that

$$T(k)\gamma_0 T^\dagger(k) = \begin{pmatrix} \cos\theta(k) & \sin\theta(k) \\ \sin\theta(k) & -\cos\theta(k) \end{pmatrix}. \qquad (9.2.49)$$

The equation (9.2.47) is equivalent to the following integral equations,

$$E(p)\cos\theta(p) = m + \frac{\gamma}{2}\!\!\!\!\fint \frac{dk}{(p-k)^2}\cos\theta(k),$$
$$E(p)\sin\theta(p) = p + \frac{\gamma}{2}\!\!\!\!\fint \frac{dk}{(p-k)^2}\sin\theta(k). \qquad (9.2.50)$$

Notice that $E(p)$ is an even function of p, while $\theta(p)$ is an odd function. By taking appropriate linear combinations of (9.2.50), one finds the following equation for $\theta(p)$,

$$p\cos\theta(p) - m\sin\theta(p) = \frac{\gamma}{2}\!\!\!\!\fint \frac{dk}{(p-k)^2}\sin\left[\theta(p) - \theta(k)\right], \qquad (9.2.51)$$

as well as

$$E(p) = m\cos\theta(p) + p\sin\theta(p) + \frac{\gamma}{2}\!\!\!\!\fint \frac{dk}{(p-k)^2}\cos\left[\theta(p) - \theta(k)\right]. \qquad (9.2.52)$$

Therefore, once $\theta(p)$ is determined by the integral equation (9.2.51), $E(p)$ is determined by (9.2.52). This solves the diagonalization problem. When $\gamma = 0$ we recover the free theory, and

$$E(p) = \sqrt{m^2 + p^2}, \qquad \theta(p) = \tan^{-1}\left(\frac{p}{m}\right). \qquad (9.2.53)$$

From the integral equations (9.2.50) we can deduce the asymptotic behavior as $p \to \infty$ of $\theta(p)$, $E(p)$, which will be useful later on (of course, the behavior as $p \to -\infty$ follows from the parity properties of these functions). The free particle solution (9.2.53) suggests the following ansatz as $p \to \infty$

$$\theta(p) \approx \frac{\pi}{2} + \frac{\theta_1}{p} + \cdots, \qquad E(p) \approx p + \frac{e_1}{p} + \cdots. \qquad (9.2.54)$$

Plugging this ansatz into the equations (9.2.50) we find immediately,

$$\theta_1 = -m, \tag{9.2.55}$$

therefore, as $p \to \infty$,

$$\theta(p) \approx \frac{\pi}{2} - \frac{m}{p}, \qquad \sin\theta(p) \approx 1 - \frac{m^2}{2p^2}. \tag{9.2.56}$$

Similarly, we find

$$e_1 = \frac{m^2}{2} + \frac{\gamma}{2} s_1, \tag{9.2.57}$$

where s_1 is defined by the large p expansion

$$\fint \frac{dk}{(p-k)^2} \sin\theta(k) = \frac{s_1}{p} + \mathcal{O}\left(\frac{1}{p^2}\right). \tag{9.2.58}$$

To calculate s_1, we use the following trick. Since

$$\lim_{p \to \pm\infty} \theta(p) = \pm\frac{\pi}{2}, \tag{9.2.59}$$

we can write

$$\sin\theta(k) = \epsilon(k) + g(k), \tag{9.2.60}$$

where $g(k)$ is a bounded function which at infinity behaves like $\approx k^{-2}$. An elementary calculation shows that

$$\fint \frac{dk}{(p-k)^2} \epsilon(k) = -\frac{2}{p}, \tag{9.2.61}$$

and by using the properties of $g(k)$ one finds that, as $p \to \infty$,

$$\fint \frac{dk}{(p-k)^2} g(k) = \mathcal{O}\left(\frac{1}{p^2}\right). \tag{9.2.62}$$

We conclude that $s_1 = -2$, for any value of m, γ, and

$$E(p) \approx p + \frac{m^2 - 2\gamma}{p}, \qquad p \to \infty. \tag{9.2.63}$$

For general values of m, p, there are no analytic solutions, and the integral equations (9.2.51), (9.2.52) have to be solved numerically.

The quadratic Hamiltonian H_2 gives the time evolution

$$b^i(p, t) = e^{iH_2 t} b^i(p) e^{-iH_2 t} = b^i(p) e^{-iE(p)t},$$
$$d^i(p, t) = e^{iH_2 t} d^i(p) e^{-iH_2 t} = d^i(p) e^{-iE(p)t}, \tag{9.2.64}$$

and the time dependent Fermi field is

$$\psi^i_\alpha(x, t) = \frac{1}{\sqrt{2\pi}} \int_{-\infty}^{\infty} dp \left[u_\alpha(p) b^i(p) e^{-ip^\mu x_\mu} + v_\alpha(p) d^{\dagger i}(p) e^{ip^\mu x_\mu} \right]. \quad (9.2.65)$$

It is now easy to compute the fermion propagator by direct calculation. We find,

$$\langle 0 | \psi^i_\alpha(x, t) \overline{\psi}_{j\beta}(y, t') | 0 \rangle = \delta^i_j \int \frac{dp}{2\pi} e^{ip(x-y) - iE(p)(t-t')} u_\alpha(p) u^*_\beta(p), \quad (9.2.66)$$

and

$$-\langle 0 | \overline{\psi}_{j\beta}(y, t') \psi^i_\alpha(x, t) | 0 \rangle = -\delta^i_j \int \frac{dp}{2\pi} e^{ip(x-y) + iE(p)(t-t')} v_\alpha(-p) v^*_\beta(-p). \quad (9.2.67)$$

As usual, we can represent this result in a compact way by using the time-ordered product and the Feynman propagator. We use the following integrals for $\epsilon > 0$,

$$\int \frac{dp^0}{2\pi} \frac{e^{-ip^0(t-t')}}{p^0 - E(p) + i\epsilon} = \begin{cases} -ie^{-iE(p)(t-t')}, & \text{if } t > t', \\ 0, & \text{if } t < t', \end{cases} \quad (9.2.68)$$

and

$$\int \frac{dp^0}{2\pi} \frac{e^{-ip^0(t-t')}}{p^0 + E(p) - i\epsilon} = \begin{cases} ie^{iE(p)(t-t')}, & \text{if } t < t', \\ 0, & \text{if } t > t', \end{cases} \quad (9.2.69)$$

to write the time-ordered product of the fields as,

$$\langle 0 | T \left[\psi^i_\alpha(x, t) \overline{\psi}_{j\beta}(y, t') \right] | 0 \rangle = \delta^i_j \int \frac{dp \, dp^0}{(2\pi)^2} e^{-ip^\mu(x-y)_\mu} S(p, p_0)_{\alpha\beta}, \quad (9.2.70)$$

where

$$p^\mu(x - y)_\mu = p^0(t - t') - p(x - y) \quad (9.2.71)$$

is the usual product in Minkowski space, and

$$\begin{aligned} S(p^\mu) &= \frac{iu(p)u^\dagger(p)}{p^0 - E(p) + i\epsilon} + \frac{iv(-p)v^\dagger(-p)}{p^0 + E(p) - i\epsilon} \\ &= i\frac{p^0\gamma^0 - \gamma^1 E(p)\sin\theta(p) + E(p)\cos\theta(p)}{(p^0 - E(p) + i\epsilon)(p^0 + E(p) - i\epsilon)}. \end{aligned} \quad (9.2.72)$$

In going from the first to the second expression on the right hand side, we have used (9.2.46) and (9.2.49). Notice that, in the free case $\gamma = 0$, we find

$$S(p^\mu) = \frac{i(\gamma^\mu p_\mu + m)}{(p^0)^2 - E(p)^2 + 2i\epsilon E(p)}. \quad (9.2.73)$$

In a free theory, $E(p)$ is always positive and the ϵ prescription appearing here leads to a propagator which is equivalent to

$$S(p^\mu) = \frac{i\left(\gamma^\mu p_\mu + m\right)}{p_\mu p^\mu - m^2 + i\epsilon}, \qquad (9.2.74)$$

which is the standard Feynman propagator for a free fermion. Notice, however, that in the interacting theory $E(p)$ is not necessarily positive, and one should use the precise ϵ prescription appearing in the original expression (9.2.72).

We should now pause and interpret the results obtained so far. Given the mass of the quarks, the equations for $\theta(p)$ and $E(p)$, (9.2.51) and (9.2.52), provide the large N corrections to the results obtained in free field theory, and they depend on the 't Hooft parameter γ. Therefore, we should interpret the propagator in (9.2.72) as a quantum-corrected propagator, where we take into account the corrections due to planar diagrams. Quantum-corrected propagators are usually parametrized in terms of the self-energy $\Sigma(p)$,

$$S(p^\mu) = \frac{i}{\gamma^\mu p_\mu - m - \Sigma(p)}. \qquad (9.2.75)$$

On the other hand, from (9.2.72) we have (we set $\epsilon = 0$)

$$S(p^\mu) = \frac{i}{p^0\gamma^0 - E(p)\sin\theta(p)\gamma^1 - E(p)\cos\theta(p)}, \qquad (9.2.76)$$

therefore

$$\Sigma(p) = E(p)\cos\theta(p) - m + (E(p)\sin\theta(p) - p)\gamma^1. \qquad (9.2.77)$$

It is easy to check that the integral equations (9.2.50) can be written as

$$\Sigma(p) = \frac{\gamma}{2\pi}\fint \frac{dk^0 dk}{(p-k)^2}\gamma^0 S(k^\mu)\gamma^0, \qquad (9.2.78)$$

where the integral over k^0 is understood as a principal value integral (so that the term $k^0\gamma^0$ in the numerator of (9.2.72) does not contribute).

The equation (9.2.78) has a natural diagrammatic interpretation, in terms of resummation of planar diagrams. Let us first recall the standard Schwinger–Dyson equation for the self-energy, which is better written in diagrammatic form as in Fig. 9.1. The thick line denotes the fully corrected quantum mechanical propagator S, in terms of the self-energy Σ and the free propagator S_0. The self-energy includes only one-particle irreducible diagrams. In the planar limit, the only diagrams that contribute to the quantum-corrected propagator are "rainbow diagrams," where we have non-intersecting gluon lines (denoted in Fig. 9.2 by dashed lines) attached to the quark propagator. The condition that lines do not intersect comes

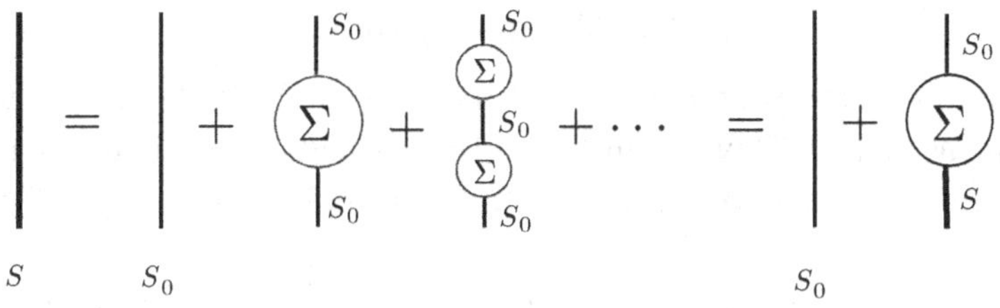

Figure 9.1 The Schwinger–Dyson equation for the quantum-corrected propagator.

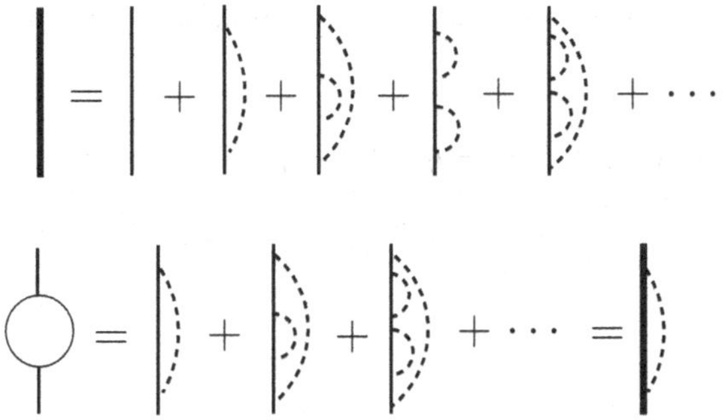

Figure 9.2 Diagrams contributing to the quantum-corrected propagator (in the first line) and the self-energy (in the second line), in the planar limit.

from the planarity condition. The one-particle irreducible, planar diagrams contributing to the self-energy are shown in the second line of Fig. 9.2. It is easy to see that these diagrams can be obtained by "capping" the full propagator S with an external gluon line. The equation (9.2.78) is the analytic expression of this diagrammatic fact.

We have seen that, at least in two dimensions, an infinite set of planar diagrams in QCD can be resummed exactly to obtain the quantum-corrected fermion propagator. This is of course due to the fact that, in two dimensions, the life of gauge fields is particularly simple. By a judicious choice of gauge, the self-interaction of the gluon field disappears and the problem of resummation can be effectively handled. Equivalently, we have been able to understand the vacuum structure of the Hamiltonian in the large N limit: the Hamiltonian becomes quadratic and can be diagonalized through a rotation in field space implemented by the operator $T(k)$ in (9.2.46). The rotation angle $\theta(p)$, as well as the resulting energies $E(p)$, depend in a non-trivial way on the 't Hooft parameter, but this dependence is encoded in a simple set of singular integral equations, (9.2.50).

9.3 Meson spectrum

Mesons are bound states of a quark and an antiquark. To study these bound states in two-dimensional QCD we should consider the quartic terms in the Hamiltonian, and to do this it is useful to consider the following color-singlet bilinear operators:

$$B(p, p') = \frac{1}{\sqrt{N}} b^{\dagger i}(p) b^i(p'), \qquad D(p, p') = \frac{1}{\sqrt{N}} d^{\dagger i}(-p) d^i(-p'),$$

$$M(p, p') = \frac{1}{\sqrt{N}} d^i(-p) b^i(p'), \qquad M^\dagger(p, p') = \frac{1}{\sqrt{N}} b^{\dagger i}(p') d^{\dagger i}(-p). \tag{9.3.1}$$

The operators M, $M^\dagger$ correspond to meson operators. They satisfy the commutation relations,

$$\left[M(p, p'), M^\dagger(q, q') \right] = \delta(p' - q') \delta(p - q)$$
$$- \frac{1}{\sqrt{N}} \left(D(q, p) \delta(p' - q') + B(q', p') \delta(p - q) \right), \tag{9.3.2}$$

while the other operators satisfy,

$$\left[B(p, p'), B(q, q') \right] = \frac{1}{\sqrt{N}} \left(B(p, q') \delta(p' - q) - B(q, p') \delta(p - q') \right),$$

$$\left[D(p, p'), D(q, q') \right] = \frac{1}{\sqrt{N}} \left(D(p, q') \delta(p' - q) - D(q, p') \delta(p - q') \right),$$

$$\left[B(p, p'), M(q, q') \right] = -\frac{1}{\sqrt{N}} M(q, p') \delta(p - q'),$$

$$\left[B(p, p'), M^\dagger(q, q') \right] = \frac{1}{\sqrt{N}} M^\dagger(q, p) \delta(p' - q'), \tag{9.3.3}$$

$$\left[D(p, p'), M(q, q') \right] = -\frac{1}{\sqrt{N}} M(p', q') \delta(p - q),$$

$$\left[D(p, p'), M^\dagger(q, q') \right] = \frac{1}{\sqrt{N}} M^\dagger(p, q') \delta(p' - q).$$

These commutators are suppressed at large N. This suggests that, at large N, the operators M, $M^\dagger$ are of order $\mathcal{O}(1)$, while the operators B, D are of order $\mathcal{O}\left(1/\sqrt{N}\right)$. Indeed, the ansatz

$$B(p, p') = \frac{1}{\sqrt{N}} \int dq \, M^\dagger(q, p) M(q, p'),$$

$$D(p, p') = \frac{1}{\sqrt{N}} \int dq \, M^\dagger(p, q) M(p', q), \tag{9.3.4}$$

satisfies the commutation relations for the B and D operators at leading order in the $1/N$ expansion, and indicates that in the large N limit the operators B and D can be neglected, and only the contributions of the M, $M^\dagger$ operators should be kept.

Remember that, after normal ordering, the Hamiltonian of the theory is given by a quadratic term H_2, and a quartic, normal-ordered term which we will denote as H_4. In terms of the bilinear operators introduced above, we see that

$$H_2 = \sqrt{N} \int dk\, E(k)\,(B(k,k) + D(k,k))$$
$$= \int dQ dp\,(E(p) + E(Q-p))\,M^\dagger(p-Q,p)M(p-Q,p). \qquad (9.3.5)$$

In order to compute the quartic term H_4 at leading order in the $1/N$ expansion, it is enough to look at the terms which lead to M, $M^\dagger$ operators. There are only four terms in the product of four fermion fields in (9.2.26) which lead to such terms. They involve the products:

$$
\begin{array}{ll}
(1) & b^{\dagger i}(k_1)b^j(k_2)d^j(-k_3)d^{\dagger i}(-k_4)\,u_\alpha^*(k_1)\,u_\alpha(k_2)v_\beta(-k_3)v_\beta^*(-k_4), \\[4pt]
(2) & b^{\dagger i}(k_1)d^{\dagger j}(-k_2)b^{\dagger j}(k_3)d^{\dagger i}(-k_4)\,u_\alpha^*(k_1)v_\alpha(-k_2)u_\beta^*(k_3)v_\beta^*(-k_4), \\[4pt]
(3) & d^i(-k_1)d^{\dagger j}(-k_2)b^{\dagger j}(k_3)b^i(k_4)\,v_\alpha^*(-k_1)v_\alpha(-k_2)u_\beta^*(k_3)u_\beta(k_4), \\[4pt]
(4) & d^i(-k_1)b^j(k_2)d^j(-k_3)b^i(k_4)\,v_\alpha^*(-k_1)u_\alpha(k_2)v_\beta^*(-k_3)u_\beta(k_4),
\end{array}
\qquad (9.3.6)
$$

which are integrated with the factor

$$
\int \frac{dx\,dy}{(2\pi)^2}\,|x-y|e^{i(k_2-k_1)x+i(k_4-k_3)y}
$$
$$
= \frac{1}{2\pi}\delta(k_2+k_4-k_1-k_3)\int dv\,e^{iv(k_2-k_1-k_4+k_3)/2}|v|. \qquad (9.3.7)
$$

Due to the overall delta function, we can parametrize the four momenta in terms of three independent momenta k, p, Q, as

$$k_1 = p, \quad k_2 = k, \quad k_3 = k - Q, \quad k_4 = p - Q, \qquad (9.3.8)$$

and by using (9.2.38) we see that (9.3.7) leads to the distribution

$$-\frac{1}{\pi}H\frac{1}{(p-k)^2}. \qquad (9.3.9)$$

The factors involving the spinors u, v can be evaluated in terms of the angle $\theta(k)$, by using (9.2.46). One finds, for (1) and (3) in (9.3.6), the factor

$$C(p,k,Q) = \cos\frac{\theta(p)-\theta(k)}{2}\cos\frac{\theta(Q-p)-\theta(Q-k)}{2}, \qquad (9.3.10)$$

while for the factors (2) and (4) one finds

$$S(p, k, Q) = \sin \frac{\theta(p) - \theta(k)}{2} \sin \frac{\theta(Q - p) - \theta(Q - k)}{2}. \tag{9.3.11}$$

The function $C(p, k, Q)$ sastisfies the symmetry properties

$$C(p, k, Q) = C(k, p, Q),$$
$$C(p - Q, k - Q, -Q) = C(p, k, Q), \tag{9.3.12}$$

and the same properties hold for $S(p, k, Q)$. The normal ordering of the operators appearing in (9.3.6) is easily computed in terms of the meson operators M, $M^\dagger$. One obtains

$$
\begin{aligned}
&(1) \quad - NM^\dagger(p - Q, p)M(k - Q, k),\\
&(2) \quad - NM^\dagger(p - Q, p)M^\dagger(k, k - Q),\\
&(3) \quad - NM^\dagger(k, k - Q)M(p, p - Q),\\
&(4) \quad - NM(p, p - Q)M(k - Q, k).
\end{aligned}
\tag{9.3.13}
$$

Using now the symmetries (9.3.12), we can change variables in the integration over Q, p, k to obtain the final expression for H_4, at large N,

$$
H_4 = -\frac{\gamma}{2} \int dQ dp \fint \frac{dk}{(p - k)^2} \Big\{ 2C(p, k, Q)M^\dagger(p - Q, p)M(k - Q, k)
$$
$$
+ S(p, k, Q)\,(M(p, p - Q)M(k - Q, k)
$$
$$
+ M^\dagger(p, p - Q)M^\dagger(k - Q, k))\Big\}. \tag{9.3.14}
$$

We would like now to calculate the spectrum of the operator $H_2 + H_4$. This is a quadratic operator in the bilocal meson fields M, $M^\dagger$, so the diagonalization process can be done with a linear transformation of the operators. Such a linear transformation is usually called a Bogoliubov transformation. Let us then define the new operators

$$m_n^\dagger(Q) = \int dq \left\{ M^\dagger(q - Q, q)\varphi_+^n(q, Q) + M(q, q - Q)\varphi_-^n(q, Q)\right\},$$
$$m_n(Q) = \int dq \left\{ M(q - Q, q)\varphi_+^n(q, Q) + M^\dagger(q, q - Q)\varphi_-^n(q, Q)\right\}, \tag{9.3.15}$$

with $n = 0, 1, \ldots$ The real functions $\varphi_+^n(p, Q)$ and $\varphi_-^n(p, Q)$ satisfy the orthonormality conditions

$$\int dp \left(\varphi_+^n(p, Q)\varphi_+^m(p, Q) - \varphi_-^n(p, Q)\varphi_-^m(p, Q)\right) = \delta_{nm},$$
$$\int dp \left(\varphi_+^n(p, Q)\varphi_-^m(p, Q) - \varphi_-^n(p, Q)\varphi_+^m(p, Q)\right) = 0, \tag{9.3.16}$$

as well as the corresponding completeness relations

$$\sum_{n=0}^{\infty} \left(\varphi_+^n(p, Q)\varphi_+^n(k, Q) - \varphi_-^n(p, Q)\varphi_-^n(k, Q) \right) = \delta(p - k),$$

$$\sum_{n=0}^{\infty} \left(\varphi_+^n(p, Q)\varphi_-^n(k, Q) - \varphi_-^n(p, Q)\varphi_+^n(k, Q) \right) = 0. \tag{9.3.17}$$

These are natural generalizations to bilocal functions of the standard conditions for Bogoliubov transformations. Using these conditions, we can write M, $M^\dagger$ in terms of the new mesonic operators m, $m^\dagger$,

$$M(k - Q, k) = \sum_n \left\{ m_n(Q)\varphi_+^n(k, Q) - m_n^\dagger(-Q)\varphi_-^n(k, Q) \right\},$$

$$M^\dagger(k - Q, k) = \sum_n \left\{ m_n^\dagger(Q)\varphi_+^n(k, Q) - m_n(-Q)\varphi_-^n(k, Q) \right\}. \tag{9.3.18}$$

The new operators $m^\dagger$, m are constructed in such a way that they obey the standard commutation relations of bosonic creation and annihilation operators,

$$\left[m_n(Q), m_p^\dagger(Q') \right] = \delta(Q - Q')\delta_{np},$$

$$\left[m_n(Q), m_p(Q') \right] = \left[m_n^\dagger(Q), m_p^\dagger(Q') \right] = 0. \tag{9.3.19}$$

We will now require the functions $\varphi_\pm^n(p, Q)$ to be solutions of the set of integral equations

$$(E(p) + E(Q - p) - E_n(Q))\, \varphi_+^n(p, Q)$$

$$= \gamma \!\!\!\!\!\!\fint \frac{dk}{(p - k)^2} \left\{ C(p, k, Q)\varphi_+^n(k, Q) - S(p, k, Q)\varphi_-^n(k, Q) \right\},$$

$$(E(p) + E(Q - p) + E_n(Q))\, \varphi_-^n(p, Q)$$

$$= \gamma \!\!\!\!\!\!\fint \frac{dk}{(p - k)^2} \left\{ C(p, k, Q)\varphi_-^n(k, Q) - S(p, k, Q)\varphi_+^n(k, Q) \right\}. \tag{9.3.20}$$

In these equations, $E_n(Q)$ are positive eigenvalues for the system of coupled integral equations, and they are even functions of Q. It is easy to see that, if $(\varphi_+^n, \varphi_-^n)$ is a solution with eigenvalue E_n, then $(\varphi_-^n, \varphi_+^n)$ is also a solution with eigenvalue $-E_n$ (see (9.3.28) below), but we will restrict ourselves to the positive eigenvalues of the system. Due to the symmetries (9.3.12), the functions $\varphi_\pm^n(p, Q)$ satisfy the symmetry property,

$$\varphi_\pm^n(p - P, -P) = \varphi_\pm^n(p, P). \tag{9.3.21}$$

It is easy to show that the orthogonality conditions (9.3.16) follow from the integral equations (9.3.20). Let us consider the space of wavefunctions defined by the pairs

$$\varphi = \begin{pmatrix} \varphi_+ \\ \varphi_- \end{pmatrix} \tag{9.3.22}$$

and endowed with the inner product

$$(\varphi, \psi) = \int dp \left(\varphi_+^*(p)\psi_+(p) - \varphi_-^*(p)\psi_-(p) \right) \tag{9.3.23}$$

and the involution

$$\varphi_I = \begin{pmatrix} \varphi_- \\ \varphi_+ \end{pmatrix} . \tag{9.3.24}$$

The equations (9.3.20) can be written as

$$\mathcal{K}\varphi^n = E_n \varphi^n, \tag{9.3.25}$$

where

$$\mathcal{K} = \begin{pmatrix} K_1 & K_2 \\ -K_2 & -K_1 \end{pmatrix}, \tag{9.3.26}$$

and the operators $K_{1,2}$ are given by

$$(K_1\psi)(p) = (E(p) + E(Q - p)) \, \psi(p) - \gamma \!\!\!\! \fint \frac{dk}{(p-k)^2} C(p, k, Q)\psi(k),$$

$$(K_2\psi)(p) = \gamma \!\!\!\! \fint \frac{dk}{(p-k)^2} S(p, k, Q)\psi(k). \tag{9.3.27}$$

Due to the first symmetry property in (9.3.12), $K_{1,2}$ are Hermitian. It follows that, with the inner product (9.3.23), the operator $\mathcal{K}$ is also Hermitian. It is immediate to check that, if (9.3.25) holds, then

$$\mathcal{K}\varphi_I^n = -E_n \varphi_I^n. \tag{9.3.28}$$

By subtracting the two equations

$$\left(\varphi^m, \mathcal{K}\varphi^n \right) = E_n \left(\varphi^m, \varphi^n \right), \qquad \left(\varphi^n, \mathcal{K}\varphi^m \right) = E_m \left(\varphi^n, \varphi^m \right), \tag{9.3.29}$$

the first orthogonality equation in (9.3.16) follows. Also, by considering

$$\left(\varphi_I^m, \mathcal{K}\varphi^n \right) = E_n \left(\varphi_I^m, \varphi^n \right) = \left(\mathcal{K}\varphi_I^m, \varphi^n \right) = -E_m \left(\varphi_I^m, \varphi^n \right) \tag{9.3.30}$$

the second line in (9.3.16) follows.

We now claim that, if (9.3.20) hold, the Hamiltonian $H_2 + H_4$ is diagonal when written in terms of the operators m, $m^\dagger$, and it can be written as

$$H_2 + H_4 = \frac{1}{2} \sum_{n \geq 0} \int dQ \, E_n \left\{ m_n^\dagger(Q) m_n(Q) + m_n(Q) m_n^\dagger(Q) \right\}, \tag{9.3.31}$$

up to a normal-ordering c-number. This is seen by direct calculation. When (9.3.18) is plugged into $H_2 + H_4$, we find an expression of the form

$$H_2 + H_4 = \sum_{n,r} \int dQ \left\{ a_{nr}(Q) m_n^\dagger(Q) m_r(Q) \right.$$

$$+ b_{nr}(Q) m_n(Q) m_r^\dagger(Q) + c_{nr}(Q) m_n^\dagger(Q) m_r^\dagger(-Q)$$

$$\left. + d_{nr}(Q) m_n(-Q) m_r(Q) \right\}. \tag{9.3.32}$$

Let us first look at $c_{nr}(Q)$. It is given by

$$\int dp \, (E(p) + E(Q - p)) \varphi_+^n(p, Q) \varphi_-^r(p, Q)$$

$$- \frac{\gamma}{2} \int dp \fint \frac{dk}{(p - k)^2} \left\{ 2C(p, k, Q) \varphi_+^n(k, Q) \varphi_-^r(k, Q) \right.$$

$$- S(p, k, Q) \varphi_-^n(p, Q) \varphi_-^r(p, Q)$$

$$\left. - S(p, k, Q) \varphi_+^n(p, Q) \varphi_+^r(p, Q) \right\}. \tag{9.3.33}$$

To write the last term, we have used the first symmetry (9.3.12) for the coefficient $S(p, k, Q)$ and the commutation relations (9.3.19). If we use the eigenvalue equations (9.3.20), we find

$$\sum_{n,r} \int c_{nr}(Q) m_n^\dagger(Q) m_r^\dagger(-Q) dQ$$

$$= \sum_{n,r} \int m_n^\dagger(Q) m_r^\dagger(-Q) \, (E_r - E_n) \, \varphi_+^n(p, Q) \varphi_-^r(p, Q) dQ \, dp$$

$$= \sum_{n,r} E_n \int dQ \, m_n^\dagger(Q) m_r^\dagger(-Q) \int \left(\varphi_+^r(p, Q) \varphi_-^n(p, Q) - (n \leftrightarrow r) \right) dp$$

$$= 0. \tag{9.3.34}$$

In going from the second to the third line, we exchanged $r \leftrightarrow n$ in the first term, used the commutation relations (9.3.19), exchanged $Q \to -Q$, used the symmetry (9.3.21), and changed variables. In the last step we used the second orthogonality condition in (9.3.16). The vanishing of d_{nr} is proved along similar lines. Let us now look at the coefficient $a_{nr}(Q)$. It is given by

$$\int dp \, (E(p) + E(Q - p))\varphi_+^n(p, Q)\varphi_+^r(p, Q)$$

$$-\frac{\gamma}{2} \int dp \fint \frac{dk}{(p-k)^2} \Big\{ 2C(p, k, Q)\varphi_+^n(k, Q)\varphi_+^r(k, Q)$$

$$- S(p, k, Q)\varphi_-^n(p, Q)\varphi_+^r(p, Q)$$

$$- S(p, k, Q)\varphi_+^n(p, Q)\varphi_-^r(p, Q) \Big\}. \qquad (9.3.35)$$

Using again the equations (9.3.20), we find that

$$a_{nr}(Q) = \frac{1}{2}(E_n + E_r) \int dp \, \varphi_+^n(p, Q)\varphi_+^r(p, Q). \qquad (9.3.36)$$

A similar calculation shows that

$$b_{nr}(Q) = -\frac{1}{2}(E_n + E_r) \int dp \, \varphi_-^n(p, Q)\varphi_-^r(p, Q). \qquad (9.3.37)$$

Since in (9.3.32) we sum over n, r, we can symmetrize the resulting terms with respect to these indices, and use the commutation relations (9.3.19). The commutator gives a volume divergence proportional to $\delta(0)$. Using the first orthonormality condition (9.3.16), we find indeed (9.3.31), up to an (infinite) additive constant or zero point energy, which is proportional to the volume divergence and to the (divergent) sum of the energies. Both divergences can be regulated appropriately.

The spectrum of the Hamiltonian (9.3.31) is now easy to compute: the new vacuum is defined by the state annihilated by the operators $m_n^\dagger(Q)$, while the operators $m_n^\dagger(Q)$ create mesons out of the vacuum, with energies given by $E_n(Q)$. These energies should be interpreted as the energies of a bound state of two quarks of the same mass, with total momentum Q, i.e. they are given by

$$E_n(Q) = \sqrt{Q^2 + M_n^2}, \qquad n = 1, 2, \ldots, \qquad (9.3.38)$$

where M_n is the rest mass of the nth meson. The equations (9.3.20) give them the mass spectrum of mesons in two-dimensional QCD.

An alternative way of deriving these equations is by a diagrammatic analysis of the bound state problem. We will sketch how this alternative derivation goes, and the interested reader can find more details in the literature cited at the end of this chapter. In the diagrammatic approach, one studies the two-point function of the quark–antiquark current. The meson bound states lead to the sum on the right hand side of (7.3.27). The factor denoted by a_n in (7.3.27) defines the form factor of the meson. More precisely, we have

$$\tilde{\Gamma}(x^\mu, y^\mu, Q^\mu) = \langle 0|T\left(\overline{\psi}_i(x^\mu)\psi^i(y^\mu)\right)|Q^\mu\rangle, \qquad (9.3.39)$$

where $|Q\rangle$ is a bound state with momentum Q^μ. After Fourier transformation, one finds a form factor $\tilde{\Gamma}(p^\mu, Q^\mu)$, which describes a meson with total momentum Q^μ, and where the quark has momentum p^μ (therefore the antiquark has momentum $Q^\mu - p^\mu$). It is more convenient to use the wavefunction

$$\Gamma(p^\mu, Q^\mu) \propto S(p)\tilde{\Gamma}(p^\mu, Q^\mu)S(Q^\mu - p^\mu), \qquad (9.3.40)$$

where $S(p^\mu)$ is the quantum-corrected propagator (9.2.72). The interactions between the two quarks, in the planar limit, are described by the exchange of an arbitrary number of non-intersecting gluon lines, so in the planar limit only "ladder diagrams" contribute to the two-point function of the current. This fact leads to an inhomogeneous integral equation for the two-point function of the current, which in turn gives a homogeneous integral equation for (9.3.40):

$$\Gamma(p^\mu, Q^\mu) = \frac{i\gamma}{2\pi} \fint \frac{dk^0 dk}{(p-k)^2} S(p^\mu)\gamma^0\Gamma(k^\mu, Q^\mu)\gamma^0 S(p^\mu - Q^\mu). \qquad (9.3.41)$$

This is the homogeneous Bethe–Salpeter equation for the bound state. Let us now define the wavefunction

$$\varphi(p, Q) = \int dp^0 \, \Gamma(p^\mu, Q^\mu). \qquad (9.3.42)$$

If we now plug into (9.3.41) the explicit expression of the propagator in (9.2.72), we can perform the p^0 integration by residue calculus, after closing the integration contour appropriately, and we find

$$\varphi(p, Q) = \gamma \fint \frac{dk^0 dk}{(p-k)^2} \left[\frac{u(p)\bar{u}(p)\gamma^0\varphi(k, Q)\gamma^0 v(Q-p)\bar{v}(Q-p)}{E(p) + E(Q-p) - Q^0} \right.$$
$$\left. + \frac{v(-p)\bar{v}(-p)\gamma^0\varphi(k, Q)\gamma^0 u(p-Q)\bar{u}(p-Q)}{E(p) + E(Q-p) + Q^0} \right].$$

$$(9.3.43)$$

We now introduce the following unitary transformation of $\varphi(p, Q)$:

$$\tilde{\varphi}(p, Q) = T^\dagger(p)\varphi(p, Q)T(Q-p), \qquad (9.3.44)$$

and we decompose it as

$$\tilde{\varphi}(p, Q) = \varphi_+(p, Q)M^+ + \varphi_-(p, Q)M^-, \qquad (9.3.45)$$

where

$$M^\pm = \frac{1}{2}(1 \pm \gamma^0)\gamma_5. \qquad (9.3.46)$$

It is now straightforward to check that the integral equation (9.3.43) decomposes in two different integral equations for $\varphi_\pm(p, Q)$, which indeed agree with (9.3.20).

In principle, we could study the integral equations (9.3.20) for different values of Q: by Lorentz invariance, they should all lead to the same spectrum of masses. It turns out that a simpler set of integral equations appears when one considers the so-called *infinite-momentum frame*. This is the limit in which $Q \to \infty$. In order to understand this limit, one should rescale all the variables appearing in (9.3.20),

$$p = xQ, \qquad k = xQ, \tag{9.3.47}$$

and take the limit $Q \to \infty$ but keeping x, y fixed. Using now (9.2.63) we find, for $0 < x < 1$,

$$E(p) + E(Q - p) - E_n(Q) = \frac{1}{2Q} \left(\frac{m^2 - 2\gamma}{x} + \frac{m^2 - 2\gamma}{1 - x} - M_n^2 \right)$$
$$+ \mathcal{O}\left(\frac{1}{Q^2} \right). \tag{9.3.48}$$

For the other values of x we find, as $Q \to \infty$,

$$E(p) + E(Q - p) - E_n(Q) = \begin{cases} -2xQ + \cdots, & \text{if } x < 0, \\ 2(x - 1)Q + \cdots, & \text{if } x > 1. \end{cases} \tag{9.3.49}$$

We also find,

$$E(p) + E(Q-p) + E_n(Q) = \begin{cases} 2Q + \cdots, & \text{if } 0 < x < 1, \\ 2(1 - x)Q + \cdots, & \text{if } x < 0, \\ 2xQ + \cdots, & \text{if } x > 1. \end{cases} \tag{9.3.50}$$

Now, the right hand side of the two equations in (9.3.20) is at most of order $1/Q$, since both $S(p, k, Q)$ and $C(p, k, Q)$ are bounded. Therefore, all the terms on the left hand side which grow like Q should vanish. In view of (9.3.50) we conclude that

$$\lim_{Q \to \infty} \varphi_-^n(xQ, Q) = 0, \tag{9.3.51}$$

for all values of x, and from (9.3.49) we find

$$\lim_{Q \to \infty} \varphi_+^n(xQ, Q) = 0, \qquad x \notin [0, 1]. \tag{9.3.52}$$

Let us now denote

$$\varphi_n(x) = \lim_{Q \to \infty} \varphi_+^n(xQ, Q), \qquad 0 \le x \le 1. \tag{9.3.53}$$

Then, the first equation in (9.3.20) becomes, in the large Q limit,

$$\left(\frac{m_\gamma^2}{x} + \frac{m_\gamma^2}{1 - x} - M_n^2 \right) \varphi_n(x) = 2\gamma \fint \frac{dy}{(x - y)^2} \varphi_n(y), \tag{9.3.54}$$

where we have denoted

$$m_\gamma^2 = m^2 - 2\gamma. \tag{9.3.55}$$

This is the famous 't Hooft integral equation determining the mass spectrum of mesons in two-dimensional QCD at large N. It was originally derived by 't Hooft by analyzing the theory in the light-cone gauge. Here, we have derived it as a particular limit of the more general set of equations (9.3.20) obtained by Bars and Green in the axial gauge. A detailed analysis of the boundary conditions appropriate for this equation indicates that the wavefunctions $\varphi_n(x)$ should vanish at $x = 0, 1$.

't Hooft's equation can be regarded as a time independent Schrödinger equation of the form,

$$H|\phi_n\rangle = M_n^2|\phi_n\rangle, \tag{9.3.56}$$

where the Hamiltonian H has the following expression in coordinate space,

$$\langle x|H|x'\rangle = \delta(x - x') \left(\frac{m_\gamma^2}{x} + \frac{m_\gamma^2}{1 - x}\right) - 2\gamma H \frac{1}{(x - x')^2}. \tag{9.3.57}$$

This Hamiltonian can be be interpreted in two different ways. In light-cone coordinates, one has

$$p_\pm = \frac{1}{\sqrt{2}} \left(p^0 \pm p^1\right), \tag{9.3.58}$$

so we have

$$2p_+ p_- = m^2. \tag{9.3.59}$$

Let us consider a quark–antiquark system with total light-cone momentum $P_- = 1$, so that $p_- = x$ for one of the particles, and $p_- = 1 - x$ for the other particle. It follows that

$$2P_+ = \frac{m_1^2}{x} + \frac{m_2^2}{1 - x}. \tag{9.3.60}$$

Therefore, (9.3.57) can be interpreted as the Hamiltonian in momentum space. The first term is $P_+/2$, and it can be interpreted as the contribution of the momentum to the total mass of the particle. The second term can be interpreted as the potential energy between the two particles. Due to (9.2.38), in position space this is a linear confining potential which confines the quark and the antiquark inside the meson.

Conversely, we can shift roles and consider x as a position coordinate. In this case, the first term in (9.3.57) is the potential energy, and the second term in (9.3.57) is the kinetic energy written down in position space. If we write it in momentum space using (9.2.38), we find that this Hamiltonian can be written as

$$H = \pi|p| + V(x), \qquad V(x) = \frac{m_\gamma^2}{x} + \frac{m_\gamma^2}{1 - x}. \tag{9.3.61}$$

Therefore, we are studying the spectrum of an ultrarelativistic particle (with kinetic energy $\propto |p|$), confined in a box (since its wavefunction satisfies $\phi(0) = \phi(1) = 0$), and in the presence of a potential $V(x)$, which can be thought of as a Coulomb potential for two repulsive charges at $x = 0, 1$. This picture leads to some useful information about the spectrum of 't Hooft's equation. For example, it is interesting to understand how the masses M_n^2 in the meson spectrum behave as n becomes large. From the point of view of the Schrödinger equation, this is the semiclassical regime of large quantum numbers and it can be treated with the Bohr–Sommerfeld quantization condition. To write down this quantization condition, one computes the volume of phase space $\text{vol}(E)$ enclosed by the curve $H = E$, or equivalently one computes the classical action

$$\text{vol}(E) = \oint p(E) dq \tag{9.3.62}$$

for a periodic trajectory with energy E. The turning points for this system are

$$x_{1,2} = \frac{1}{2} \pm \frac{1}{2} \sqrt{1 - \frac{4m_\gamma^2}{E}}, \tag{9.3.63}$$

and one easily finds

$$\text{vol}(E) = \frac{2}{\pi} \left\{ E (x_2 - x_1) + m_\gamma^2 \log \left[\frac{x_1 (1 - x_2)}{x_2 (1 - x_1)} \right] \right\}. \tag{9.3.64}$$

At large E, we have

$$\text{vol}(E) \approx \frac{2}{\pi} \left\{ E - 2m_\gamma^2 \log(E) + \cdots \right\}. \tag{9.3.65}$$

The Bohr–Sommerfeld quantization condition reads, for n large,

$$\text{vol}(E) = 2\pi n, \tag{9.3.66}$$

since here we are setting $\hbar = 1$. This determines the spectrum of masses for large quantum numbers,

$$M_n^2 \approx \pi^2 n + 2m_\gamma^2 \log(n), \qquad n \gg 1. \tag{9.3.67}$$

We then have an infinite number of mesons, as expected from the general arguments in Section 7.3. Their masses fall asymptotically on a straight line of slope π^2, which is sometimes called a Regge trajectory.

9.4 Quantum Chromodynamics in two dimensions and the Hartree–Fock approximation

In the last section we showed that the Hamiltonian of QCD in two dimensions, at large N, becomes a quartic Hamiltonian in N fermion fields. The diagonalization of this Hamiltonian can be regarded as an N-body problem, which we solved at large N. We will now show that this solution is nothing but the Hartree–Fock or mean-field approximation to the N-body problem, which becomes exact at large N.

The mean-field method in many-body physics can be formulated in many ways, but an elegant and convenient one is based on the path integral formulation of the many-body propagator. In this language, the mean-field method consists in writing the path integral in terms of an auxiliary scalar field, which roughly speaking is a bilinear field in the fermions. The mean-field approximation is the stationary phase approximation applied to the scalar field path integral, and the resulting saddle-point equations are the Hartree–Fock equations of the original theory. The path integral derivation makes it possible to construct corrections to the Hartree–Fock solution in a systematic way, by just calculating corrections to the saddle-point approximation. In general, the mean-field approximation is not justified by any small parameter playing the role of $\hbar$, and one needs additional physical arguments to justify it. In the case at hand, however, we will see that there is an explicit $\hbar$ parameter in the problem, namely $1/N$, and the large N limit is the semiclassical limit of the mean-field path integral.

Let us consider a theory involving Grassmannian fields ζ_A and their conjugates $\bar{\zeta}_A$. Here, the index A is a generic index labelling spacetime dependence, spin or color. The path integral defining the partition function of the theory is

$$
Z = \int \mathcal{D}\bar{\zeta}\,\mathcal{D}\zeta\,\exp\left[i\bar{\zeta}_A K_{AB}\zeta_B - \frac{i}{2}\bar{\zeta}_A \zeta_C v_{ABCD}\bar{\zeta}_B \zeta_D \right]. \tag{9.4.1}
$$

Here, K_{AB} is a kinetic operator, and v_{ABCD} is the quartic interaction term. We can assume, without lack of generality, that

$$
v_{ABCD} = v_{BADC}. \tag{9.4.2}
$$

We now introduce an auxiliary scalar field, or mean field, with two indices, σ_{AB}. By using elementary Gaussian integration, we can rewrite the above path integral, up to an overall factor, as

$$
Z = \int \mathcal{D}\sigma\,\mathcal{D}\bar{\zeta}\,\mathcal{D}\zeta\,\exp\left[\frac{i}{2}\sigma_{AC} v_{ABCD}\sigma_{BD} + i\bar{\zeta}_A (K_{AC} - \Sigma_{AC})\,\zeta_C \right], \tag{9.4.3}
$$

where we have denoted

$$
\Sigma_{AC} = (v\sigma)_{AC} = v_{ABCD}\sigma_{BD}. \tag{9.4.4}
$$

Of course, if we integrate out the auxiliary field, we recover the original path integral (9.4.1). But in the new path integral (9.4.3), the dependence on the fermion fields is now quadratic, and we can integrate them out immediately to obtain a path integral for the mean field

$$Z = \int \mathcal{D}\sigma \, \exp\left[\frac{i}{2}\sigma_{AC}v_{ABCD}\sigma_{BD} + \log\det(K - \Sigma)\right]. \tag{9.4.5}$$

This transformation, in which a path integral with a quartic fermionic term is written in terms of an equivalent path integral involving a scalar field, is usually called a *Hubbard–Stratonovich* transformation. We can now study the resulting path integral for σ in the saddle-point approximation. By varying the new action functional with respect to σ we derive the saddle-point equation,

$$i\Sigma_{AC} = \left(\frac{1}{K - \Sigma}\right)_{BD} v_{DABC} \tag{9.4.6}$$

where we have used (5.4.5). If the path integral above describes a non-relativistic interacting Fermi gas, the saddle-point equation (9.4.6) leads to the Hartree–Fock approximation. More precisely, it leads to a mean-field theory which includes the Hartree term or the Fock term, depending on the way one writes the quartic interaction term (this choice is sometimes referred to as the choice of "channel").

Let us now apply this formalism to the problem of calculating the path integral of two-dimensional QCD in the Coulomb gauge. For simplicity, we will only consider the partition function, although correlators can easily be included. The action of the theory is, at leading order in N,

$$S = \int \bar{\psi}_i \left(i\partial\!\!\!/ - m\right) \psi^i \, dx \, dt + \frac{g_0^2}{8} \int \bar{\psi}_i \gamma^0 \psi^j(x,t)|x - y|\bar{\psi}_j \gamma^0 \psi^i(y,t) dx \, dy \, dt, \tag{9.4.7}$$

where we have used (9.2.24) and kept the first term. The partition function

$$Z = \int \mathcal{D}\bar{\psi} \, \mathcal{D}\psi \, e^{iS} \tag{9.4.8}$$

can be put in the form (9.4.1). The index $A = (x^\mu, i, \alpha)$ runs over spacetime position $x^\mu = (t, x)$, color and spinor indices. The kinetic operator K is given by

$$K_{AB} = \delta_{i_A i_B}(i\partial\!\!\!/ - m)\delta(x_A^\mu - x_B^\mu), \tag{9.4.9}$$

where we have written the spacetime indices as $x_A^\mu = (t_A, x_A)$, $x_B^\mu, \ldots$, and the color indices as $i_A, i_B, \ldots$. The spinor indices corresponding to A, B, C, D are $\alpha, \beta, \gamma, \delta$, respectively. The quartic interaction can be written as

$$v_{ABCD} = \frac{g_0^2}{4} \gamma^0_{\alpha\delta} \gamma^0_{\beta\gamma} |x_A - x_B| \delta(x_A^\mu - x_D^\mu)\delta(x_B^\mu - x_C^\mu)\delta(t_A - t_B)\delta_{i_A i_C}\delta_{i_B i_D}. \tag{9.4.10}$$

Let us analyze the resulting path integral at large N. First of all, notice that both K_{AB} and $\Sigma_{AB} = (v\sigma)_{AB}$ are diagonal in color space: due to their index structure, they are both proportional to $\delta_{i_A i_B}$, and we can write

$$(K - \Sigma)_{AB} = \delta_{i_A i_B} (K - \Sigma)_{\alpha\beta} (x_A^\mu, x_B^\mu) \tag{9.4.11}$$

where the operator of the right hand side acts only on spacetime and spinor indices. Therefore,

$$\log \det (K - \Sigma) = N \log \det (K - \Sigma)_{\alpha\beta} (x_A^\mu, x_B^\mu). \tag{9.4.12}$$

On the other hand, in the 't Hooft limit, $v \propto 1/N$. Let us extract this explicit factor of $1/N$ in v, and redefine $\sigma \to N\sigma$. The path integral reads

$$Z = \int \mathcal{D}\sigma \, \exp\left[\frac{i}{2} N \sigma_{AC} v_{ABCD} \sigma_{BD} + N \log \det (K - \Sigma)_{\alpha\beta} (x_A^\mu, x_B^\mu)\right]. \tag{9.4.13}$$

We see now that, in the 't Hooft limit, N plays the role of $\hbar$, and at large N the saddle point approximation is exact. The equation for the saddle point is

$$i\Sigma_{\alpha\gamma}(x_A^\mu, x_C^\mu) = \frac{g_0^2 N}{4} \left(\gamma^0 \frac{1}{i\slashed{\partial} - m - \Sigma} \gamma^0\right)_{\alpha\gamma} (x_A^\mu, x_C^\mu)|x_A - x_C|\delta(t_A - t_C). \tag{9.4.14}$$

This is precisely the equation (9.2.78) for the quantum-corrected propagator in the planar limit, in position space.

The above analysis shows that, at large N, the Hartree–Fock approximation is exact, and leads to the same results that we obtained before by using more conventional quantum field theoretical methods. In particular, we showed that the mean-field equation leads to the equation (9.2.78) for the quantum propagator. It can be seen that in our problem the Hartree term is absent, and there is only a Fock exchange term. The choice of (9.4.10) is indeed the standard choice for the "exchange" channel, in which the saddle-point approximation includes only the Fock term.

Although we have restricted ourselves to a re-derivation of the quantum propagator using path integral methods, one can also recover the meson spectrum that we obtained in the last section by analyzing the quadratic fluctuations around the saddle point. In many-body theory, the incorporation of these fluctuations is usually called the random phase approximation (RPA). The meson resonances turn out to correspond to the so-called RPA modes.

9.5 Bibliographical notes

The solution of two-dimensional QCD in the planar limit was found by 't Hooft in the classic paper [174] and is reviewed in [50]. 't Hooft worked in the light-cone gauge, while here we have followed very closely the analysis of Bars and Green in the Coulomb gauge [21]. A very useful review of two-dimensional QCD in the Coulomb gauge can be found in [117], as well as in the book [167]. The integral equations for the meson spectrum (9.3.20) were obtained by Bars and Green using diagrammatic methods, and they generalize the original equations obtained by 't Hooft in [174]. They were studied numerically in [124], where it was shown that they lead to the same spectrum as in [174]. The Hamiltonian approach to the meson spectrum was first developed by Lenz *et al.* [123] in the light-cone gauge. The extension of this approach to the Coulomb gauge was done in [118], which we have followed in our exposition. Many aspects of 't Hooft's equation (including the use of the Bohr–Sommerfeld quantization condition) are developed in [41].

The analogy between the large N solution of two-dimensional QCD and the Hartree–Fock approximation was pointed out by Witten in [189], and we have presented his argument in the standard language of many-body theory, as developed for example in [9, 146, 147]. For a many-body approach to the meson spectrum, see for example [108, 123].

10

Instantons at large N

10.1 Introduction

In this final chapter we will analyze the interplay between instantons and the large N expansion. As we will see, in the same way that the "classical" large N approximation involves a non-trivial resummation of perturbation theory, an instanton configuration at large N is much more complicated that its standard semiclassical counterpart. In some cases, they involve a resummation of Feynman diagrams in the background of a conventional instanton. However, large N instantons play an important conceptual role. As we have seen in Chapter 3, in theories without renormalons, instantons can be "discovered" by looking at the large order behavior of the perturbative expansion. One might suspect that a similar phenomenon would occur for the $1/N$ expansion, and indeed, in many examples, large N instantons are connected to the large order behavior of the $1/N$ expansion.

In this chapter, we will first discuss some general properties of the $1/N$ expansion and large N instantons. We will then show that, in some of the toy models considered in this book, such as matrix models and $\mathbb{CP}^{N-1}$ models, large N instantons can be analyzed in detail.

10.2 Analyticity and the $1/N$ expansion

As we discussed in Chapter 3, standard perturbation theory (even in the absence of renormalons) is divergent due to the factorial growth of the number of Feynman diagrams. In the $1/N$ expansion, however, we have to consider fatgraphs with a fixed topology. For example, each genus g free energy (7.3.3) is given by a sum of fatgraphs of genus g. It is therefore interesting to know how the number of such fatgraphs grows with the number of loops, since this will determine to a large extent the growth of the coefficients $a_{g,h}$ in (7.3.3), as h grows large.

It turns out that the number of planar diagrams (and, more generally, the number of fatgraphs with a fixed genus) grows only *exponentially* with the number of loops. This is much slower growth than the factorial growth of usual Feynman diagrams. Such a phenomenon can already be observed when we compare the number of graphs in the standard quartic oscillator and the number of planar graphs for the matrix quartic oscillator of Section 8.4: if we look at the numbers involved in Table 1.1 and at the numbers involved in Table 8.1, we see that, by restricting ourselves to planar diagrams, the symmetry factors are drastically reduced.

Another simple example where we can see in a precise way this difference of growth rates is the fatgraph version of the calculation (1.2.22). In this calculation, one counts the number of possible contractions in a vertex with $2k$ legs, and this number grows factorially. However, if we now regard the vertex as a fatgraph, and we consider the possible contractions, we will of course get planar and non-planar diagrams. The total number of contractions remains the same, and given by (1.2.22), but if one considers contractions that lead only to *planar* diagrams, as shown in Fig. 10.1 the resulting number is much smaller. One way to derive this number is to notice that the planar diagrams have a "petal" structure, in which the petals are either juxtaposed or included in one another (with no edge crossings). The counting of these petal diagrams is a standard problem in combinatorics which might be solved by using a recursion relation. Let us imagine that we want to obtain a petal diagram starting with a vertex with $2k$ edges, and let c_k be the number of such diagrams. We first fix one edge (say at position 1), and then we consider the edges which can be contracted with the first one. These edges are at positions $2j$, where $j = 1, 2, \ldots, k$ (other positions will lead to crossing edges, which are forbidden due to the planarity condition). The diagram obtained after contracting the edge $2j$ with the first one has two halves, with $2(j - 1)$ edges in one half and $2(k - j)$ edges in the other half, and therefore leads to $c_{j-1} \cdot c_{k-j}$ different petals. Summing over all the possible positions of the edge at $2j$ gives the recursion relation

Figure 10.1 Wick contractions in a vertex with $2k$ legs can be represented as "flower" diagrams. In the case of standard Feynman diagrams, petals can cross (left). If we consider instead planar fatgraphs, petals should not overlap (right).

$$c_k = \sum_{j=1}^{k} c_{j-1} c_{k-j}, \qquad c_0 = 1, \tag{10.2.1}$$

which is solved by the Catalan numbers

$$c_k = \frac{(2k)!}{(k+1)!k!}. \tag{10.2.2}$$

This number, in contrast to (1.2.22), grows only exponentially with k

$$c_k \sim 4^k, \qquad k \gg 1. \tag{10.2.3}$$

The fact that the number of planar diagrams, and more generally, the number of fatgraphs of a fixed genus, grows only exponentially, has important consequences for the analytic structure of the observables. In order to understand this more concretely, let us compare the ground state energy of the standard quartic oscillator (1.2.18) with its matrix version, equations (8.4.8) and (8.4.9). As we explained at the end of Section 3.4, the expression (1.2.19) for the coefficients in the perturbative series gives a divergent series due to the factorial growth of the symmetry factors s_Γ with the number of vertices, since the Feynman integrals $\mathcal{I}_\Gamma$ grow only exponentially. On the other hand, the expression (8.4.10) involves symmetry factors for fatgraphs $s_{\Gamma_{g,h}}$. As suggested by the example of the planar petals, these grow only exponentially with h, for a *fixed* genus g. This means that the genus g energies $\mathcal{E}_g(g_4)$ are *analytic* at the origin $g_4 = 0$, and as power series they have a finite radius of convergence. We can verify this by looking at the explicit solution (8.4.53) in terms of elliptic functions (note that in order to calculate $e_F(g)$ we have to invert an analytic function, and this preserves analyticity). The radius of convergence of the expansion (8.4.55) can be calculated by finding the singularity in the g_4-plane which is closest to the origin. This can be found by looking at the expressions for the turning points (8.4.49): the singularity occurs at the branch cut of the square root, since this is where analyticity breaks down. The value of e which appears in the expression (8.4.53) for the ground state energy is the Fermi energy $e_F(g_4)$. Therefore, the singular value of the coupling, g_4^c, satisfies

$$e_F(g_4^c) = -\frac{1}{4 g_4^c}. \tag{10.2.4}$$

For this value of the coupling, the modulus (8.4.50) becomes $-\infty$. Let us consider the limit of (8.4.51) as we approach this point. As $k^2 \to -\infty$, $K(k)$ vanishes, while

$$E(k) \approx \sqrt{-k^2}, \qquad k^2 \to -\infty, \tag{10.2.5}$$

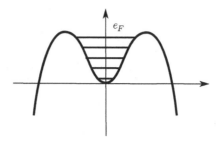

Figure 10.2 The Fermi level e_F in the quartic potential with negative coupling $g_4 < 0$. The singular value of g_4 is such that the Fermi energy e_F reaches the maximum of the potential.

and one easily deduces the value

$$g_4^c = -2\frac{\sqrt{2}}{3\pi}. \tag{10.2.6}$$

This has a nice interpretation in terms of the fermion picture: since g_4^c is negative, we have an inverted quartic potential, and at the critical value of g_4, the Fermi sea reaches the maximum of the potential, see Fig. 10.2. This indicates the onset of an instability.

Similar arguments can be used to show the following. Let us consider an observable in a gauge theory, and let us assume that the Feynman integrals only grow exponentially with the number of loops (i.e. renormalons are absent). Then, the contribution to the observable due to fatgraphs with a fixed topology and more than one loop is *analytic* at the origin, in the complex plane of the 't Hooft parameter (as we mentioned in Chapter 7, the tree level and one-loop contributions are somewhat special and might lead to non-analytic behavior). For example, in the expression (7.3.7) for the coefficients of the genus g free energy, both the coefficients $p_{g,h}(\Gamma)$ and the number of diagrams involved in the sum grow only exponentially with h. Given our assumption about the behavior of Feynman integrals, one concludes that each genus g free energy (7.3.3) is analytic at $t_0 = 0$. This assumption is *not* valid in theories with renormalon singularities (like YM theory and QCD), since in those cases the Feynman integrals diverge factorially, and even after restricting to diagrams with a fixed topology, one finds divergent series. In some gauge theories with special finiteness properties, like Chern–Simons theory or maximally super-symmetric YM theory, there are no renormalon singularities, and indeed explicit calculations show that planar amplitudes are analytic. It is also possible to show, by using estimates on the exponential growth, that when the genus g free energies $F_g(t_0)$ are analytic, they have a *finite*, *common* radius of convergence ρ. Typically we find a singularity in the complex plane of the 't Hooft parameter t_0^c, with

$$|t_0^c| = \rho. \tag{10.2.7}$$

This is what we observed in the example of MQM, where g_4 plays the role of the 't Hooft parameter, and there is a singularity (10.2.6) in the negative real axis of g_4 which sets the radius of convergence of the series (8.4.55). It can be shown that the higher genus ground state energies $\mathcal{E}_g(g_4)$ have the same radius of convergence and they are singular at the same value (10.2.6).

A final question we should ask is: assuming that a given theory is free of renormalon singularities, does the full $1/N$ expansion lead to analytic answers for the observables? In fact, this is not the case. We have argued that, after restricting to a fixed topology, we obtain analytic functions. But physical observables, for example the full ground state energy (8.4.8) or the full free energy (7.3.2), are themselves, for a fixed value of the 't Hooft couplings, formal power series in $1/N$. It turns out that this series is again factorially divergent. This is due to the fact that the number of fatgraphs of genus g grows with the genus as $(2g)!$. For example, in the case of MQM, one finds that, at fixed g_4, the genus g ground state energies behave like

$$\mathcal{E}_g(g_4) \sim (2g)!(A(g_4))^{-2g}, \tag{10.2.8}$$

where $A(g_4)$ is a function of g_4. Similarly, in gauge theories with no renormalons, the free energies are expected to grow as

$$F_g(t_0) \sim (2g)!(A(t_0))^{-2g}. \tag{10.2.9}$$

This is very similar to the behavior (3.2.5) typical of quantum perturbative series. Therefore, the $1/N$ expansion has in general a zero radius of convergence, and to make sense of it one has to use the techniques developed in Chapter 3.

10.3 Large N instantons

What happens to instantons in large N theories? Let us consider a YM theory with bare coupling constant g_0^2, and an instanton solution with action

$$S_{\text{inst}} = \frac{c_0}{g_0^2}. \tag{10.3.1}$$

Let us assume that the action of the instanton is of order one at large N, i.e. $c_0 \sim \mathcal{O}(1)$. This is generally the case, since as we saw in Chapter 4, we can build an instanton by using just an $SU(2)$ subgroup of $SU(N)$. We can now do perturbation theory around this instanton configuration. The one-loop fluctuations give a term with the generic form (at large N)

$$\left(\frac{c}{g_0^2}\right)^{c_1 N} \tag{10.3.2}$$

where $c_1 N$ is the number of zero modes, or collective coordinates of the instanton, at large N (for example, in the calculation of the one-instanton partition function for YM theory on $\mathbb{S}^4$, we have $c_1 = 4$, as shown in (4.5.80) and (4.5.91)). Putting both things together and, expressing everything in terms of g_0^2 and the 't Hooft parameter t_0, we find

$$\left(\frac{c}{g_0^2}\right)^{c_1 N} e^{-c_0/g_0^2} \approx \exp\left(-\frac{A(t_0)}{g_0^2}\right) \tag{10.3.3}$$

where

$$A(t_0) = c_0 - c_1 t_0 \log\left(\frac{c}{t_0}\right) + \mathcal{O}(t_0), \tag{10.3.4}$$

is called the large N instanton action. It is given by a series in t_0 which incorporates, on top of the classical action and the one-loop fluctuations which we have written down explicitly, all vacuum, connected bubble planar diagrams (at all loops) in the background of the classical instanton. To see how these appear, let us focus for simplicity on the interaction given by the cubic vertex (see Fig. 7.4). Let us consider fluctuations of the gauge connection $\mathcal{A}$ around an instanton solution $\overline{\mathcal{A}}$ (we use the notation $\mathcal{A}$ for the gauge connection to avoid confusion with the notation for the instanton action):

$$\mathcal{A} = \overline{\mathcal{A}} + \mathcal{A}'. \tag{10.3.5}$$

The action for the fluctuations will include a vertex of the form

$$\sum_{i,j,k} (\overline{\mathcal{A}}_\mu)^i{}_j (\mathcal{A}_\nu)^j{}_k (\mathcal{A}_\rho)^k{}_i \tag{10.3.6}$$

and involve the instanton background. We can represent this vertex in the double-line notation as in Fig. 10.3, where the lines ending on the blob correspond to the instanton background. This gives a factor of g_0^2, but only the interior line gives a factor of N after tracing over color indices. A simple example of a diagram contributing to the instanton action is the one depicted on the right hand side of Fig. 10.3. The closed lines in the interior of the diagram give a factor of N^3, and the diagram is proportional to

$$\mathrm{Tr}\left(\overline{\mathcal{A}}^3\right) N^3 g_0^4 = \frac{1}{g_0^2} t_0^3 \mathrm{Tr}\left(\overline{\mathcal{A}}^3\right), \tag{10.3.7}$$

since there are nine edges $E = 9$ and seven vertices $V = 7$, so the power of $g_0^{2(E-V)}$ is four. This diagram gives a correction of order t_0^3 to $A(t_0)$. It transpires from this example that the calculation of large N instanton actions is a difficult matter, since we have to sum up an infinite number of planar diagrams (in the same

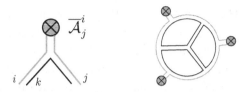

Figure 10.3 The instanton vertex (10.3.6) (left) and a planar diagram contributing to the large N instanton action $A(t_0)$ a term of order t_0^3 (right).

way that calculating the planar free energy involves adding up an infinite number of diagrams at all loops).

An alternative, more general way to think about large N instantons is in terms of large N effective actions. The idea of the master field suggests that ordinary theories at large N can be reformulated in terms of a "large N effective action" with coupling constant $1/N$. In this effective theory, correlation functions at large N are obtained by solving the classical EOM of the effective action in the presence of sources. A large N instanton is simply an instanton solution of this large N effective theory, i.e. a solution of the Euclidean EOM with finite action. This is in general different from the usual instanton configurations, which are saddle points of the *classical* action. In some cases, large N instantons can be thought of as deformations of classical instantons, where the deformation parameter is the 't Hooft parameter. This can easily be seen in the diagrammatic argument presented above: it is clear from (10.3.4) that when $t_0 \to 0$ we recover the gauge theory instanton action.

In the same way that the standard factorial growth of the perturbative expansion is related to "standard" instantons, the factorial growth of the $1/N$ expansion, as noted in (10.2.9), is related to large N instantons. In particular, the function $A(t_0)$ can be identified in many cases with the action of a large N instanton. Notice that, in the calculation of observables, large N instantons are weighted by the factor

$$\exp\left(-A(t_0)/g_0^2\right) = \exp\left(-NA(t_0)/t_0\right), \tag{10.3.8}$$

where $A(t_0)$, the large N instanton action, is in general a non-trivial function of the 't Hooft parameter. If $\mathrm{Re}(A(t_0)/g_0^2) > 0$, large N instantons are suppressed exponentially at large N. This might lead one to think that instanton corrections are unimportant at large N, but this is not necessarily the case. It might happen for example that $A(t_0)$ vanishes at a particular value of t_0. In this case, instantons are no longer suppressed and their contribution becomes as important as the $1/N$ expansion itself. When this is the case, we have typically a phase transition in which the $1/N$ expansion changes discontinuously. The value of the 't Hooft parameter for which $A(t_0)$ vanishes very often signals a *large N phase transition*, or a critical point, in the theory. The critical value of the 't Hooft parameter is also, in many cases, the first singularity t_0^c in the t_0-plane which we found in (10.2.7). This is natural, since this singularity signals the breakdown of the 't Hooft expansion.

It has been pointed out that sometimes instanton methods and large N methods might lead to contradictory results. For example, a naive instanton analysis, like the one leading to (4.5.104), seems to indicate that the topological susceptibility of YM theory or the $\mathbb{CP}^{N-1}$ model is an instanton effect, therefore it should be exponentially suppressed at large N, as in (10.3.8). However, the $1/N$ analysis of (7.4.5) and (7.4.7) does not indicate such an exponential suppression, and indeed, our explicit calculation in the $\mathbb{CP}^{N-1}$ model at large N agrees with the scaling (7.4.7). The resolution of this apparent paradox is simply that, in the absence of an IR cutoff, semiclassical instanton analyses are not justified, and one should not trust them. It has been verified in many concrete models that, when both large N methods and instanton methods are valid, they lead to perfectly compatible results. We will see an example of this in our study of large N instantons in the $\mathbb{CP}^{N-1}$ model on the sphere, in Section 10.5.

The analysis in this section has focused on the instantons of YM theory, but it is easy to see that, as long as the action of an instanton is of order one at large N, many of the results we have obtained generalize to other models. In the next sections we will consider concrete examples of large N instantons in simple models where one can obtain analytic results and make concrete the general considerations of this section.

10.4 Large N instantons in matrix models and matrix Quantum Mechanics

Let us first consider Hermitian matrix models of the form (8.2.4), which can be written in terms of eigenvalues as in (8.2.35). The analogue of the classical EOM is (8.2.42). As we have seen, in the one-cut solution to the matrix model, the solution to these equations is a distribution of eigenvalues filling an interval, and along this interval the effective potential $V_{\text{eff}}(\lambda)$ introduced in (8.2.54) is constant. Let us now assume that $V_{\text{eff}}(\lambda)$ has critical points outside the cut where the eigenvalues condense. In order to find such critical points, we can proceed as follows. Consider (8.2.54) and suppose for example that $x > a$, where a is the endpoint of the cut. Then, by the definition of planar resolvent (8.2.60), we find

$$V_{\text{eff}}(x) - V_{\text{eff}}(a) = \int_a^x y(p)\mathrm{d}p, \qquad (10.4.1)$$

where $y(p)$ is defined in (8.2.77). Therefore,

$$V'_{\text{eff}}(x) = y(x), \qquad (10.4.2)$$

and we have a critical point x_0 if the moment function $M(p)$ defined in (8.2.75) has a zero at x_0, i.e.

$$M(x_0) = 0. \tag{10.4.3}$$

In general, an instanton configuration must also solve the EOM, but the instanton solution must be different from the perturbative background. In our case, the perturbative background corresponds to the situation in which all eigenvalues condense at the cut. We can obtain another solution to the EOM if we take a single eigenvalue, out of the N eigenvalues, and we put it in a critical point x_0 of $V_{\text{eff}}(x)$ which is outside the cut. Such a configuration gives a contribution to the matrix model partition function with a weight,

$$\exp\left(-A/g_s\right). \tag{10.4.4}$$

Here, g_s is the coupling constant introduced in (8.2.4), and the action of the instanton is given by the difference between the effective potential at x_0 and the effective potential at the cut,

$$A = V_{\text{eff}}(x_0) - V_{\text{eff}}(a). \tag{10.4.5}$$

The effect (10.4.4) is clearly non-perturbative in g_s.

This picture of a large N instanton in the matrix model is sometimes called *eigenvalue tunneling*, since a one-instanton effect can be interpreted as the effect of moving one eigenvalue from the cut where the eigenvalues condense, to a non-trivial saddle point (see Fig. 10.4). The action of the instanton is just the cost of moving the eigenvalue, as measured by the effective potential. Clearly, the ℓ-instanton sector is obtained by considering the tunneling of ℓ eigenvalues instead of one. The action of an ℓ-instanton is just given by ℓA. Notice that the action of an ℓ-instanton in which $\ell \ll N$ is of order one at large N, as in the analysis of large N instantons in gauge theories.

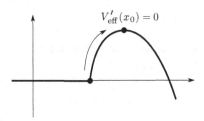

Figure 10.4 A large N instanton in the one-cut matrix model can be obtained by "tunneling" one eigenvalue from the support of the density of eigenvalues, to a critical point of the effective potential outside the cut.

The instantons of the matrix model can be regarded as "deformations" of the non-trivial saddle points appearing in the study of the one-dimensional integral

$$I(g_s) = \int e^{-V(x)/g_s} dx, \qquad (10.4.6)$$

where $V(x)$ is the potential of the matrix model. Indeed, if we assume that $x = 0$ is a minimum of the potential, we can do a saddle-point calculation of the integral (10.4.6) around this point. However, there might be other saddle points x_*, given by critical values of $V(x)$,

$$V'(x_*) = 0. \qquad (10.4.7)$$

The integral around this saddle point is of order $\exp(-V(x_*)/g_s)$. In the limit in which the 't Hooft parameter goes to zero, the effective potential of the matrix model $V_{\text{eff}}(x)$ becomes just $V(x)$, therefore

$$A \rightarrow V(x_*) - V(0) \qquad (10.4.8)$$

for $t \rightarrow 0$. This is easy to understand, since in the limit of zero 't Hooft coupling, the interaction between the eigenvalues is turned off, and the matrix integral becomes a product of one-dimensional integrals of the form (10.4.6). This is a simple example of the phenomenon noted in the previous section, when discussing large N instantons in general: in many cases, large N instantons can be regarded as deformations of conventional instantons, and in particular their action becomes the classical instanton action when $t \rightarrow 0$. This is precisely the content of (10.4.8).

Example 10.1   *Large N instanton in the quartic matrix model.* Let us consider the quartic matrix model, defined by (8.2.87), and let us set $g = -\lambda/48$ to simplify the resulting expressions. With this notation, the quartic potential has three critical values:

$$x = 0, \qquad x_* = \pm \frac{2\sqrt{3}}{\sqrt{\lambda}}. \qquad (10.4.9)$$

The moment function can be read off from (8.2.92). It is given by

$$M(x) = 1 - \frac{\lambda}{6}a^2 - \frac{\lambda}{12}x^2. \qquad (10.4.10)$$

The moment function has two zeros which give two non-trivial saddle points of the effective potential, namely $\pm x_0$ with

$$x_0^2 = \frac{12}{\lambda} - 2a^2. \qquad (10.4.11)$$

As $t \rightarrow 0$, $\alpha \rightarrow 0$ and they become the two non-trivial critical points of the potential given in (10.4.9). Since the potential is symmetric, there are *two* one-instanton solutions, corresponding to one eigenvalue tunneling from the cut to the two saddles $\pm x_0$. Both instantons have the same action, which is given by (10.4.5). The effective potential can be obtained by integrating the equation for the spectral curve (8.2.92). One finds,

$$
A(t) = \frac{3}{\lambda}\sqrt{\left(1 - \frac{a^2\lambda}{2}\right)\left(1 - \frac{a^2\lambda}{6}\right)}
$$
$$
- \frac{1}{4}a^2\left(a^2\lambda - 4\right)\left(\log\left(\frac{a^2\lambda}{3}\right) - 2\log\left(\sqrt{1 - \frac{a^2\lambda}{2}} + \sqrt{1 - \frac{a^2\lambda}{6}}\right)\right).
$$

$$(10.4.12)$$

This function has the small t expansion

$$
A(t) = \frac{3}{\lambda} + t\left(-1 + \log\left(\frac{\lambda t}{12}\right)\right) + \mathcal{O}(t^2). \tag{10.4.13}
$$

The first term (i.e. the instanton action as $t \to 0$) is given by

$$
V\left(\frac{2\sqrt{3}}{\sqrt{\lambda}}\right) \tag{10.4.14}
$$

in agreement with (10.4.8). The structure of the next-to-leading correction is in agreement with the general expectation (10.3.4). □

The same techniques can be used to study the large N instantons in the Gross–Witten–Wadia unitary matrix model, in the weakly coupled or gapped phase, in which the eigenvalues condense in the cut $[-\alpha_c, \alpha_c]$, with $\alpha_c < \pi$. In this case, the potential (which is minus the action $S(\phi)$) is of the form

$$
V(\phi) = -\beta \cos \phi. \tag{10.4.15}
$$

It has a minimum at $\phi = 0$, and a saddle point (maximum) at $\phi = \pi$. In complete analogy with what we did in the case of Hermitian matrix models, we can define an effective potential which satisfies

$$
V'_{\text{eff}}(\phi) = V'(\phi) - 2\omega_0(\phi), \tag{10.4.16}
$$

where $\omega_0(\phi)$ is the planar resolvent written down in (8.3.34). We then obtain

$$
V_{\text{eff}}(\phi) - V_{\text{eff}}(\alpha_c) = 2\beta \int_{\alpha_c}^{\phi} \cos\frac{z}{2}\sqrt{\sin^2\frac{z}{2} - \sin^2\frac{\alpha_c}{2}}\,dz = 2\Phi\left(\frac{\sin\frac{\phi}{2}}{\sin\frac{\alpha_c}{2}}\right),
$$

$$(10.4.17)$$

where

$$
\Phi(x) = x\sqrt{x^2 - 1} - \cosh^{-1}(x), \tag{10.4.18}
$$

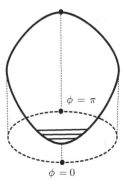

$\phi = \pi$

$\phi = 0$

Figure 10.5 A large N instanton in the weakly coupled phase of the Gross–Witten–Wadia model is also obtained by eigenvalue tunneling from the cut where the eigenvalues condense, to the saddle point at $\phi = \pi$. In this figure we have depicted schematically the effective potential of the model, which is defined on the circle where the eigenvalues live and is given by (10.4.17).

and we have used (8.3.37). The point $\phi = \pi$ is also a saddle point of the effective potential, therefore an instanton configuration corresponds to an eigenvalue which tunnels from the cut to $\phi = \pi$, see Fig. 10.5. Its action is given by

$$A(\beta) = V_{\text{eff}}(\pi) - V_{\text{eff}}(\alpha_c) = 2\Phi(\beta^{1/2}). \tag{10.4.19}$$

It can be seen that this instanton configuration, as well as the multi-instanton generalizations in which one tunnels ℓ eigenvalues, lead to corrections to the partition function with a weight of the form

$$\exp\left(-N\ell A(\beta)\right). \tag{10.4.20}$$

In principle, one might think that these corrections would be suppressed at large N. However, as we pointed out above, one has to be careful since the instanton action might vanish for some value of the 't Hooft parameter. Indeed, when $\beta = 1$, the instanton action (10.4.19) *vanishes*, and corrections are no longer suppressed. On the other hand, $\beta = 1$ is the value of the inverse 't Hooft coupling at which the Gross–Witten–Wadia phase transition takes place. We can now interpret this phase transition as triggered by instantons: at $\beta = 1$, the exponentially small corrections to the $1/N$ expansion are no longer suppressed, and this expansion must change. This is in agreement with our explicit analysis in Section 8.3, where we showed that the planar free energies are different in each phase. This is very similar to the Stokes phenomenon of classical asymptotic theory. The Stokes phenomenon takes place when the asymptotic form of a function changes in a discontinuous way as we change the parameters or the direction in which we expand the function. This is typically due to saddle points which used to be exponentially suppressed and become suddenly of order one. Of course, the situation in the Gross–Witten–Wadia matrix model is more complicated than in the standard examples of the Stokes phenomenon, but the underlying principle is the same.

It turns out that large N instanton calculations are relevant for the study of the statistics of eigenvalues in the theory of random matrices. In this theory, the integrand of the partition function (8.2.4) is regarded as a (non-normalized) probability distribution for a Hermitian matrix M. It is natural in that context to study the statistical properties of the largest or extreme eigenvalue of the random matrix. This is of course a fluctuating variable, and its average value at large N, in a one-cut distribution, is the endpoint of the cut a. Its typical or small fluctuations are well described, in a certain scaling limit, by the so-called Tracy–Widom distribution. However, one is also interested in the statistical properties of large deviation, which in certain cases are accesible to experimental measurement.

A particularly important case in the study of maximum eigenvalue statistics are Wishart random matrices. A Wishart matrix is defined as the product

$$W = X^\dagger X, \tag{10.4.21}$$

where X is an $M \times N$ random matrix. We will set $N \le M$. The probability distribution for X is a Gaussian distribution

$$P(X) \propto \exp\left[-\frac{\beta}{2}\mathrm{Tr}\left(X^\dagger X\right)\right]. \tag{10.4.22}$$

We will focus here on the case $\beta = 2$, which corresponds to complex X. One can now integrate over "angular variables," as in the case of Hermitian matrices, in order to find the probability distribution for the eigenvalues. This is similar to the Faddeev–Popov calculation we did in Section 8.2, and one finds

$$P(\lambda_i) \propto \prod_{i=1}^{N} d\lambda_i e^{-\frac{1}{g_s}V(\lambda_i)} \prod_{i<j}(\lambda_i - \lambda_j)^2, \tag{10.4.23}$$

where

$$V(\lambda) = \lambda - \xi \log \lambda, \tag{10.4.24}$$

and

$$\xi = \frac{1}{c} - 1, \qquad c = \frac{N}{M} \le 1. \tag{10.4.25}$$

The eigenvalues λ_i live on the positive real axis. We have introduced a parameter g_s which can be set to $1/N$ at the end of the calculation. This corresponds to setting $t = 1$, where $t = g_s N$ is as usual the 't Hooft parameter.

The above probability distribution suggests considering the partition function

$$Z \propto \int \prod_{i=1}^{N} d\lambda_i e^{-\frac{1}{g_s} V(\lambda_i)} \prod_{i<j} (\lambda_i - \lambda_j)^2, \qquad (10.4.26)$$

which can be be studied with the techniques developed in Section 8.2. The potential (10.4.24) has a minimum at $\lambda = \xi$. We can use the formulae (8.2.74) and (8.2.75) to obtain the resolvent. Although these formulae were derived in the case in which $V(\lambda)$ is a polynomial potential, it can easily be seen that the derivation goes through whenever $V'(\lambda)$ is a rational function. One finds,

$$\omega_0(p) = \frac{1}{2t}\left(1 - \frac{\xi}{p}\right) - \frac{\xi}{2t\sqrt{ab}} \frac{\sqrt{(p-a)(p-b)}}{p}. \qquad (10.4.27)$$

The conditions (8.2.73) give for the endpoints,

$$a = \left((t+\xi)^{1/2} + t^{1/2}\right)^2, \qquad b = \left((t+\xi)^{1/2} - t^{1/2}\right)^2, \qquad (10.4.28)$$

and the density of eigenvalues is

$$\rho(\lambda) = \frac{1}{2\pi t \lambda}\sqrt{(\lambda - a)(b - \lambda)}. \qquad (10.4.29)$$

After setting $t = 1$, in order to recover the standard results for Wishart random matrices, one finds

$$a = \left(c^{-1/2} + 1\right)^2, \qquad b = \left(c^{-1/2} - 1\right)^2. \qquad (10.4.30)$$

The density of eigenvalues (10.4.29) for the Wishart ensemble, together with the endpoints of the cut (10.4.30), is usually called the *Marcenko–Pastur law*. The value of a gives the average value for the maximal eigenvalue of W, at large N.

The effective potential of this model $V_{\text{eff}}(z)$ turns out to calculate the so-called *right deviation function* for the maximal eigenvalue statistics, which measures the distribution function of large deviations for the top eigenvalue of the matrix W. This is because this function studies the effect of pushing one eigenvalue out of the cut to $z > a$. We will denote it by $\Phi_+(z)$, and it is given by our general formula (10.4.1),

$$\Phi_+(z) = \frac{1}{2} \int_a^z y(p) dp. \qquad (10.4.31)$$

The factor of $1/2$ is introduced in order to agree with the conventions in the literature on extreme statistics. The spectral curve $y(p)$ is given in our case by

$$y(p) = \frac{\sqrt{(p-a)(p-b)}}{p}. \qquad (10.4.32)$$

The primitive of $y(z)$ is elementary and can be computed in closed form:

$$\int^z y(z')dz' = \sqrt{(z-a)(z-b)} + 2\sqrt{ab}\tanh^{-1}\left[\sqrt{\frac{b(z-a)}{a(z-b)}}\right]$$

$$- (a+b)\log\left[2\sqrt{z-a} + 2\sqrt{z-b}\right]. \tag{10.4.33}$$

After using this formula, and the explicit values of a, b as a function of ξ, we obtain

$$\Phi_+(z) = -\xi\log\left(\sqrt{z-a} + \sqrt{z-b}\right) + \xi\log\left(\sqrt{b(z-a)} + \sqrt{a(z-b)}\right)$$

$$+ \frac{1}{2}\sqrt{(z-a)(z-b)} - 2\log\left(\sqrt{z-a} + \sqrt{z-b}\right)$$

$$+ \log(b-a) - \frac{1}{2}\xi\log(z). \tag{10.4.34}$$

This function is known as the Majumdar–Vergassola right deviation function for the Wishart ensemble. It has been measured experimentally in systems of coupled lasers, which then provide an experimental implementation of some of the ideas of large N instantons.

Finally, let us study instanton effects in MQM, focusing for simplicity on the quartic model (8.4.5). When $g_4 = -\kappa < 0$ we have a metastable vacuum at the origin and we should expect the existence of an instanton configuration mediating vacuum decay. This instanton solution can be found by tunneling one single eigenvalue of the matrix M, which has an action of order $\mathcal{O}(1)$, i.e. it is of the form,

$$M_c(t) = \text{diag}(0, \ldots, 0, q_c(t), 0, \ldots, 0), \tag{10.4.35}$$

where $q_c(t)$ is the bounce (1.4.18). In principle, one could consider the path integral of MQM around this configuration and compute quantum planar fluctuations to determine the large N instanton action. However, the fermion picture, which gave us a compact way of computing the planar ground state energy, should also give us an efficient way to compute the large N instanton action in a single strike. In this picture, the ground state is given by a filled Fermi level. As in any Fermi system, tunneling effects will first affect fermions which are near the Fermi surface. The instanton action of such a fermion is just given by the standard WKB action,

$$\frac{A(\kappa)}{\lambda} = N(2\kappa)^{1/2}\int_a^b\sqrt{(\lambda^2 - a^2)(b^2 - \lambda^2)}, \tag{10.4.36}$$

where $\lambda = \kappa/N$ is the original coupling of the theory, a, b are the turning points associated to the Fermi energy e_F, and they can be obtained from the expressions (8.4.49) by setting $g_4 = -\kappa$. There is an extra factor of 2 due to the symmetry of the problem, and the factor of N on the right hand side is due to the fact that $1/N$

is the effective Planck constant of this problem. The integral in (10.4.36) can be explicitly computed by using elliptic functions, and the final result is

$$A(\kappa) = \frac{1}{3}(2\kappa^3)^{1/2}b\left[(a^2 + b^2)E(k) - 2a^2 K(k)\right], \tag{10.4.37}$$

where the elliptic modulus k is given by

$$k^2 = \frac{b^2 - a^2}{b^2}. \tag{10.4.38}$$

The instanton action has the following expansion around $\kappa = 0$,

$$A(\kappa) = \frac{4}{3} - \kappa \log\left(\frac{16e}{\kappa}\right) + \frac{17\kappa^2}{16} + \frac{125\kappa^3}{128} + \cdots. \tag{10.4.39}$$

This is precisely the expected structure for a large N instanton action, as we discussed in (10.3.4): the leading term is the action for an instanton (1.7.9) in the $N = 1$ quantum mechanical problem. The log term is a one-loop factor in disguise, and the rest of the series is a sum of loop corrections in the background of the "classical" instanton. An interesting property of $A(\kappa)$ is that it vanishes at the critical value

$$\kappa = -g_4^c, \tag{10.4.40}$$

where g_4^c is given in (10.2.6). As we explained in Section 10.2, this is the first singularity in the complex g_4-plane. It corresponds to the critical point in which the Fermi sea reaches the local maximum. At this value of κ, instanton tunneling is no longer exponentially suppressed and the theory becomes unstable, even in the planar limit.

10.5 Large N instantons in the $\mathbb{CP}^{N-1}$ model

An interesting and highly non-trivial example of large N instantons occurs in the $\mathbb{CP}^{N-1}$ model on the two-sphere. In this case, since there is an IR cutoff due to the finite volume of the two-sphere, instanton calculus is well defined, and one can in principle compute various quantities by using a semiclassical expansion in each instanton sector. For example, one can calculate the probability of finding a field configuration in the topological sector with charge k by using instanton methods (this is the quantity P_k that we introduced in the context of YM theory in (4.3.75), but of course we can also define it for the $\mathbb{CP}^{N-1}$ model). This calculation can be done for finite N and low instanton number, but also at large N and arbitrary instanton number. This is a generalization of the large N calculation in Section 6.3 to a general instanton sector on the two-sphere. Such a calculation was done by Münster, building on the work of Berg and Lüscher on instantons in the $\mathbb{CP}^{N-1}$

model, and it is a nice example of how large N methods make it possible to calculate a non-trivial instanton action as a function of the (renormalized) 't Hooft parameter.

We consider then the $\mathbb{CP}^{N-1}$ model on the two-sphere. Since we want to analyze non-trivial instanton sectors, we use the formulation in terms of $SU(2)$ developed in Section 6.3. For later convenience, we will normalize the fields ξ defined in (6.3.51) as follows,

$$\|\xi_j(g)\|^2 = \frac{N}{t_0}, \tag{10.5.1}$$

where t_0 is the bare 't Hooft coupling introduced in (6.2.4). Let us denote by $\mathcal{H}_k$ the subset of field configurations with topological charge k. The partition function in the sector $\mathcal{H}_k$ is given by the Euclidean functional integral,

$$Z^{(k)} = \int_{\mathcal{H}_k} \mathcal{D}\bar{\xi}\mathcal{D}\xi\, \delta\left(\|\xi\|^2 - \frac{N}{t_0}\right) e^{-S}, \tag{10.5.2}$$

where the action is the one given in (6.3.76), after the change of normalization of ξ_j:

$$S = 4\pi \sum_{a=1}^{2} \int_{SU(2)} dg\, \|J_a\xi\|^2, \tag{10.5.3}$$

and the operator J_a reads, with this normalization,

$$J_a = L_a - \frac{t_0}{N}\bar{\xi} \cdot L_a\xi. \tag{10.5.4}$$

Therefore,

$$\|J_a\xi\|^2 = \|L_a\xi\|^2 - \frac{t_0}{N}\|\bar{\xi} \cdot L_a\xi\|^2. \tag{10.5.5}$$

The second term is quartic in the fields. We can now perform a Hubbard–Stratonovich transformation, similar to what we did in (9.4.3), i.e. we can introduce two real, scalar auxiliary fields λ_a, which generate this term after integrating them out. Let us then consider the path integral

$$\int \mathcal{D}\lambda_a \exp\left[4\pi \int dg\left(-\frac{1}{N}\sum_{a=1}^{2}\lambda_a^2\|\xi\|^2 + \frac{2}{\sqrt{N}}\sum_{a=1}^{2}\lambda_a \mathrm{Re}\left(\bar{\xi} \cdot L_a\xi\right)\right)\right]. \tag{10.5.6}$$

The λ_a do not propagate, since they have no kinetic term. Hence, the path integral (10.5.6) can be calculated, up to an overall normalization factor, by simply replacing the field λ_a by the solution to the algebraic EOM,

$$\lambda_a = \frac{\sqrt{N}}{\|\xi\|^2}\mathrm{Re}\left(\bar{\xi} \cdot L_a\xi\right). \tag{10.5.7}$$

The path integral (10.5.6) is then proportional to

$$
\exp\left[4\pi \sum_{a=1}^{2} \int dg \frac{\left(\mathrm{Re}\left(\bar{\xi} \cdot L_a \xi \right) \right)^2}{\| \xi \|^2} \right] = \exp\left[\frac{4\pi t_0}{N} \sum_{a=1}^{2} \int dg \, \| \bar{\xi} \cdot L_a \xi \|^2 \right],
$$

(10.5.8)

where we took into account (6.3.69). This is the quartic term appearing in $-S$. We also introduce a Lagrange multiplier α imposing the delta function constraint in the path integral (10.5.2),

$$
\delta\left(\| \xi \|^2 - \frac{N}{t_0} \right) = \int \mathcal{D}\alpha \exp\left[4\pi i \int dg \, \alpha \left(\| \xi \|^2 - \frac{N}{t_0} \right) \right].
$$

(10.5.9)

We conclude that the partition function can be represented as

$$
Z^{(k)} = \int_{\mathcal{H}_k} \mathcal{D}\lambda_a \mathcal{D}\alpha \mathcal{D}\bar{\xi} \mathcal{D}\xi \, e^{-S'},
$$

(10.5.10)

where

$$
S' = 4\pi \int_{SU(2)} dg \left(\bar{\xi} \cdot \Delta_k \xi + \frac{iN}{t_0} \alpha \right),
$$

(10.5.11)

and

$$
\Delta_k = \sum_{a=1}^{2} \left(L_a - \frac{1}{\sqrt{N}} \lambda_a \right)^2 - i\alpha.
$$

(10.5.12)

Note that the operator whose square appears here becomes J_a once we replace λ_a by the value (10.5.7). The subindex k in Δ_k means that we are considering fields with topological charge k. We can now integrate out ξ, $\bar{\xi}$ to obtain an effective action for λ_a and α, of the form

$$
S_{\mathrm{eff}} = N \mathrm{Tr} \log \Delta_k + \frac{4\pi iN}{t_0} \int_{SU(2)} dg \, \alpha.
$$

(10.5.13)

This is just the generalization of (6.3.81) to the path integral on the two-sphere, and to an arbitrary instanton sector. Notice that the scalar fields λ_a play the role of the gauge field A_μ in flat space.

We now follow the same steps as in our previous analysis of the $\mathbb{CP}^{N-1}$ model: at large N the path integral can be computed by a saddle-point approximation. We assume, as before, that the saddle point occurs at

$$
\lambda_a = 0, \qquad \alpha = im_k^2 R^2,
$$

(10.5.14)

where R is the radius of $\mathbb{S}^2$, and m_k^2 is a quantity to be determined by the saddle-point condition, as in the previous analysis of sigma models at large N. This condition reads, as in (6.2.15),

$$\Delta_k^{-1}(g, g) = \frac{4\pi}{t_0},$$

(10.5.15)

where

$$\Delta_k = \sum_{a=1}^{2} L_a^2 + m_k^2 R^2.$$

(10.5.16)

The action (10.5.13), evaluated at this saddle point, is of the form $N\widetilde{\omega}_k$, where

$$\widetilde{\omega}_k = \text{Tr} \log \Delta_k - \frac{4\pi}{t_0} m_k^2 R^2.$$

(10.5.17)

This is the action of the large N instanton, and it can be computed as a function of the 't Hooft parameter: indeed, the condition (10.5.15) gives the value of m_k^2 as a function of t_0, which we then plug into (10.5.17).

In order to calculate $\widetilde{\omega}_k$ explicitly, we have to be more explicit about the solution of (10.5.15) and about the trace of the operator $\text{Tr} \log \Delta_k$. In flat space, we can evaluate such a trace by going to momentum space and using Feynman diagrams. Here this approach is no longer possible, since we do not have translation invariance on $\mathbb{S}^2$. However, we can use heat kernel techniques, as explained in Appendix B.4. The only information needed for this calculation is the spectrum of the operator Δ_k. In studying the path integral in the sector with topological charge k, we have to consider fields which are eigenfunctions of L_3 with eigenvalue $-k/2$, as explained in (6.3.64). This is the only constraint on the fields. Let us now denote by $|j, \ell\rangle$ a field which is an eigenfunction of

$$\mathbf{L}^2 = \sum_{a=1}^{3} L_a^2$$

(10.5.18)

with eigenvalue $j(j + 1)$, and an eigenfunction of L_3 with eigenvalue

$$\ell = -j, -j + 1, \ldots, j - 1, j.$$

(10.5.19)

The set of such fields forms a basis for the harmonic analysis on $\mathbb{S}^3$. The constraint that L_3 has eigenvalue $-k/2$ means that j should be of the form

$$j = k/2 + n, \qquad n = 0, 1, \ldots,$$

(10.5.20)

and the eigenvalues of Δ_k are then

$$\lambda_n = \left(n + \frac{k}{2} + 1\right)\left(n + \frac{k}{2}\right) - \frac{k^2}{4} + m_k^2 R^2, \qquad n = 0, 1, \ldots,$$

(10.5.21)

with degeneracies

$$d_n = 2j + 1 = 2n + k + 1.$$

(10.5.22)

The relevant quantities appearing in (10.5.15) and (10.5.17) are computed in Appendix B.4, using dimensional regularization, i.e. by doing the calculation in $d = 2 + \epsilon$ dimensions. The extra ϵ dimensions will be taken to have the geometry of a torus $(\mathbb{S}^1)^\epsilon$, where $\mathbb{S}^1$ has length L. The operator Δ_k is then extended to an operator $\Delta_{k,\epsilon}$ which acts on the extra dimensions, as indicated in (B.114). In order to state the result of these computations, we will denote

$$a = \frac{k+1}{2}, \qquad c^2 = \frac{k^2+1}{4} - m_k^2 R^2. \qquad (10.5.23)$$

We first analyze (10.5.15). Using the explicit result (B.141) we find the equation

$$-\frac{2}{\epsilon} + \log\left(4\pi R^2\right) + F_k(m_k^2 R^2) + \mathcal{O}(\epsilon) = \frac{4\pi}{t_0}, \qquad (10.5.24)$$

where (see (B.138))

$$F_k(m_k^2 R^2) = -\psi(a+c) - \psi(a-c) - \gamma_E, \qquad (10.5.25)$$

and $\psi(z)$ is the digamma function. The equation (10.5.24) is very similar to the equation (6.2.18) for the non-linear sigma model and for the $\mathbb{CP}^{N-1}$ model in flat space. It follows from (6.2.20) that the structure of the UV divergences is the same, as should be expected, since they only depend on the short-distance structure of the theory. As we did in the examples in flat space, we introduce a renormalized 't Hooft parameter t and a renormalization scale μ through the equation (6.2.21). After doing this, equation (10.5.24) becomes,

$$\frac{4\pi}{t(\mu)} = \log\left(4\pi R^2 \mu^2\right) + F_k(m_k^2 R^2). \qquad (10.5.26)$$

This equation can be written in terms of the dynamically generated scale of the theory in flat space, Λ, which we introduced in (6.2.27). We finally obtain:

$$F_k(m_k^2 R^2) + \log\left(4\pi R^2 \Lambda^2\right) = 0. \qquad (10.5.27)$$

This equation determines the value of $m_k^2 R^2$, the solution of the saddle-point equation, as a function of the dimensionless, renormalization group invariant quantity characterizing the theory, ΛR. Note however that this quantity is determined by the large N beta function, which is exact at one-loop, and we can regard it as a renormalized 't Hooft parameter.

Let us now compute the effective action $\tilde{\omega}_k$. After dimensional regularization, the second term in (10.5.17) picks an extra factor due to the volume of the extra dimensions, and one finds

$$\tilde{\omega}_k(\epsilon) = \mathrm{Tr} \log \Delta_{k,\epsilon} - \frac{4\pi}{t_0} m_k^2 R^2 L^\epsilon, \qquad (10.5.28)$$

where L^ϵ is the volume of the regulating manifold $\left(\mathbb{S}^1\right)^\epsilon$. By using the result (B.146), as well as (6.2.21), we obtain

$$\widetilde{\omega}_k(\epsilon) = \frac{2}{3\epsilon} + \frac{1}{3}\left(\gamma_E - \log\left(4\pi R^2/L^2\right)\right) - m_k^2 R^2\left(\gamma_E - \log\left(4\pi R^2\Lambda^2\right)\right)$$
$$- \zeta'_{\Delta_k}(0) + \mathcal{O}(\epsilon). \tag{10.5.29}$$

This quantity is divergent as $\epsilon \to 0$, and it depends on L. The physically relevant quantities are actually quotients of partition functions, which lead to differences of effective actions. We therefore introduce

$$\omega_k(\epsilon) = \widetilde{\omega}_k(\epsilon) - \widetilde{\omega}_0(\epsilon), \tag{10.5.30}$$

whose limit as $\epsilon \to 0$ is finite and independent of L. We will denote this limit simply as ω_k, and we find

$$\omega_k = -\zeta'_{\Delta_k}(0) + \zeta'_{\Delta_0}(0) - \left(m_k^2 - m_0^2\right) R^2 \left(\gamma_E - \log\left(4\pi R^2\Lambda^2\right)\right). \tag{10.5.31}$$

The derivative of the zeta function can be computed explicitly. The calculation is done in Appendix B.4 and the result is given in (B.144). The result (10.5.31) gives the appropriately renormalized effective action of an instanton at large N. Once the relationship (10.5.27) is taken into account, this computes the effective action of an instanton in the large N 't Hooft limit, as a function of ΛR. Of course, the dependence on ΛR is complicated, since it involves transcendental functions appearing in the explicit computation of the functional determinants. However, one can evaluate the instanton action numerically. In Fig. 10.6 we plot the instanton action ω_k as a function of ΛR, for $k = 1$ (bottom) and $k = 2$ (top).

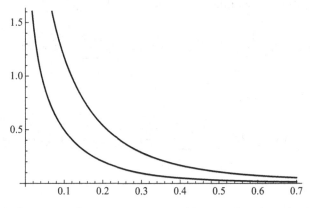

Figure 10.6 The large N instanton action (10.5.31) as a function of ΛR, for $k = 1$ (bottom) and $k = 2$ (top).

There are two natural limits to consider in this theory: the small volume limit, and the large volume limit. In the small volume limit, the running coupling constant is small, and one should make contact with the perturbative expansion. In the large volume limit we should make contact with the solution on flat space worked out in Section 6.3.

Let us first study the above solution in the limit of small volume $R \to 0$. In this limit, the parameter

$$F \equiv -\gamma_E - \log\left(4\pi R^2 \Lambda^2\right) \tag{10.5.32}$$

diverges logarithmically and is positive. If we want to solve (10.5.27) in this limit, i.e. if we want to solve

$$F_k(m_k^2 R^2) = F + \gamma_E, \tag{10.5.33}$$

the value of $m_k^2 R^2$ should lead to a logarithmic growth. By looking at (10.5.25), we see that this is achieved if $a \approx c$, since in this case the digamma function has a singularity. This occurs when the parameter

$$\xi_k \equiv m_k^2 R^2 + \frac{k}{2} \tag{10.5.34}$$

is small: $\xi_k \ll 1$. In this regime, we can solve (10.5.33) as a series expansion, and we find

$$\frac{1}{\xi_k} = \frac{F}{1+k} + \frac{H_{k+1}}{1+k} + \mathcal{O}\left(\frac{1}{F}\right), \tag{10.5.35}$$

where

$$H_n = \sum_{i=1}^{n} \frac{1}{i} \tag{10.5.36}$$

is the harmonic number. Similarly, we can expand $-\zeta'_{\Delta_k}(0)$,

$$-\zeta'_{\Delta_k}(0) = (1+k)\log\xi_k + \frac{1}{2}(k+1)^2 - 4\zeta'(-1) - \sum_{i=1}^{k+1}(2i-k-1)\log(i) + \mathcal{O}(\xi_k). \tag{10.5.37}$$

By using (10.5.35), we find

$$-\zeta'_{\Delta_k}(0) + \zeta'_{\Delta_0}(0) = -k\log F - \sum_{i=1}^{k}(2i-k-1)\log(i) + \frac{1}{2}k(k+1) + \mathcal{O}\left(\frac{1}{F}\right). \tag{10.5.38}$$

On the other hand,

$$-\left(m_k^2 - m_0^2\right) R^2 \left(\gamma_E - \log\left(4\pi R^2 \Lambda^2\right)\right) = \frac{kF}{2} + (\gamma_E - 1)k + \mathcal{O}\left(\frac{1}{F}\right), \tag{10.5.39}$$

and we finally obtain

$$\omega_k = \frac{kF}{2} - k\log F + \frac{k^2}{2} + \gamma_E k - \sum_{i=1}^{k}(2i - k - 1)\log(i) + \mathcal{O}\left(\frac{1}{F}\right). \quad (10.5.40)$$

This should be related to an instanton calculation at weak coupling but large N. To see this, we should take into account that the natural coupling appearing in such a calculation is the running coupling constant at the scale R, as we discussed in Section 4.5 in the context of YM theory on $\mathbb{S}^4$. In this case, by using (6.2.27), we find

$$\bar{t} = t\,(1/R) = -\frac{4\pi}{\log \Lambda^2 R^2}. \quad (10.5.41)$$

In terms of this coupling, we find

$$\omega_k = \frac{2\pi k}{\bar{t}} - k\log \frac{4\pi}{\bar{t}} + r_k + \mathcal{O}(\bar{t}), \quad (10.5.42)$$

where

$$r_k = \frac{1}{2}k\,(k + \gamma_E - \log(4\pi)) - \sum_{i=1}^{k}(2i - k - 1)\log(i). \quad (10.5.43)$$

The partition function in the k-instanton sector is then given by

$$e^{-N\omega_k} = C_k^N \left(\frac{4\pi}{\bar{t}}\right)^{Nk} \exp\left(-\frac{2\pi k N}{\bar{t}}\right)(1 + \mathcal{O}(\bar{t}))^N \quad (10.5.44)$$

where

$$C_k = e^{-r_k}. \quad (10.5.45)$$

This is precisely the structure we expect for the contribution of an instanton at weak coupling: the exponential term can be written as

$$\exp\left(-\frac{2\pi k}{g^2(1/R)}\right) \quad (10.5.46)$$

which is precisely the action of an instanton of charge k, evaluated with the running coupling constant at the scale R. The factor

$$\left(\frac{4\pi}{\bar{t}}\right)^{Nk} \quad (10.5.47)$$

is also what is expected from the presence of zero modes, as we pointed out in (10.3.2). It turns out that the one-instanton and two-instanton contributions for the

$\mathbb{CP}^{N-1}$ model on $\mathbb{S}^2$ have been computed at next-to-leading order in the semiclassical expansion, using the ideas that we presented in Section 4.5. The result for the renormalized partition function in the kth instanton sector has the form,

$$C_k(N) \left(\frac{4\pi}{\bar{t}}\right)^{Nk} \exp\left(-\frac{2\pi k N}{\bar{t}}\right) \left(1 + \mathcal{O}(\bar{g}^2)\right),\tag{10.5.48}$$

where $C_k(N)$ is a numerical coefficient with the property that

$$\lim_{N\to\infty} (C_k(N))^{1/N} = C_k \tag{10.5.49}$$

for $k = 1, 2$. Therefore, in the weak coupling (or small volume) limit, the instanton action ω_k incorporates and resums all the quantum corrections around the instanton, at large N.

Let us now consider the opposite regime: the large volume limit. This corresponds to the regime in which $m_k^2 R^2$ is large. One finds, by expanding the relevant functions in (10.5.25),

$$F_k(m_k^2 R^2) = -\gamma_{\rm E} - \log(m_k^2 R^2) + \frac{1}{3}(m_k^2 R^2)^{-1}$$
$$+ \left(\frac{1}{15} - \frac{k^2}{24}\right)(m_k^2 R^2)^{-2} + \cdots,\tag{10.5.50}$$

and (10.5.27) gives the expansion

$$m_k^2 R^2 = m^2 R^2 + \frac{1}{3} + \left(\frac{1}{90} - \frac{k^2}{24}\right)(m^2 R^2)^{-1} + \cdots,\tag{10.5.51}$$

where m^2 is the squared mass appearing in (6.2.28). Similarly, by expanding (B.144), we find

$$\zeta'_{\Delta_k}(0) = m_k^2 R^2 \log(m_k^2 R^2) - m_k^2 R^2 - \frac{1}{3}\log(m_k^2 R^2)$$
$$+ \left(\frac{1}{15} - \frac{k^2}{24}\right)(m_k^2 R^2)^{-2} + \cdots,\tag{10.5.52}$$

and we finally obtain

$$\omega_k = \frac{k^2}{24}(m^2 R^2)^{-1} + \cdots.\tag{10.5.53}$$

What is the interpretation of this result? Remember that $P_k \approx \exp(-N\omega_k)$ can be interpreted as the probability of finding a field configuration with topological charge k. Using that the volume of the two-dimensional, spherical spacetime is $V = 4\pi R^2$, we obtain

$$P_k \approx \exp\left(-\frac{k^2}{2V \chi_t^{\rm large\ N}}\right),\tag{10.5.54}$$

where $\chi_t^{\text{large } N}$ is the topological susceptibility of the $\mathbb{CP}^{N-1}$ model on flat space, i.e. in the limit $R \to \infty$, which we computed in (6.3.101). Therefore, at large volume, the probabilities P_k follow a Gaussian distribution, and the second moment of the probability distribution is given by

$$V \chi_t^{\text{large } N}. \tag{10.5.55}$$

This is in precise agreement with the large N limit of (4.3.76). The Gaussian character of the probability distribution P_k at large volume has been argued by Lüscher on general grounds: the topological charge can be computed by an integral of a local density, and therefore it is additive. At large volume, the different contributions to the charge at different points of spacetime are uncorrelated. Therefore, by the central limit theorem of probability theory, we should expect the distribution of the total charge to be Gaussian. The above calculation confirms this general expectation.

Another consequence of the above calculation is the following: at *fixed, finite* volume, when instanton calculus is reliable, the probabilities P_k are exponentially suppressed as $N \to \infty$. In particular, the topological susceptibility at finite volume (4.3.76) is also exponentially suppressed. In this case, and as can be seen in Fig. 10.6, the actions ω_k are non-vanishing, for any value of ΛR, and there are no phase transitions like those found in the Gross–Witten–Wadia model. However, if we take first the limit of infinite volume, the topological susceptibility is no longer exponentially suppressed, but has a $1/N$ expansion whose first term is (6.3.101). Therefore, we find an interpolation between two very different behaviors at large N as we go from small to large volume. In particular, the large N limit and the large volume limit *do not* commute.

10.6 Bibliographical notes

The analyticity of the fixed genus amplitudes in the $1/N$ expansion was first pointed out in [121], and it has been analyzed in detail in many other models since then. The example of the petal diagrams can be found in [68]. Unfortunately, there are very few discussions of general aspects of large N instantons in the literature. The material presented here comes from the review [137], and precious insights can be found in [101, 149, 150]. The apparent incompatibility between some naive instanton predictions and the large N expansion was pointed out by Witten in [187], and it was addressed in explicit calculations in [2, 144, 145] in the context of the $\mathbb{CP}^{N-1}$ model. The connection between vanishing large N instanton actions and large N phase transitions was proposed by Neuberger in [149, 150]. Large N instantons in matrix models and MQM were first discussed in [64, 65, 95, 166], and a detailed analysis (together with a complete list of references) can be found in

[138, 139] as well as in the review [137]. The large N instanton in the weakly coupled phase of the Gross–Witten–Wadia model and in MQM with a quartic potential, as well as their connection to the large order behavior of the $1/N$ expansion, are discussed in detail in [136] and [138], respectively. The Majumdar–Vergassola right deviation formula for the Wishart ensemble was obtained in [133]. A beautiful review of extreme value statistics and random matrices can be found in [132].

The large N instanton of the $\mathbb{CP}^{N-1}$ model on the two-sphere was analyzed by Münster in [144, 145]. His analysis is further clarified and extended in [5]. A detailed calculation of the one-instanton and two-instanton partition functions of the $\mathbb{CP}^{N-1}$ model at arbitrary N, building on the results in [33], is given in [161, 162].

Appendix A
Harmonic analysis on $\mathbb{S}^3$

In this appendix, we collect some useful facts on harmonic analysis on $\mathbb{S}^3$ which are used throughout the book.

The elements of $SU(2)$ are of the form

$$g = \begin{pmatrix} \alpha & \beta \\ -\bar{\beta} & \bar{\alpha} \end{pmatrix}, \qquad |\alpha|^2 + |\beta|^2 = 1, \tag{A.1}$$

which is the equation for $\mathbb{S}^3$ as a submanifold of $\mathbb{C}^2 = \mathbb{R}^4$. Therefore, we can identify $SU(2) = \mathbb{S}^3$. We parametrize this element as

$$\begin{aligned} |\alpha| &= \cos\frac{t_1}{2}, & |\beta| &= \sin\frac{t_1}{2}, \\ \operatorname{Arg}\alpha &= \frac{t_2 + t_3}{2}, & \operatorname{Arg}\beta &= \frac{t_2 - t_3 + \pi}{2}, \end{aligned} \tag{A.2}$$

where t_i are the Euler angles and span the range

$$0 \le t_1 < \pi, \qquad 0 \le t_2 < 2\pi, \qquad -2\pi \le t_3 < 2\pi. \tag{A.3}$$

The general element of $SU(2)$ will then be given by

$$\begin{aligned} g = u(t_1, t_2, t_3) &= \begin{pmatrix} \cos(t_1/2)e^{i(t_2+t_3)/2} & i\sin(t_1/2)e^{i(t_2-t_3)/2} \\ i\sin(t_1/2)e^{i(-t_2+t_3)/2} & \cos(t_1/2)e^{-i(t_2+t_3)/2} \end{pmatrix} \\ &= u(0, t_2, 0)u(t_1, 0, 0)u(0, 0, t_3), \end{aligned} \tag{A.4}$$

where

$$u(0, 0, t_3) = e^{it_3\sigma_3/2}. \tag{A.5}$$

The Maurer–Cartan forms for $SU(2)$ are defined by the equation

$$g^{-1}dg = \frac{i}{2}\sum_{a=1}^{3}\sigma_a\omega^a, \tag{A.6}$$

and they are expressed in the coordinates introduced above as

$$
\begin{aligned}
\omega^1 &= \cos t_3 dt_1 + \sin t_3 \sin t_1 dt_2, \\
\omega^2 &= \sin t_3 dt_1 - \cos t_3 \sin t_1 dt_2, \\
\omega^3 &= \cos t_1 dt_2 + dt_3.
\end{aligned}
\tag{A.7}
$$

They satisfy the Cartan equations,

$$
d\omega^a = \frac{1}{2}\epsilon^{abc}\,\omega^b \wedge \omega^c,
\tag{A.8}
$$

where ϵ^{abc} denotes, as usual, the totally antisymmetric Levi-Civita symbol.

The metric $G_{\mu\nu}$ on $SU(2) = \mathbb{S}^3$ is induced from its embedding in $\mathbb{R}^4$, and it reads

$$
ds^2 = G_{\mu\nu}dt_\mu dt_\nu = \frac{r^2}{4}\left(dt_1^2 + dt_2^2 + dt_3^2 + 2\cos t_1\, dt_2 dt_3\right),
\tag{A.9}
$$

where r is the radius of the sphere. The volume element on $\mathbb{S}^3$ is, in these coordinates,

$$
\sqrt{G}dt_1\, dt_2\, dt_3 = \frac{r^3}{8}\sin(t_1)dt_1\, dt_2\, dt_3.
\tag{A.10}
$$

It is convenient to use (A.10) to define a normalized volume element in $SU(2)$,

$$
dg = \frac{1}{\mathrm{vol}(\mathbb{S}^3)}\sqrt{G}dt_1\, dt_2\, dt_3 = \frac{1}{16\pi^2}\sin(t_1)dt_1\, dt_2\, dt_3,
\tag{A.11}
$$

which coincides with the normalized Haar measure in $SU(2)$. Using this volume element, we define a natural scalar product for complex-valued functions on $SU(2)$ as follows,

$$
(f_1, f_2) = \int_{SU(2)} dg\,\overline{f_1(g)}f_2(g).
\tag{A.12}
$$

We can use the Maurer–Cartan forms to analyze the differential geometry of $\mathbb{S}^3$. The vierbein of $\mathbb{S}^3$ satisfies

$$
e_\mu^a e_\nu^b \delta_{ab} = G_{\mu\nu}.
\tag{A.13}
$$

It can be seen to be proportional to ω^a:

$$
e_\mu^a = \frac{r}{2}\omega_\mu^a.
\tag{A.14}
$$

The inverse vierbein is defined by

$$
E_a^\mu = \delta_{ab}G^{\mu\nu}e_\mu^b,
\tag{A.15}
$$

which can be used to define left invariant vector fields

$$\ell_a = E_a^\mu \frac{\partial}{\partial x^\mu}. \tag{A.16}$$

Their explicit expressions are the following:

$$\ell_1 = \frac{2}{r} \left(\cos t_3 \frac{\partial}{\partial t_1} + \frac{\sin t_3}{\sin t_1} \frac{\partial}{\partial t_2} - \sin t_3 \cot t_1 \frac{\partial}{\partial t_3} \right),$$

$$\ell_2 = \frac{2}{r} \left(\sin t_3 \frac{\partial}{\partial t_1} - \frac{\cos t_3}{\sin t_1} \frac{\partial}{\partial t_2} + \cos t_3 \cot t_1 \frac{\partial}{\partial t_3} \right), \tag{A.17}$$

$$\ell_3 = \frac{2}{r} \frac{\partial}{\partial t_3}.$$

Of course, they obey

$$e^a(\ell_b) = \delta_b^a, \tag{A.18}$$

as well as the following commutation relations:

$$[\ell_a, \ell_b] = -\frac{2}{r} \epsilon_{abc} \ell_c. \tag{A.19}$$

This can be checked by direct computation. If we now introduce the operators L_a through

$$L_a = \frac{r}{2i} \ell_a, \tag{A.20}$$

we see that they satisfy the standard commutation relations of the $SU(2)$ angular momentum operators:

$$[L_a, L_b] = i\epsilon_{abc} L_c. \tag{A.21}$$

These operators are Hermitian with respect to the following standard inner product on $\mathbb{S}^3$:

$$(f, g) = \int_{\mathbb{S}^3} d^3t \sqrt{G} \overline{f(t_i)} g(t_i). \tag{A.22}$$

Appendix B

Heat kernel and zeta functions

B.1 Heat kernel and heat kernel expansion

Let M be a Riemannian manifold M of dimension d, and let A be a differential operator acting on fields on M with "internal" indices, which can be spinor indices, color indices, etc. In a more precise language, A acts on sections of a bundle on M. We will assume that it has a discrete set of eigenvalues and eigenfunctions λ_n, ϕ_n. A typical example is the scalar Laplacian on a compact manifold M. For simplicity, we will assume that the λ_n are non-negative. The *heat kernel* of the operator A is defined as the operator

$$K_A(\tau) = e^{-\tau A} = \sum_n e^{-\lambda_n \tau} |\phi_n\rangle \langle \phi_n|. \tag{B.1}$$

Here, we sum over all possible eigenvalues, including the vanishing ones (i.e. we include the so-called zero modes of the operator). If A is the Hamiltonian, the heat kernel can be interpreted as the unnormalized canonical density matrix, and τ is then the inverse temperature. The trace of the heat kernel is given by

$$\operatorname{Tr} K_A(\tau) = \sum_n e^{-\lambda_n \tau}. \tag{B.2}$$

One important property of K_A is that it satisfies an ordinary differential equation with respect to τ

$$\frac{dK_A}{d\tau} = -A K_A, \tag{B.3}$$

with initial condition

$$K_A(0) = 1. \tag{B.4}$$

This equation can be written as a heat-diffusion equation in coordinate space. If we define

$$K_A(x, x'; \tau) = \langle x|K_A(\tau)|x'\rangle, \tag{B.5}$$

we find that

$$\frac{dK_A(x, x'; \tau)}{d\tau} = -\int dx'' \langle x|A|x''\rangle K_A(x'', x'; \tau). \tag{B.6}$$

For example, if $A = -\partial_x^2$ is the Laplacian in one dimension, we have that

$$\langle x|A|x''\rangle = -\partial_x^2 \delta(x - x''), \tag{B.7}$$

and the differential equation (B.3) reads

$$\frac{dK_A(x, x'; \tau)}{d\tau} = \frac{\partial^2 K_A(x'', x'; \tau)}{\partial x^2}, \tag{B.8}$$

which is precisely the heat equation in $\mathbb{R}$, and has the solution

$$K_A(x, x'; \tau) = \frac{1}{\sqrt{4\pi\tau}} \exp\left[-\frac{(x - x')^2}{4\tau}\right]. \tag{B.9}$$

Let us denote by $g_{\mu\nu}$ the Riemannian metric of M. Our conventions for the Riemann tensor and the Ricci tensor are as in [179], i.e.

$$R^\lambda_{\rho\mu\nu} = \partial_\nu \Gamma^\lambda_{\rho\mu} - \partial_\mu \Gamma^\lambda_{\rho\nu} + \Gamma^\lambda_{\nu\delta} \Gamma^\delta_{\mu\rho} - \Gamma^\lambda_{\mu\delta} \Gamma^\delta_{\nu\rho},$$
$$R_{\mu\nu} = R^\sigma_{\mu\nu\sigma}. \tag{B.10}$$

The heat kernel of the operator A has an asymptotic expansion as $\tau \to 0$ called the *heat kernel* or *Seeley–De Witt expansion*. We will present a simplified version of this expansion, in which we take a trace over the internal indices of the fields. We will denote such a trace as tr, to be distinguished from Tr in which we also take the trace over the coordinates of M. The heat kernel expansion has the structure

$$\text{tr} K_A(x, x'; \tau) = \frac{1}{(4\pi\tau)^{d/2}} \sum_{n\geq 0} a_n(x, x')\tau^n, \tag{B.11}$$

where $a_n(x, x')$ depend on the type of field and on the differential operator. They are called the heat kernel or Seeley–De Witt coefficients. Notice that

$$\text{Tr} K_A(\tau) = \int d^d x \sqrt{g}\, \text{tr} K_A(x, x; \tau), \tag{B.12}$$

where $g = \det(g_{\mu\nu})$. By using the heat kernel expansion we find

$$\text{Tr} K_A(\tau) = \frac{1}{(4\pi)^{d/2}} \sum_{n=0}^\infty \tau^{n-d/2} \int d^d x \sqrt{g}\, a_n(x), \tag{B.13}$$

which involves the integrated versions of the Seeley–De Witt coefficients along the diagonal $a_n(x) \equiv a_n(x, x)$.

The coefficients of the expansion (B.11) contain a lot of information, and they are useful in the calculation of many quantities, such as anomalies, beta functions, etc. It is then important to have a recipe to compute them in detail. A lot of effort has been spent on such computations, and an excellent exposition can be found in [179], to which we refer for more details. We will now present formulae for these coefficients which are useful for the computations needed in this book, as well as some relevant examples. We will also briefly present a method to derive these expressions in the simpler case of Euclidean, flat space $\mathbb{R}^d$.

Let us write the operator A in the form

$$A = - \left(g^{\mu\nu} \nabla_\mu \nabla_\nu + E \right), \tag{B.14}$$

where E is a matrix-valued function, acting on the "internal" indices, and the covariant derivative is written as

$$\nabla_\mu = \partial_\mu + \omega_\mu, \tag{B.15}$$

where ω_μ is a connection which incorporates both the Levi-Civita connection and the connection acting on the internal indices. Let us define the field strength of the connection ω_μ as

$$\Omega_{\mu\nu} = \partial_\mu \omega_\nu - \partial_\nu \omega_\mu + [\omega_\mu, \omega_\nu], \tag{B.16}$$

where the commutator refers to the matrix structure of ω_μ. In terms of these data, the first three Seeley–De Witt coefficients are given by (up to total derivatives)

$$a_0(x) = \text{tr}[I],$$

$$a_1(x) = \frac{1}{6}\text{tr}[6E + I\,R],$$

$$a_2(x) = \frac{1}{360}\text{tr}\left[60RE + 180E^2 + 5I\,R^2 - 2I\,R_{\mu\nu}R^{\mu\nu} \right.$$

$$\left. + 2I\,R_{\mu\nu\rho\sigma}R^{\mu\nu\rho\sigma} + 30\Omega_{\mu\nu}\Omega^{\mu\nu} \right]. \tag{B.17}$$

In these equations, I is the identity matrix acting on the internal indices.

We will now work out some important examples of operators which are used in the book. Let us first consider a massless scalar field in a representation r of a gauge group G, and let A be the operator,

$$A = -g^{\mu\nu}\nabla_\mu\nabla_\nu + \xi R, \tag{B.18}$$

where ∇_μ is the covariant derivative including both the gauge and the Levi-Civita connections. Notice that we have a non-minimal coupling to gravity given by the parameter ξ. In this case, the term E is

$$E_{kl} = -\xi R\delta_{kl}, \qquad k, l = 1, \ldots, d(r), \tag{B.19}$$

and the connection is

$$\left(\omega_\mu\right)_{kl} = -iA_\mu^a \left(T_a^r\right)_{kl}, \tag{B.20}$$

with field strength

$$\left(\Omega_{\mu\nu}\right)_{kl} = -iF_{\mu\nu}^a \left(T_a^r\right)_{kl}. \tag{B.21}$$

One easily finds,

$$\mathrm{tr}\, E = -d(r)\xi R, \qquad \mathrm{tr}\, E^2 = d(r)\xi^2 R^2, \qquad \mathrm{tr}\left(\Omega_{\mu\nu}\Omega^{\mu\nu}\right) = -C(r)F_{\mu\nu}^a F^{\mu\nu a}, \tag{B.22}$$

where $d(r)$ is the dimension of r, $C(r)$ is defined in (4.2.11), and

$$a_0(x) = d(r),$$

$$a_1(x) = d(r)\left(\frac{1}{6} - \xi\right)R,$$

$$a_2(x) = d(r)\left\{\frac{1}{180}R_{\mu\nu\rho\sigma}R^{\mu\nu\rho\sigma} - \frac{1}{180}R_{\mu\nu}R^{\mu\nu} + \frac{1}{2}\left(\frac{1}{6} - \xi\right)^2 R^2\right\}$$

$$- \frac{1}{12}C(r)F_{\mu\nu}^a F^{\mu\nu a}. \tag{B.23}$$

Our next example, relevant to the quantization of YM theory, is the operator Δ_1' introduced in (4.5.76), which acts on one-forms with values in the Lie algebra. We find,

$$E_{\nu b}^{a\rho} = -R_\nu^\rho \delta_b^a + 2f_{abc}F_\nu^{\rho c}, \tag{B.24}$$

while the connection is

$$\left(\omega_\mu\right)_{\nu b}^{a\rho} = f_{acb}A_\mu^c \delta_\nu^\rho - \Gamma_{\mu\nu}^\rho \delta_b^a. \tag{B.25}$$

It follows that the field strength is

$$\left(\Omega_{\mu\nu}\right)_{\rho b}^{a\sigma} = R^\sigma{}_{\rho\mu\nu}\delta_a^b + f_{acb}F_{\mu\nu}^c \delta_\rho^\sigma. \tag{B.26}$$

One calculates,

$$\mathrm{tr}\, E = -Rd(G),$$
$$\mathrm{tr}\, E^2 = d(G)R_{\rho\sigma}R^{\rho\sigma} + 4c_2(G)F_{\mu\nu}^a F^{\mu\nu a}, \tag{B.27}$$
$$\mathrm{tr}\left(\Omega_{\mu\nu}\Omega^{\mu\nu}\right) = -d(G)R_{\mu\nu\rho\sigma}R^{\mu\nu\rho\sigma} - dC_2(G)F_{\mu\nu}^a F^{\mu\nu a},$$

where $C_2(G)$ is the quadratic Casimir of the Lie algebra, see (4.2.11) and (4.2.13). One finds,

$$a_0(x) = d(G),$$

$$a_1(x) = -\frac{5d(G)}{6}R,$$

$$a_2(x) = \frac{d(G)}{360}\left[(5d-60)R^2 + (180-2d)R_{\mu\nu}R^{\mu\nu} + (2d-30)R_{\mu\nu\rho\sigma}R^{\mu\nu\rho\sigma}\right]$$
$$+ \frac{C_2(G)}{12}(24-d)F_{\mu\nu}^a F^{\mu\nu a}. \tag{B.28}$$

In $d = 4$, we find for $a_2(x)$,

$$a_2 = d(G)\left[-\frac{1}{9}R^2 + \frac{43}{90}R_{\mu\nu}R^{\mu\nu} - \frac{11}{180}R_{\mu\nu\rho\sigma}R^{\mu\nu\rho\sigma}\right] + \frac{5}{3}C_2(G)F_{\mu\nu}^a F^{\mu\nu a}. \tag{B.29}$$

In the case of flat space, the results for the heat kernel expansion can easily be obtained by using the following trick [74, 148, 179]. After inserting a complete set of plane waves, we can write

$$K_A(x, x'; \tau) = \int \frac{d^d k}{(2\pi)^d} e^{-ikx'} e^{-\tau A_x} e^{ikx}. \tag{B.30}$$

It is immediate to verify that

$$e^{-ikx}\nabla_\mu e^{ikx} = ik_\mu + \nabla_\mu, \tag{B.31}$$

therefore

$$e^{-ikx}\left(-\nabla^\mu\nabla_\mu\right)e^{ikx} = k^2 - 2ik_\mu\nabla^\mu - \nabla^\mu\nabla_\mu, \tag{B.32}$$

and (B.30) reads, after rescaling $k \to k/\tau^{1/2}$,

$$K_A(x, x'; \tau) = \frac{1}{\tau^{d/2}}\int \frac{d^d k}{(2\pi)^d} e^{-k^2} e^{2i\tau^{1/2}k_\mu\nabla^\mu - \tau A_x}$$
$$= \sum_{n=0}^{\infty}\frac{1}{(4\pi\tau)^{d/2}}\int \frac{d^d k}{\pi^{d/2}} e^{-k^2}\left(2i\tau^{1/2}k_\mu\nabla^\mu - \tau A_x\right)^n. \tag{B.33}$$

This leads to a systematic expansion in powers of τ, by performing the integrals over the momenta. Since these are Gaussian integrals, they can be computed with Wick's theorem. If we denote by

$$\langle f(k)\rangle = \int \frac{d^d k}{\pi^{d/2}} e^{-k^2} f(k), \tag{B.34}$$

the "propagator" is given by

$$\langle k_\mu k_\nu \rangle = \frac{1}{2}\delta_{\mu\nu}, \tag{B.35}$$

and any other correlator is computed by Wick's theorem. For example,

$$\langle k_\mu k_\nu k_\alpha k_\beta \rangle = \langle k_\mu k_\nu \rangle \langle k_\alpha k_\beta \rangle + \langle k_\mu k_\alpha \rangle \langle k_\nu k_\beta \rangle + \langle k_\mu k_\beta \rangle \langle k_\nu k_\alpha \rangle. \tag{B.36}$$

As an example of this, we can compute the coefficient a_1 of the heat kernel expansion. It is given by

$$a_1(x) = \mathrm{tr}\left(-2\langle k_\mu k_\nu \rangle \nabla^\mu \nabla^\nu - A_x\right) = \mathrm{tr}(E), \tag{B.37}$$

which is the result quoted above in the case of $M = \mathbb{R}^d$. It is straightforward to compute the coefficient a_2, see for example [74, 148, 179] for the details.

B.2 Zeta function and determinants

In QFT we often have to compute determinants of differential operators. These determinants are generically UV divergent. To regularize them, one can use *zeta function regularization*. The zeta function of the operator A is defined as

$$\zeta_A(s) = {\sum_n}' \lambda_n^{-s}, \tag{B.38}$$

where the $'$ indicates that we are only including strictly positive eigenvalues. This function is well defined if $\mathrm{Re}(s)$ is large enough, and it can be extended to a mero-morphic function on the s-plane. The zeta function can sometimes be computed directly, as the following example shows.

Example B.1 Let us consider the one-dimensional Laplacian

$$A = -\partial_x^2 \tag{B.39}$$

on a circle of length L. The eigenfunctions must be periodic under $x \rightarrow x + L$, and they are given by

$$\phi_n(x) = e^{\frac{2\pi i n x}{L}}, \qquad n \in \mathbb{Z}. \tag{B.40}$$

The eigenvalues of A are

$$\lambda_n = \left(\frac{2\pi n}{L}\right)^2, \qquad n \in \mathbb{Z}. \tag{B.41}$$

Therefore, the zeta function is given by

$$\zeta_A(s) = 2\sum_{n=1}^{\infty} \lambda_n^{-s} = \frac{2L^{2s}}{(2\pi)^{2s}} \sum_{n=1}^{\infty} \frac{1}{n^{2s}} = \frac{2L^{2s}}{(2\pi)^{2s}} \zeta(2s), \tag{B.42}$$

where

$$\zeta(z) = \sum_{n=1}^{\infty} \frac{1}{n^z} \tag{B.43}$$

is Riemann's zeta function. We then find,

$$-\zeta'_A(0) = -4 \log\left(\frac{L}{2\pi}\right) \zeta(0) - 4\zeta'(0) = -2 \log(L), \tag{B.44}$$

where we used

$$\zeta(0) = -\frac{1}{2}, \qquad \zeta'(0) = -\frac{1}{2}\log(2\pi). \tag{B.45}$$

$\square$

The zeta function of the operator A is closely related to the heat kernel, since it is given by a Mellin transform of the trace of the heat kernel (after zero modes have been removed). If s is large enough, so that the sum over eigenvalues defining the zeta function is well defined, we can write

$$\zeta_A(s) = \frac{1}{\Gamma(s)} \int_0^{\infty} \left(\sum_n{}' e^{-\lambda_n \tau}\right) \tau^{s-1} d\tau$$

$$= \frac{1}{\Gamma(s)} \int_0^{\infty} \left(\mathrm{Tr} K_A(\tau) - n_A^0\right) \tau^{s-1} d\tau, \tag{B.46}$$

where n_A^0 is the number of zero modes of A. The resulting expression is then extended to the complex s-plane by analytic continuation. We can now study $\zeta_A(s)$ by using the heat kernel, and in particular its Seeley–De Witt coefficients. Let us write (B.46) as

$$\zeta_A(s) = \frac{1}{\Gamma(s)} \Bigg[\int_0^1 d\tau\, \tau^{s-1} \left(\mathrm{Tr} K_A(\tau) - n_A^0\right)$$

$$+ \int_1^{\infty} d\tau\, \tau^{s-1} \left(\mathrm{Tr} K_A(\tau) - n_A^0\right)\Bigg]. \tag{B.47}$$

The second integral is convergent for all values of s: there is no singularity at short distances due to the cutoff at $\tau = 1$, and at large distances we have

$$\mathrm{Tr} K(\tau) \sim e^{-\lambda_1 \tau}, \tag{B.48}$$

where λ_1 is the smallest non-zero eigenvalue of A. It can easily be seen that this defines an analytic function of s which we will denote by $F(s)$. The first integral in (B.47) is now evaluated for

$$\mathrm{Re}(s) \geq \frac{d}{2}, \tag{B.49}$$

and then extended by analytic continuation to the full complex s-plane. One finds, after using (B.13),

$$\frac{1}{\Gamma(s)} \int_0^1 d\tau\, \tau^{s-1} \left(\mathrm{Tr} K_A(\tau) - n_A^0 \right)$$

$$= \frac{1}{(4\pi)^{d/2}\Gamma(s)} \sum_{n \geq 0} \frac{1}{s - d/2 + n} \int d^d x \sqrt{g}\, a_n(x) - \frac{n_A^0}{s\Gamma(s)}. \qquad (B.50)$$

We then obtain the following representation for the zeta function,

$$\zeta_A(s) = \frac{1}{(4\pi)^{d/2}\Gamma(s)} \sum_{n \geq 0} \frac{1}{s - d/2 + n} \int d^d x \sqrt{g}\, a_n(x) + \frac{F(s)}{\Gamma(s)} - \frac{n_A^0}{s\Gamma(s)}, \qquad (B.51)$$

where $F(s)$ is analytic. We can now deduce some important analyticity properties of $\zeta_A(s)$. First of all, we see that $\zeta_A(s)$ is analytic in the region (B.49), and its only singularities are simple poles. These are located in principle at the poles appearing in the sum in (B.51)

$$s = d/2 - n, \qquad n \geq 0. \qquad (B.52)$$

However, for $n \geq d/2$, the possible poles at non-positive values of $s = -k$ are cancelled by the poles in the Γ function, since

$$\Gamma(s) = \frac{(-1)^k}{k!} \frac{1}{s + k} + \cdots, \qquad (B.53)$$

and we find the useful formula

$$\zeta_A(-k) = \frac{(-1)^k k!}{(4\pi)^{d/2}} \int d^d x \sqrt{g}\, a_{k+d/2}(x), \qquad k = 1, 2, \ldots. \qquad (B.54)$$

For $s = 0$ there is also a contribution due to the number of zero modes, and one obtains

$$n_A^0 + \zeta_A(0) = \frac{1}{(4\pi)^{d/2}} \int d^d x \sqrt{g}\, a_{d/2}(x). \qquad (B.55)$$

The only poles in (B.52) then occur when

$$n = 0, 1, \ldots, d/2 - 1, \qquad (B.56)$$

i.e. at positive values of s,

$$s = 1, 2, \ldots, \frac{d}{2}. \qquad (B.57)$$

The residue at these poles is given by

$$\frac{1}{(4\pi)^{d/2}} \frac{1}{\Gamma(s)} \int d^d x \sqrt{g}\, a_{d/2-s}(x). \qquad (B.58)$$

Let us now define the zeta function regularization of the determinant of A. By taking the derivative of (B.38) we find,

$$- \frac{d\zeta_A}{ds} = {\sum_n}' \lambda_n^{-s} \log \lambda_n \qquad (B.59)$$

which is again well defined if $\mathrm{Re}(s)$ is large enough. As $s \to 0$, the left hand side becomes $-\zeta_A'(0)$, while the right hand side becomes

$$ {\sum_n}' \log \lambda_n. \qquad (B.60)$$

Of course, this sum is not well defined, but the equation (B.59) suggests that we *define* this sum as $-\zeta_A'(0)$. This leads to the zeta function regularization of the determinant of the operator A,

$$ \det{}'(A) = e^{-\zeta_A'(0)}, \qquad (B.61)$$

which computes the determinant of A once zero modes have been removed. As we have seen, $\zeta_A(s)$ is analytic at $s = 0$, and (B.61) is well defined. By taking the logarithm of this expression, we find

$$ \mathrm{Tr} \log{}'(A) = -\zeta_A'(0). \qquad (B.62)$$

B.3 Zeta function regularization in Quantum Mechanics

We will now compute, by using zeta function regularization, the determinant of the one-dimensional operator

$$ A - \lambda = -\frac{d^2}{dt^2} + u(t) - \lambda \qquad (B.63)$$

on the interval $[-\beta/2, \beta/2]$ with periodic boundary conditions, i.e. we consider eigenfunctions which satisfy

$$ \psi(-\beta/2) = \psi(\beta/2), \qquad \psi'(-\beta/2) = \psi'(\beta/2). \qquad (B.64)$$

Our derivation follows the treatment in [172]. We start with the expression (B.46), take a derivative with respect to λ and another derivative with respect to s, and then we evaluate the result at $s = 0$. We find

$$ \frac{d^2}{ds d\lambda} \zeta_{A-\lambda}(0) = \int_0^\infty \mathrm{Tr} \, e^{-(A-\lambda)\tau} \, d\tau, \qquad (B.65)$$

i.e.

$$ \frac{d}{d\lambda} \log \det (A - \lambda) = -\mathrm{Tr} \, R_\lambda, \qquad (B.66)$$

where the resolvent R_λ of the operator A is defined as

$$R_\lambda = \frac{1}{A - \lambda}. \tag{B.67}$$

This is nothing but the Green function of the operator $A - \lambda$. In the case of one-dimensional Schrödinger operators, the Green function can be calculated directly by using for example the method of variation of parameters. It can be expressed in terms of the fundamental solutions defined just before (1.6.2) and appearing in (1.6.3): it is given by the symmetric function in t, t' which, for $t \leq t'$, satisfies,

$$\mathrm{Tr}\,(T(\lambda) - 1)\,R_\lambda(t, t') = \psi_\lambda^1(t)\left\{\psi_\lambda^2(\beta/2)\psi_\lambda^1(t') - \left(\psi_\lambda^1(\beta/2) - 1\right)\psi_\lambda^2(t')\right\}$$
$$+ \psi_\lambda^2(t)\left\{\left(\dot\psi_\lambda^2(\beta/2) - 1\right)\psi_\lambda^1(t') - \dot\psi_\lambda^2(\beta/2)\psi_\lambda^2(t')\right\}. \tag{B.68}$$

In this equation, $T(\lambda)$ is the monodromy matrix introduced in (1.6.5). It can easily be checked that this function satisfies the defining properties of a Green's function with the appropriate boundary conditions:

1. $R_\lambda(\beta/2, t') = R_\lambda(-\beta/2, t')$, $\partial_t R_\lambda(\beta/2, t') = \partial_t R_\lambda(-\beta/2, t')$,
2. $(-\partial_t^2 + u(t) - \lambda)R_\lambda(t, t') = 0$, $t \neq t'$,
3. $\lim_{t' \to t}\left(\partial_t R_\lambda(t, t')|_{t > t'} - \partial_t R_\lambda(t, t')|_{t < t'}\right) = -1$.

We can write, for $t \leq t'$,

$$R_\lambda(t, t') = -\mathrm{Tr}\left\{(T(\lambda) - 1)^{-1}\,T(\lambda)Z(t, t'; \lambda)\right\}, \tag{B.69}$$

where

$$Z(t, t'; \lambda) = \begin{pmatrix} \psi_\lambda^1(t)\psi_\lambda^2(t') & \psi_\lambda^2(t)\psi_\lambda^2(t') \\ -\psi_\lambda^1(t)\psi_\lambda^1(t') & -\psi_\lambda^2(t)\psi_\lambda^1(t') \end{pmatrix}. \tag{B.70}$$

Let us now consider two solutions $a_\lambda(t)$, $b_\lambda(t)$ of the equation $(A - \lambda)y(t) = 0$. It is easy to deduce that

$$a_\lambda(t)b_\lambda(t) = \frac{d}{dt}W\left[\partial_\lambda a_\lambda(t), b_\lambda(t)\right], \tag{B.71}$$

where W is the Wronskian; therefore

$$\int_{-\beta/2}^{\beta/2} a_\lambda(t)b_\lambda(t)dt = W\left[\partial_\lambda a_\lambda(t), b_\lambda(t)\right]\Big|_{-\beta/2}^{\beta/2}. \tag{B.72}$$

We can then calculate

$$\int_{-\beta/2}^{\beta/2} Z(t, t; \lambda)dt = T(\lambda)^{-1}\frac{dT(\lambda)}{d\lambda}, \tag{B.73}$$

to obtain

$$-\operatorname{Tr} R_\lambda = \int_{-\beta/2}^{\beta/2} R_\lambda(t,t)\,dt = \operatorname{Tr}\left\{ (T(\lambda)-1)^{-1}\frac{dT(\lambda)}{d\lambda} \right\}$$
$$= \frac{d}{d\lambda}\log\det\left(T(\lambda)-1 \right). \tag{B.74}$$

From (B.66) we conclude that

$$\det(A-\lambda) = C\det\left(T(\lambda)-1 \right), \tag{B.75}$$

where C is a constant, independent of λ. This constant can be evaluated by looking at the limiting behavior of both sides when $\lambda = -\mu$, $\mu \to \infty$. In order to study $\det(A+\mu)$ for large μ, we use (B.47). Since $A+\mu$ does not have zero modes, we can write

$$\zeta_{A+\mu}(s) = \frac{1}{\Gamma(s)}\left[\int_0^1 d\tau\, \tau^{s-1}e^{-\mu\tau}\operatorname{Tr}K_A(\tau) + \int_1^\infty d\tau\, \tau^{s-1}e^{-\mu\tau}\operatorname{Tr}K_A(\tau) \right]. \tag{B.76}$$

The derivative at $s = 0$ of the term involving the second integral is of order $e^{-\mu}$. To compute the first integral, we use the heat kernel expansion for $\operatorname{Tr}K_A(\tau)$, which reads

$$\operatorname{Tr}K_A(\tau) = \frac{\beta}{\sqrt{4\pi\tau}} + \mathcal{O}\left(\sqrt{\tau}\right), \tag{B.77}$$

since in this case $a_0(x) = 1$, and β comes from the integration on spacetime. The first term in the expansion gives, when plugged into the first integral of (B.76),

$$\frac{1}{\Gamma(s)}\int_0^1 d\tau\, \tau^{s-3/2}e^{-\mu\tau} = \frac{\mu^{1/2-s}}{\Gamma(s)}\left\{ \Gamma(s-1/2) - \int_\mu^\infty du\, u^{s-3/2}e^{-u} \right\}. \tag{B.78}$$

When taking the derivative with respect to s at $s = 0$, the first term gives

$$-2\sqrt{\pi\mu} \tag{B.79}$$

while the second term is again of order $e^{-\mu}$. Finally, the higher order terms in the heat kernel expansion can be easily estimated and they are of order $\mathcal{O}\left(\mu^{-1/2}\right)$. We conclude that

$$\det(A+\mu) = e^{\sqrt{\mu}\beta}\left(1 + \mathcal{O}\left(\mu^{-1/2}\right) \right). \tag{B.80}$$

We now calculate $T(-\mu)$ for large μ. In this case, the fundamental matrix (1.6.3) at large μ is obtained by solving (1.6.1) with $\lambda = -\mu$ and $u(t) = 0$. We have

$$M_{-\mu}(t) \approx \begin{pmatrix} \cosh\left(\sqrt{\mu}(t+\beta/2)\right) & \frac{1}{\sqrt{\mu}}\sinh\left(\sqrt{\mu}(t+\beta/2)\right) \\ \sqrt{\mu}\sinh\left(\sqrt{\mu}(t+\beta/2)\right) & \cosh\left(\sqrt{\mu}(t+\beta/2)\right) \end{pmatrix}, \tag{B.81}$$

and

$$\det{(T(-\mu) - 1)} \approx -e^{\sqrt{\mu}\beta}. \tag{B.82}$$

We then conclude that $C = -1$, and we finally obtain

$$\det(A - \lambda) = -\det{(T(\lambda) - 1)} = \text{Tr}\,(T(\lambda) - 1), \tag{B.83}$$

which is the result (1.6.10).

B.4 Relation to dimensional regularization

In QFT on flat space, UV divergences are often regularized with dimensional regularization, which leads to finite physical amplitudes after using renormalized coulings. It is then useful to understand what is the relationship between zeta function regularization, which also gives a finite result for the functional determinant, and dimensional regularization. A natural way to define dimensional regularization on a curved space was proposed by Lüscher in [127], and we will follow his exposition. Other approaches can be found in [109, 168], for example.

Let us assume that we have an operator A defined on a certain space M. We now introduce ϵ extra dimensions, compactified on a manifold Y_ϵ of dimension ϵ, i.e we consider the manifold $M \times Y_\epsilon$, and we extend the relevant operator A defined on M to the full manifold Y_ϵ. It can be seen that, if ϵ is sufficiently large, this procedure leads to manifestly finite quantities which depend on ϵ, and which can be extended to meromorphic functions on the complex ϵ-plane. When $\epsilon \to 0$, one finds divergent terms which can be absorbed in a redefinition of the couplings, and the remaining finite parts can be related to the zeta function regularization. Notice that, when using zeta function regularization, the divergent part is somewhat implicit, and therefore it is not obvious how to introduce renormalized couplings in this scheme. By combining zeta function regularization with dimensional regularization, we have control on the structure of the divergences and we also have a well-defined and powerful method to compute the finite parts. We will illustrate this method in two examples relevant to calculations in Chapter 4 and Chapter 10, in four and two dimensions, respectively.

An example in four dimensions

Let us consider an operator A defined on a four-dimensional manifold M, and let n_A^0 be the number of its zero modes. Given the heat kernel K_A of the operator A, a sensible definition of $\log(\det'(A)) = \text{Tr}\log'(A)$ should start from

$$-\int_0^\infty \frac{d\tau}{\tau} \text{Tr} K_A(\tau), \tag{B.84}$$

since this is what one obtains, for example, by naively computing $-\zeta'_A(0)$ from (B.46). This expression has two kinds of divergence: there is a UV divergence at $\tau \to 0$, due to the short-distance behavior of the heat kernel, and there is an IR divergence at $\tau \to \infty$, due to the presence of zero modes. The way the zeta function regularization deals with these issues is clear in (B.46): we regulate the UV divergence by introducing the variable s and performing an analytic continuation with the help of the Gamma function. For the IR divergence, we simply subtract the zero mode. We will now propose a different regularization based on analytic continuation in the number of dimensions, for the UV divergences, and the introduction of a mass regulator, for the IR divergence. Let us consider the IR divergence first. We will define

$$\mathrm{Tr}\log'(A) = \lim_{m^2 \to 0} \left\{ \mathrm{Tr}\log(A + m^2) - n_A^0 \log(m^2) \right\}. \tag{B.85}$$

This is free of IR divergences, so we can write

$$\mathrm{Tr}\log(A + m^2) = -\int_0^\infty \frac{d\tau}{\tau} e^{-\tau m^2} \mathrm{Tr}K_A(\tau)$$

$$= -\int_1^\infty \frac{d\tau}{\tau} e^{-\tau m^2} \left(\mathrm{Tr}K_A(\tau) - n_A^0 \right) - \int_0^1 \frac{d\tau}{\tau} e^{-\tau m^2} \mathrm{Tr}K_A(\tau)$$

$$- n_A^0 \int_1^\infty \frac{d\tau}{\tau} e^{-\tau m^2}. \tag{B.86}$$

The first two terms in the last line do not have IR divergences as we take the limit $m^2 \to 0$. The last term has an IR divergence, indeed,

$$\int_1^\infty \frac{d\tau}{\tau} e^{-\tau m^2} = \Gamma(0, m^2) = -\log(m^2) - \gamma_E + \mathcal{O}(m^2), \tag{B.87}$$

where $\Gamma(z, a)$ is the incomplete Gamma function introduced in (1.9.53). The divergence is cancelled by the subtraction in (B.85). We can now write

$$\mathrm{Tr}\log'(A) = -\int_1^\infty \frac{d\tau}{\tau} \left(\mathrm{Tr}K_A(\tau) - n_A^0 \right) - \int_0^1 \frac{d\tau}{\tau} \mathrm{Tr}K_A(\tau) + n_A^0 \gamma_E. \tag{B.88}$$

Let us now consider the UV divergences. In the spirit of dimensional regularization, we will consider a suitable extension of the operator A on M to an operator A_p on the manifold $X = M \times Y_p$ (we will be completely specific when considering particular examples below). In the four-dimensional examples worked out in this book, we will take $Y_p = (\mathbb{S}^2)^p$, where $\mathbb{S}^2$ is a two-sphere of radius r. The two-sphere is a particularly useful example since the analysis of the usual operators on this space is particularly simple. We can now consider the definition (B.88) for the operator A_p. This is a function of p, and as we will see in examples, after an analytic continuation to negative $p = -\epsilon$, (B.88) is well defined for $\mathrm{Re}(\epsilon) > 2$. As we take the limit $\epsilon \to 0$, we will find divergences which have to be absorbed in a renormalization of the coupling.

In order to make this procedure concrete, we will consider the operators appearing in the analysis of the fluctuations around the instanton solution of Euclidean YM theory. Let us start with the Laplace operator Δ_0 acting on scalars taking values in the Lie algebra. This is easily extended to arbitrary manifolds, and in particular to the product manifold $X = M \times Y_p$, and we will denote the extension by Δ_0^X (if no superscript is indicated, the corresponding operator is simply the operator on M). We have

$$\mathrm{Tr} K_{\Delta_0^X}(\tau) = \left(\mathrm{Tr} K_{\Delta_0}(\tau)\right) \left(\mathrm{Tr} K_{\Delta_0^{Y_p}}(\tau)\right). \tag{B.89}$$

In addition, since Y_p is a Cartesian product, the heat kernel factorizes and

$$\mathrm{Tr} K_{\Delta_0^{Y_p}}(\tau) = \left(\mathrm{Tr} K_{\Delta_0^{\mathbb{S}^2}}\right)^p. \tag{B.90}$$

The operator Δ_0 on the four-dimensional manifold M has the Seeley–De Witt expansion

$$\mathrm{Tr} K_{\Delta_0}(\tau) = \frac{1}{\tau^2}\alpha_0(\Delta_0) + \frac{1}{\tau}\alpha_1(\Delta_0) + \alpha_2(\Delta_0) + \mathcal{O}(\tau), \tag{B.91}$$

where $\alpha_n(\Delta_0)$, $n \geq 0$, are the coefficients appearing in the expansion (B.13). Their explicit form can easily be obtained by integrating the corresponding coefficients in (B.23), for $r = G$ the adjoint representation. On the other hand, we have

$$\mathrm{Tr} K_{\Delta_0^{\mathbb{S}^2}} = \frac{r^2}{\tau} + \alpha_1(\Delta_0^{\mathbb{S}^2}) + \frac{\tau}{r^2}\alpha_2(\Delta_0^{\mathbb{S}^2}) + \mathcal{O}(\tau^2), \tag{B.92}$$

where we recall that r is the radius of the two-sphere. We have written down these factors explicitly so that the coefficients $\alpha_n(\Delta_0^{\mathbb{S}^2})$ are adimensional. Of course, these coefficients can be computed explicitly by using (B.13) and the formulae (B.23), and we have used that $a_0(x) = 1$, so that the first coefficient in the expansion (B.92) is the one we have written down. We can now split

$$\int_0^1 \frac{d\tau}{\tau}\mathrm{Tr} K_{\Delta_0^X}(\tau) = \int_0^1 \frac{d\tau}{\tau}\left(\mathrm{Tr} K_{\Delta_0^{\mathbb{S}^2}}\right)^p \left(\frac{1}{\tau^2}\alpha_0(\Delta_0) + \frac{1}{\tau}\alpha_1(\Delta_0) + \alpha_2(\Delta_0)\right)$$
$$+ \int_0^1 \frac{d\tau}{\tau}\left(\mathrm{Tr} K_{\Delta_0^{\mathbb{S}^2}}\right)^p \left(\mathrm{Tr} K_{\Delta_0}(\tau) - \frac{1}{\tau^2}\alpha_0(\Delta_0) - \frac{1}{\tau}\alpha_1(\Delta_0) - \alpha_2(\Delta_0)\right). \tag{B.93}$$

The integral in the second line is finite when $p \to 0$, since we have subtracted the terms in the Seeley–De Witt expansion which lead to UV divergences. Let us now do the first integral for $p = -\epsilon$:

$$\int_0^1 \frac{d\tau}{\tau} \left(\text{Tr}K_{\Delta_0^{S^2}}\right)^{-\epsilon} \left(\frac{1}{\tau^2}\alpha_0(\Delta_0) + \frac{1}{\tau}\alpha_1(\Delta_0) + \alpha_2(\Delta_0)\right)$$

$$= r^{-2\epsilon} \int_0^1 \frac{d\tau}{\tau} \tau^{\epsilon-1} \left\{\frac{1}{\tau^2}\alpha_0(\Delta_0) + \frac{1}{\tau}\left(\alpha_1(\Delta_0) - \frac{\epsilon b_1}{r^2}\alpha_0(\Delta_0)\right)\right.$$

$$\left. + \left(\alpha_2(\Delta_0) - \frac{\epsilon b_1}{r^2}\alpha_1(\Delta_0) - \frac{\epsilon b_2}{r^4}\alpha_0(\Delta_0)\right) + \mathcal{O}(\epsilon^2)\right\},$$

$$\text{(B.94)}$$

where b_1, b_2 are constants that can be computed from the $\alpha_n(\Delta_0^{S^2})$ in (B.92). Assuming first that $\text{Re}(\epsilon) > 2$, we find

$$\int_0^1 \tau^{\epsilon-k} \, d\tau = \frac{1}{\epsilon - k + 1}, \quad k \le 3. \tag{B.95}$$

We finally obtain, as $\epsilon \to 0$,

$$-\text{Tr}\log'\left(\Delta_0^X\right) = \frac{1}{\epsilon} r^{-2\epsilon} \alpha_2(\Delta_0) - n_A^0 \gamma_E - \frac{b_1}{r^2}\alpha_1(\Delta_0) - \frac{b_2}{r^4}\alpha_0(\Delta_0)$$

$$+ \int_1^\infty \frac{d\tau}{\tau}\left(\text{Tr}K_{\Delta_0}(\tau) - n_A^0\right) - \frac{\alpha_0(\Delta_0)}{2} - \alpha_1(\Delta_0)$$

$$+ \int_0^1 \frac{d\tau}{\tau}\left(\text{Tr}K_{\Delta_0}(\tau) - \frac{1}{\tau^2}\alpha_0(\Delta_0) - \frac{1}{\tau}\alpha_1(\Delta_0) - \alpha_2(\Delta_0)\right)$$

$$+ \mathcal{O}(\epsilon).$$

$$\text{(B.96)}$$

This is the definition of the determinant of the operator Δ_0 based on dimensional regularization. It has a pole at $\epsilon = 0$, and its finite part is free of both UV and IR divergences. What remains to be done now is to relate this finite part to the part obtained by zeta function regularization. To this aim, we start from (B.47) for a generic operator A defined on a four-dimensional manifold, and write the zeta function as

$$\zeta_A(s) = \frac{1}{\Gamma(s)} \int_0^1 d\tau \, \tau^{s-1} \left(\text{Tr}K(\tau) - \frac{1}{\tau^2}\alpha_0(A) - \frac{1}{\tau}\alpha_1(A) - \alpha_2(A)\right)$$

$$+ \frac{\alpha_2(A) - n_A^0}{s\Gamma(s)} + \frac{\alpha_1(A)}{(s-1)\Gamma(s)} + \frac{\alpha_0(A)}{(s-2)\Gamma(s)}$$

$$+ \frac{1}{\Gamma(s)} \int_1^\infty d\tau \, \tau^{s-1} \left(\text{Tr}K(\tau) - n_A^0\right). \tag{B.97}$$

We then find the manifestly finite expression,

$$
\zeta_A'(0) = \int_0^1 \frac{d\tau}{\tau} \left(\operatorname{Tr} K(\tau) - \frac{1}{\tau^2} \alpha_0(A) - \frac{1}{\tau} \alpha_1(A) - \alpha_2(A) \right)
$$
$$
+ \int_1^\infty \frac{d\tau}{\tau} \left(\operatorname{Tr} K(\tau) - n_A^0 \right) + \left(\alpha_2(A) - n_A^0 \right) \gamma_E
$$
$$
- \alpha_1(A) - \frac{\alpha_0(A)}{2}. \tag{B.98}
$$

Comparing this with (B.96) we finally obtain the sought-for result:

$$
-\operatorname{Tr} \log' \left(\Delta_0^X \right) = \alpha_2(\Delta_0) \left(\frac{1}{\epsilon} r^{-2\epsilon} - \gamma_E \right) - \frac{b_1}{r^2} \alpha_1(\Delta_0)
$$
$$
- \frac{b_2}{r^4} \alpha_0(\Delta_0) + \zeta_{\Delta_0}'(0) + \mathcal{O}(\epsilon). \tag{B.99}
$$

This expression has all the virtues we need: it displays the structure of the divergence in four dimensions through the pole at $\epsilon = 0$, and its finite part can be computed in terms of a zeta function. It can then be combined with the standard renormalization of the coupling constant and produce manifestly finite results at one-loop.

We will now consider the operator Δ_1^t defined in (4.5.76), which is an elliptic operator acting on one-forms with values in the Lie algebra. This operator can be defined on any Riemannian manifold, in particular it can be defined on the product of two Riemannian manifolds $X = M \times Y$, but the relations between the traces of the heat kernels are slightly more complicated. Indeed, if $A_m(x, y)$ is a one-form on X, where $m = 1, \ldots, \dim(X)$ and x, y are local coordinates on M and Y, respectively, we have the natural splitting (also known as dimensional reduction)

$$
A_m(x, y) = \begin{pmatrix} A_\mu(x) f(y) \\ A_i(y) g(x) \end{pmatrix}, \qquad \mu = 1, \ldots, \dim(M), \quad i = 1, \ldots, \dim(Y). \tag{B.100}
$$

We also have that

$$
\left(\Delta_1^{t,X} A \right)_m (x, y) = \begin{pmatrix} (\Delta_1^{t,M} A)_\mu(x) f(y) + A_\mu(x)(\Delta_0^Y f)(y) \\ (\Delta_1^{t,Y} A)_i(y) g(x) + A_i(x)(\Delta_0^M g)(x) \end{pmatrix}. \tag{B.101}
$$

It follows that, in order to find an eigenfunction of the operator Δ_1^t on X, we have to take $A_\mu(x)$ and $g(x)$ to be eigenfunctions of $\Delta_1^{t,M}$ and Δ_0^M, and $A_i(y)$ and $f(y)$ to be eigenfunctions of $\Delta_1^{t,Y}$ and Δ_0^Y, respectively. We conclude from the action of $\Delta_1^{t,X}$ that

$$
\operatorname{Tr} K_{\Delta_1^{t,X}}(\tau) = \operatorname{Tr} K_{\Delta_1^{t,M}}(\tau) \operatorname{Tr} K_{\Delta_0^Y}(\tau) + \operatorname{Tr} K_{\Delta_0^M}(\tau) \operatorname{Tr} K_{\Delta_1^{t,Y}}(\tau). \tag{B.102}
$$

The above result is valid for any product of manifolds. Let us now apply this to our case, in which $X = M \times Y_p$ and $Y_p = (\mathbb{S}^2)^p$ (notice that here the one-form $A_\mu(x)$ on M takes values in the Lie algebra, but $A_i(y)$ has no color indices). By iterating (B.102) we also find

$$\text{Tr}K_{\Delta_1^{t,Y_p}}(\tau) = p\left(\text{Tr}K_{\Delta_0^{\mathbb{S}^2}}(\tau)\right)^{p-1}\left(\text{Tr}K_{\Delta_1^{t,\mathbb{S}^2}}(\tau)\right), \tag{B.103}$$

and we conclude that

$$\text{Tr}K_{\Delta_1^{t,X}}(\tau) = \text{Tr}K_{\Delta_1^{t}}(\tau)\left(\text{Tr}K_{\Delta_0^{\mathbb{S}^2}}(\tau)\right)^{p}$$
$$+ p\text{Tr}K_{\Delta_0}(\tau)\left(\text{Tr}K_{\Delta_0^{\mathbb{S}^2}}(\tau)\right)^{p-1}\left(\text{Tr}K_{\Delta_1^{t,\mathbb{S}^2}}(\tau)\right). \tag{B.104}$$

We recall that, when no superscript is written explicitly, the corresponding quantity corresponds by default to the four-manifold M. Notice that, since one-forms on $\mathbb{S}^2$ have two components, we have

$$\text{Tr}K_{\Delta_1^{t,\mathbb{S}^2}}(\tau) = \frac{2r^2}{\tau} + \mathcal{O}(1). \tag{B.105}$$

We are now ready to compute $\text{Tr}\log'(\Delta_1^{t,X})$. The calculation is identical to the one we did for Δ_0^X, but we have an extra contribution due to the term proportional to $p = -\epsilon$ in (B.104). A simple calculation shows that this extra term is, as $\epsilon \to 0$,

$$- 2\alpha_2(\Delta_0). \tag{B.106}$$

We then find, after putting everything together,

$$-\text{Tr}\log'\left(\Delta_1^{t,X}\right) = \alpha_2(\Delta_1^t)\left(\frac{1}{\epsilon}r^{-2\epsilon} - \gamma_{\text{E}}\right) - \frac{b_1}{r^2}\alpha_1(\Delta_1^t) - \frac{b_2}{r^4}\alpha_0(\Delta_1^t)$$
$$+ \zeta'_{\Delta_1^t}(0) - 2\alpha_2(\Delta_0) + \mathcal{O}(\epsilon), \tag{B.107}$$

where the coefficients $\alpha_n(\Delta_1^t)$ can be obtained explicitly by integrating (B.28) in four dimensions.

We can now evaluate the above quantities (B.99) and (B.107) in the background of an instanton, after subtracting the contribution of the trivial connection, i.e. we want to evaluate

$$- \left(\text{Tr}\log'(\Delta_0) - \text{Tr}\log'\left(\Delta_0^{(0)}\right)\right),$$
$$- \left(\text{Tr}\log'(\Delta_1^t) - \text{Tr}\log'\left(\Delta_1^{t,(0)}\right)\right), \tag{B.108}$$

where we have removed the superscript X for simplicity of notation. We need to compute

$$\alpha_n(\Delta_0) - \alpha_n\left(\Delta_0^{(0)}\right), \quad n = 0, 1, 2, \tag{B.109}$$

and the same quantity for Δ_1^t. By looking at the explicit expressions (B.23) and (B.28), we see that the gravitational parts, depending solely on the Riemann tensor, cancel, and only the contribution depending on the gauge field survives. For $n = 0, 1$, since there is no contribution of the gauge field, the above differences vanish. We then find

$$-\alpha_2(\Delta_0) + \alpha_2\left(\Delta_0^{(0)}\right) = \frac{C_2(G)}{12}\frac{1}{16\pi^2}\int d^4x\sqrt{g}F^a_{\mu\nu}F^{\mu\nu a}$$
$$= \frac{C_2(G)\nu}{6}, \tag{B.110}$$

where ν is the instanton charge, we used the value of the YM action in (4.4.5), and the fact that $C(G) = C_2(G)$. Similarly, we find

$$-\alpha_2\left(\Delta_1^t\right) + \alpha_2\left(\Delta_1^{t,(0)}\right) = -\frac{5}{3}C_2(G)\frac{1}{16\pi^2}\int d^4x\sqrt{g}F^a_{\mu\nu}F^{\mu\nu a}$$
$$= -\frac{10C_2(G)\nu}{3}. \tag{B.111}$$

We conclude that

$$\text{Tr}\log'(\Delta_0) - \text{Tr}\log'\left(\Delta_0^{(0)}\right) = \frac{C_2(G)\nu}{6}\left(\frac{1}{\epsilon}r^{-2\epsilon} - \gamma_E\right)$$
$$- \zeta'_{\Delta_0}(0) + \zeta'_{\Delta_0^{(0)}}(0) + \mathcal{O}(\epsilon),$$

$$\text{Tr}\log'\left(\Delta_1^t\right) - \text{Tr}\log'\left(\Delta_1^{t,(0)}\right) = -\frac{10C_2(G)\nu}{3}\left(\frac{1}{\epsilon}r^{-2\epsilon} - \gamma_E\right) - \frac{C_2(G)\nu}{3}$$
$$- \zeta'_{\Delta_1^t}(0) + \zeta'_{\Delta_1^{t,(0)}}(0) + \mathcal{O}(\epsilon). \tag{B.112}$$

An example in two dimensions

Another possible way of performing dimensional regularization is by considering the manifold $X = M \times (\mathbb{S}^1)^\epsilon$, i.e. we take the extra space to be a torus of dimension ϵ. The coordinates on $(\mathbb{S}^1)^\epsilon$ are denoted by

$$0 \le y^\mu \le L, \qquad \mu = 1, \ldots, \epsilon, \tag{B.113}$$

where L is the length of $\mathbb{S}^1$. We now consider the operator

$$A_\epsilon = A + \Delta_\epsilon, \qquad \Delta_\epsilon = R^2\sum_{\mu=1}^\epsilon \partial_\mu^2. \tag{B.114}$$

Here, we assume that A is normalized in such a way that it is dimensionless, and R is the characteristic scale of the space M (for example, if M is a sphere $\mathbb{S}^n$, then R would be its radius).

In order to compare the two regularizations, we will consider an operator A on a two-dimensional manifold M, with no zero modes, and we will consider the quantities $\text{Tr}A^{-1}$ and $\text{Tr}\log(A)$. This is the situation relevant for the calculation in Section 10.5, but the calculation can easily be generalized to any situation.

Let us first consider the quantity $\text{Tr}A_\epsilon^{-1}$. In terms of the heat kernel

$$K_{A_\epsilon}(\tau) = e^{-\tau A_\epsilon} = K_A(\tau)K_{\Delta_\epsilon}(\tau), \tag{B.115}$$

this can be calculated as

$$\text{Tr}A_\epsilon^{-1} = \int_0^\infty \text{Tr}K_A(\tau)\,\text{Tr}K_{\Delta_\epsilon}(\tau)\,d\tau. \tag{B.116}$$

The Seeley–De Witt expansion of these heat kernels is of the form

$$\text{Tr}K_{\Delta_\epsilon}(\tau) = \frac{1}{\left(4\pi(R/L)^2\tau\right)^{\epsilon/2}}\left(1+\mathcal{O}(\tau)\right),$$

$$\text{Tr}K_A(\tau) = \frac{1}{\tau}\left(\hat{a}_0 + \hat{a}_1\tau + \mathcal{O}(\tau^2)\right), \tag{B.117}$$

where we took into account that we are working on a two-dimensional manifold, therefore the Seeley–De Witt expansion starts with a term of order τ^{-1}. We also use that, for $\epsilon = 1$, i.e. when we have one extra dimension with the geometry of a circle $\mathbb{S}^1$, of length L, the Seeley–De Witt expansion reads

$$\text{Tr}K_{\Delta_1}(\tau) = \frac{L}{(4\pi R^2\tau)^{1/2}}\left(1+\mathcal{O}(\tau)\right), \tag{B.118}$$

and for a product of ϵ circles we simply consider the ϵ power of this expression. We have,

$$\text{Tr}K_{A_\epsilon}(\tau) = \frac{c_0}{\tau^{\epsilon/2+1}}\left(1+\mathcal{O}(\tau)\right), \qquad c_0 = \frac{\hat{a}_0}{\left(4\pi(R/L)^2\right)^{\epsilon/2}}. \tag{B.119}$$

We can then write,

$$\text{Tr}\,A_\epsilon^{-1} = c_0\int_0^1 \frac{d\tau}{\tau^{\epsilon/2+1}} + \int_0^1 \left(\text{Tr}K_{A_\epsilon}(\tau) - \frac{c_0}{\tau^{\epsilon/2+1}}\right)d\tau$$
$$+ \int_1^\infty \text{Tr}K_{A_\epsilon}(\tau)\,d\tau. \tag{B.120}$$

Notice that the last two integrals are regular when $\epsilon \to 0$, while the first integral has, in that limit, a UV divergence due to the behavior at $\tau = 0$. We can now use dimensional regularization to extract the divergence of this integral: we assume that

$\epsilon > 0$, so that the integral is regular at $\tau = 0$, and then we analytically continue the result of the integration to arbitrary ϵ. We find,

$$c_0 \int_0^1 \frac{d\tau}{\tau^{\epsilon/2+1}} = -\frac{2}{\epsilon} \frac{\hat{a}_0}{(4\pi (R/L)^2 \tau)^{\epsilon/2}}$$

$$= \left[-\frac{2}{\epsilon} + \log \left(4\pi R^2/L^2 \right) \right] \hat{a}_0 + \mathcal{O}(\epsilon). \tag{B.121}$$

Therefore,

$$\mathrm{Tr} A_\epsilon^{-1} = \left[-\frac{2}{\epsilon} + \log \left(4\pi R^2/L^2 \right) \right] \hat{a}_0$$

$$+ \int_0^\infty \left(\mathrm{Tr} K_A(\tau) - \theta(1-\tau) \frac{\hat{a}_0}{\tau} \right) + \mathcal{O}(\epsilon), \tag{B.122}$$

where $\theta(\tau)$ is the Heaviside function. Similarly, for the calculation of $\mathrm{Tr} \log(A_\epsilon)$, one finds, after using the representation (B.88),

$$\mathrm{Tr} \log(A_\epsilon) = \frac{2\hat{a}_1}{\epsilon} + \hat{a}_0 - \hat{a}_1 \log \left(4\pi R^2/L^2 \right)$$

$$- \int_0^\infty \frac{d\tau}{\tau} \mathrm{Tr} K_A(\tau) \left\{ 1 - \theta(1-\tau) \left(\frac{\hat{a}_0}{\tau} + \hat{a}_1 \right) \right\} + \mathcal{O}(\epsilon). \tag{B.123}$$

The integrals appearing in (B.122) and (B.123) are convergent. We will now relate them to zeta function regularization. To do this, we consider the expression (B.46) and we split it as follows:

$$\zeta_A(s) = \frac{\hat{a}_0}{\Gamma(s)} \int_0^1 \tau^{s-2} d\tau + \int_0^\infty \tau^{s-1} \left\{ \mathrm{Tr} K_A(\tau) - \theta(1-\tau) \frac{\hat{a}_0}{\tau} \right\}. \tag{B.124}$$

We can integrate the first term by assuming that $s > 1$, and then analytically continue the result to arbitrary s. We find,

$$\zeta_A(s) = \frac{\hat{a}_0}{(s-1)\Gamma(s)} + \int_0^\infty \tau^{s-1} \left\{ \mathrm{Tr} K_A(\tau) - \theta(1-\tau) \frac{\hat{a}_0}{\tau} \right\}. \tag{B.125}$$

The first term appearing here is nothing but the pole at $s = 1$ of the zeta function in two dimensions (see (B.57)). The remaining, finite part is precisely what appears in (B.122). We conclude that,

$$\mathrm{Tr} A_\epsilon^{-1} = \left[-\frac{2}{\epsilon} + \log \left(4\pi R^2/L^2 \right) \right] \hat{a}_0$$

$$+ \lim_{s \to 1} \left(\zeta_A(s) - \frac{\hat{a}_0}{(s-1)\Gamma(s)} \right) + \mathcal{O}(\epsilon). \tag{B.126}$$

This gives the precise relation between the dimensional regularization of this quantity, and the zeta function regularization: the UV divergences appear as poles in ϵ, and the finite part can be computed via zeta function regularization. In order to recover the finite part in (B.123), we make a similar splitting of (B.46),

$$
\zeta_A(s) = \frac{\hat{a}_0}{(s-1)\Gamma(s)} + \frac{\hat{a}_1}{s\Gamma(s)}
$$
$$
+ \int_0^\infty \tau^{s-1} \left\{ \mathrm{Tr} K_A(\tau) - \theta(1-\tau) \left(\frac{\hat{a}_0}{\tau} + \hat{a}_1 \right) \right\} d\tau. \tag{B.127}
$$

If we now expand this expression in power series around $s = 0$, we find

$$
\zeta_A'(0) = -\hat{a}_0 + \gamma_E \hat{a}_1 + \int_0^\infty \frac{d\tau}{\tau} \left\{ \mathrm{Tr} K_A(\tau) - \theta(1-\tau) \left(\frac{\hat{a}_0}{\tau} + \hat{a}_1 \right) \right\}, \tag{B.128}
$$

where γ_E is the Euler–Mascheroni constant. We conclude that

$$
\mathrm{Tr} \log(A_\epsilon) = \frac{2\hat{a}_1}{\epsilon} + \hat{a}_1 \left[\gamma_E - \log \left(4\pi R^2/L^2 \right) \right] - \zeta_A'(0) + \mathcal{O}(\epsilon). \tag{B.129}
$$

We will now study, as an example of these general considerations, the operator (10.5.16) relevant for the analysis of large N instantons in $\mathbb{CP}^{N-1}$ models. The spectrum and the degeneracies of this operator are given in (10.5.21) and (10.5.22). We can write

$$
\lambda_n = (n+a)^2 - c^2, \tag{B.130}
$$

where a, c are defined in (10.5.23). The zeta function reads

$$
\zeta_{\Delta_k}(s) = 2 \sum_{n=0}^\infty \frac{n+a}{\left((n+a)^2 - c^2 \right)^s}. \tag{B.131}
$$

In order to express this in terms of known functions, we first define $R(m, s)$ by the decomposition

$$
\frac{m}{(m^2 - c^2)^s} = \frac{1}{2} \left\{ \frac{m}{(m+c)^s} + \frac{m}{(m-c)^{2s}} \right\} - \frac{2s^2 c^2}{m^{2s+1}} + R(m, s). \tag{B.132}
$$

The interest of this decomposition is that, for m large,

$$
R(m, s) \sim \frac{1}{m^{2s+3}}, \tag{B.133}
$$

therefore it leads to a series which is absolutely convergent for $s \geq 0$, and it can be evaluated directly for concrete values of s, by using for example the polygamma function and its derivatives, since

$$
\sum_{n=0}^\infty \frac{1}{(n+z)^{l+1}} = \frac{(-1)^{l+1}}{l!} \psi^{(l)}(z). \tag{B.134}
$$

The sums of the first terms in (B.132) can be found in terms of the Hurwitz zeta function, which is defined as

$$\zeta_H(s, z) = \sum_{n=0}^{\infty} \frac{1}{(n+z)^s}. \tag{B.135}$$

We find, by using the above decomposition,

$$\zeta_{\Delta_k}(s) = \zeta_H(2s - 1, a - c) + \zeta_H(2s - 1, a + c)$$
$$+ c\left(\zeta_H(2s, a - c) - \zeta_H(2s, a + c)\right) - 4s^2c^2\zeta_H(2s + 1, a)$$
$$+ 2\sum_{n=0}^{\infty} R(n + a, s). \tag{B.136}$$

Let us now study this function near $s = 1$. We find,

$$\zeta_{\Delta_k}(s) = \frac{1}{s - 1} - \psi(a + c) - \psi(a - c) + \mathcal{O}(s - 1). \tag{B.137}$$

From this we conclude that, in particular, $\hat{a}_0 = 1$ (since this is the residue of the pole at $s = 1$), and also that

$$F_k(m_k^2 R^2) \equiv \lim_{s \to 1} \left(\zeta_{\Delta_k}(s) - \frac{\hat{a}_0}{(s - 1)\Gamma(s)} \right)$$
$$= -\psi(a + c) - \psi(a - c) - \gamma_E. \tag{B.138}$$

Finally, we obtain, in dimensional regularization,

$$\mathrm{Tr}\,\Delta_{k,\epsilon}^{-1} = -\frac{2}{\epsilon} + \log\left(4\pi^2 R^2/L^2\right) + F_k(m_k^2 R^2) + \mathcal{O}(\epsilon). \tag{B.139}$$

The equation of motion (10.5.15) involves the diagonal value $\Delta_{k,\epsilon}^{-1}(g, g)$, rather than $\mathrm{Tr}\,\Delta_{k,\epsilon}^{-1}$. However, in spaces of constant curvature like the one we are considering, they differ just in the overall volume of the manifold (this is because all the coefficients $a_n(x)$ appearing in the Seeley–De Witt expansion are actually constant). In particular, $\Delta^{-1}(g, g)$ is a constant, and equals

$$\Delta_{k,\epsilon}^{-1}(g, g) = (\mathrm{vol}(\mathbb{S}^1))^{-\epsilon}\mathrm{Tr}\,\Delta_{k,\epsilon}^{-1} = L^{-\epsilon}\mathrm{Tr}\,\Delta_{k,\epsilon}^{-1}, \tag{B.140}$$

where we used the fact that the volume of $SU(2)$ is normalized to 1. We then find,

$$\Delta_{k,\epsilon}^{-1}(g, g) = -\frac{2}{\epsilon} + \log\left(4\pi^2 R^2\right) + F_k(m_k^2 R^2) + \mathcal{O}(\epsilon). \tag{B.141}$$

In order to compute $\mathrm{Tr}\log\Delta_{k,\epsilon}$, we need the value of $\hat{a}_1$. Since we have the relationship (B.55), this can be obtained by studying the value at $s = 0$ of (B.136). To do this, we need the results

$$\zeta_H(0, z) = \frac{1}{2} - z, \qquad \zeta_H(-1, z) = -\frac{1}{2}\left(z^2 - z + \frac{1}{6}\right). \tag{B.142}$$

We find,

$$\hat{a}_1 = \zeta_{\Delta_k}(0) = c^2 - a^2 + a - \frac{1}{6} = \frac{1}{3} - m_k^2 R^2. \tag{B.143}$$

Finally, we need the value of $\zeta'_{\Delta_k}(0)$. This can be obtained immediately from (B.136). The sum involving the derivatives of $R(m + a, s)$ does not contribute for $s = 0$, and we obtain

$$\zeta'_{\Delta_k}(0) = 2\zeta'_{\mathrm{H}}(-1, a - c) + 2\zeta'_{\mathrm{H}}(-1, a + c) + 2c \log\left(\frac{\Gamma(a - c)}{\Gamma(a + c)}\right) - 2c^2, \tag{B.144}$$

where the ′ in the Hurwitz zeta function denotes the derivative with respect to the first argument, and we used that

$$\zeta'_{\mathrm{H}}(0, z) = \log \Gamma(z) - \frac{1}{2} \log(2\pi). \tag{B.145}$$

Using now (B.129), we conclude that

$$\mathrm{Tr} \log \Delta_{k,\epsilon} = \frac{2}{\epsilon}\left(\frac{1}{3} - m_k^2 R^2\right) + \left(\frac{1}{3} - m_k^2 R^2\right)\left(\gamma_{\mathrm{E}} - \log(4\pi R^2/L^2)\right)$$

$$- \zeta'_{\Delta_k}(0) + \mathcal{O}(\epsilon). \tag{B.146}$$

Appendix C

Effective action for large N sigma models

In this appendix we compute the large N, effective propagators (6.2.39) and (6.3.87).

We start by calculating (6.2.39). Using the standard trick of introducing Feynman parameters (see for example [154], p. 189), we find

$$\int \frac{d^2q}{(2\pi)^2} \frac{1}{(q^2 + m^2)\left((p+q)^2 + m^2\right)}$$

$$= \int \frac{d^2q}{(2\pi)^2} \int_0^1 dx \frac{1}{\left[x(m^2 + q^2) + (1-x)((p+q)^2 + m^2)\right]^2}$$

$$= \int \frac{d^2\ell}{(2\pi)^2} \int_0^1 dx \frac{1}{\left[m^2 + \ell^2 + x(1-x)p^2\right]^2} \qquad (C.1)$$

where we have introduced

$$\ell = q + xp. \qquad (C.2)$$

We can now use the standard formula for the evaluation of Feynman integrals in dimensional regularization,

$$\int \frac{d^d\ell}{(2\pi)^d} \frac{1}{(\ell^2 + \Delta)^n} = \frac{1}{(4\pi)^{d/2}} \frac{\Gamma(n - d/2)}{\Gamma(n)} \Delta^{d/2-n}. \qquad (C.3)$$

In our case,

$$\Delta = m^2 + x(1-x)p^2. \qquad (C.4)$$

Since $d = n = 2$, the integral is convergent. We conclude that the integral (C.1) is given by,

$$\tilde{\Gamma}^\alpha(p) = \frac{1}{4\pi} \int_0^1 \frac{dx}{\Delta} = \frac{1}{4\pi} \int_0^1 \frac{dx}{m^2 + x(1-x)p^2}. \qquad (C.5)$$

The final integration over x is elementary and we finally obtain the result shown in (6.2.40).

We now compute $\tilde{\Gamma}^A_{\mu\nu}(p)$, given in (6.3.87). Both integrals appearing in (6.3.87) are divergent, but their divergences cancel. This is easily seen in dimensional regularization. We first consider the second integral, after doing the change of variables (C.2). We find,

$$
\int \frac{d^2 q}{(2\pi)^2} \frac{(p_\mu + 2q_\mu)(p_\nu + 2q_\nu)}{(q^2 + m^2)((p+q)^2 + m^2)}
$$

$$
= p_\mu p_\nu \int_0^1 dx \int \frac{d^2\ell}{(2\pi)^2} \frac{(1-2x)^2}{\left[m^2 + \ell^2 + x(1-x)p^2\right]^2}
$$

$$
+ \frac{4\delta_{\mu\nu}}{d} \int_0^1 dx \int \frac{d^2\ell}{(2\pi)^2} \frac{\ell^2}{\left[m^2 + \ell^2 + x(1-x)p^2\right]^2}. \tag{C.6}
$$

As usual, by rotational invariance, linear terms in ℓ_μ vanish after integration, and we can set

$$
\ell_\mu \ell_\nu \to \frac{\delta_{\mu\nu}}{d} \ell^2. \tag{C.7}
$$

We conclude that (6.3.87) has two contributions. The first one has the tensorial structure of $\delta_{\mu\nu}$, with coefficient

$$
2 \int \frac{d^2 q}{(2\pi)^2} \frac{1}{(q^2 + m^2)} - \frac{4}{d} \int_0^1 dx \int \frac{d^2\ell}{(2\pi)^2} \frac{\ell^2}{\left[m^2 + \ell^2 + x(1-x)p^2\right]^2}, \tag{C.8}
$$

while the second one has the tensorial structure of $p_\mu p_\nu$, and coefficient

$$
- \int_0^1 dx \int \frac{d^2\ell}{(2\pi)^2} \frac{(1-2x)^2}{\left[m^2 + \ell^2 + x(1-x)p^2\right]^2}. \tag{C.9}
$$

Let us compute (C.8). Using the formula (C.3) for $n = 1$ and the integral

$$
\int \frac{d^d\ell}{(2\pi)^d} \frac{\ell^2}{(\ell^2 + \Delta)^2} = \frac{1}{(4\pi)^{d/2}} \frac{d}{2} \Gamma(1 - d/2) \Delta^{d/2-1}, \tag{C.10}
$$

we obtain

$$
\frac{2\Gamma(1 - d/2)}{(4\pi)^{d/2}} \int_0^1 dx \left[\Delta_1^{d/2-1} - \Delta_2^{d/2-1} \right], \tag{C.11}
$$

where

$$
\Delta_1 = m^2, \qquad \Delta_2 = m^2 + x(1-x)p^2. \tag{C.12}
$$

We now expand around $\epsilon = 0$, after setting

$$
d = 2 - \epsilon. \tag{C.13}
$$

Since

$$\Delta_1^{d/2-1} - \Delta_2^{d/2-1} = e^{-\epsilon \log \Delta_1/2} - e^{-\epsilon \log \Delta_2/2} = \frac{\epsilon}{2} \log \frac{\Delta_2}{\Delta_1} + \mathcal{O}(\epsilon^2), \qquad (C.14)$$

and

$$\Gamma(\epsilon/2) = \frac{2}{\epsilon} - \gamma + \mathcal{O}(\epsilon), \qquad (C.15)$$

the total result is finite as $\epsilon \to 0$ and is given by

$$\frac{1}{2\pi} \int_0^1 dx \, \log \left[1 + x(1-x)\frac{p^2}{m^2} \right] = (p^2 + 4m^2) f(p) - \frac{1}{\pi}, \qquad (C.16)$$

where $f(p)$ is the function given in (6.2.40). On the other hand, the integral (C.9) is equal to

$$-\frac{1}{4\pi} \int_0^1 dx \frac{(1-2x)^2}{m^2 + x(1-x)p^2} = -\frac{1}{p^2}\left((p^2 + 4m^2) f(p) - \frac{1}{\pi} \right). \qquad (C.17)$$

Putting both results together, we obtain (6.3.88).

References

[1] Adams, D. 1996. Perturbative expansion in gauge theories on compact manifolds. hep-th/9602078.

[2] Affleck, I. 1980. Testing the instanton method. *Phys. Lett.*, B **92**, 149–152.

[3] Affleck, I. 1981. Quantum statistical metastability. *Phys. Rev. Lett.*, **46**, 388–391.

[4] Affleck, I. 1981. Mesons in the large N collective field method. *Nucl. Phys.*, B **185**, 346–364.

[5] Aguado, M. and Asorey, M. 2011. Theta-vacuum and large N limit in $\mathbb{CP}^{N-1}$ sigma models. *Nucl. Phys.*, B **844**, 243–265.

[6] Aharony, O., Gubser, S. S., Maldacena, J. M., Ooguri, H. and Oz, Y. 1990. Large N field theories, string theory and gravity. *Phys. Rep.*, **323**, 183–386.

[7] Akemann, G., Baik, J. and Di Francesco, P. (eds.) 2011. *The Oxford Handbook of Random Matrix Theory*. Oxford University Press.

[8] Akhiezer, N. I. 1990. *Elements of the Theory of Elliptic Functions*. American Mathematical Society.

[9] Altland, A. and Simons, B. 2006. *Condensed Matter Field Theory*. Cambridge University Press.

[10] Álvarez, G. 1988. Coupling-constant behavior of the resonances of the cubic anharmonic oscillator. *Phys. Rev.*, A **37**, 4079–4083.

[11] Álvarez, G. 2004. Langer–Cherry derivation of the multi-instanton expansion for the symmetric double well. *J. Math. Phys.*, **45**, 3095–3108.

[12] Álvarez-Gaumé, L. and Vázquez-Mozo, M. A. 2012. *An Invitation to Quantum Field Theory*. Springer-Verlag.

[13] Ambjorn, J., Chekhov, L., Kristjansen, C. F. and Makeenko, Y. 1993 Matrix model calculations beyond the spherical limit. *Nucl. Phys.*, B **404**, 127–172.

[14] Aniceto, I., Schiappa, R. and Vonk, M. 2012. The resurgence of instantons in string theory. *Commun. Num. Theor. Phys.*, **6**, 339–496.

[15] Appelquist, T. and Chodos, A. 1983. The quantum dynamics of Kaluza–Klein theories. *Phys. Rev.*, D **28**, 772–784.

[16] Atiyah, M. F., Hitchin, N. J., Drinfeld V. G. and Manin, Y. I. 1978. Construction of instantons. *Phys. Lett.*, A **65**, 185–187.

[17] Baacke, J. and Lavrelashvili, G. 2004. One-loop corrections to the metastable vacuum decay. *Phys. Rev.*, D **69**, 025009.

[18] Balian, R., Parisi, G. and Voros, A. 1978. Discrepancies from asymptotic series and their relation to complex classical trajectories. *Phys. Rev. Lett.*, **41**, 1141–1144.

[19] Balian, R., Parisi, G. and Voros, A. 1979. Quartic oscillator. In: *Feynman Path Integrals*, Lecture Notes in Physics **106**, pp. 337–360, Springer-Verlag.

[20] Bar-Natan, D. 1995. On the Vassiliev knot invariants. *Topology*, **34**, 423–472.

[21] Bars, I. and Green, M. B. 1978. Poincare and gauge invariant two-dimensional QCD. *Phys. Rev.*, D **17**, 537–545.

[22] Basar, G., Dunne, G. V. and Unsal, M. 2013. Resurgence theory, ghost-instantons, and analytic continuation of path integrals. *JHEP*, **1310**, 041.

[23] Bauer, C., Bali, G. S. and Pineda, A. 2012. Compelling evidence of renormalons in QCD from high order perturbative expansions. *Phys. Rev. Lett.*, **108**, 242002.

[24] Belavin, A. A. and Polyakov, A. M. 1977. Quantum fluctuations of pseudoparticles. *Nucl. Phys.*, B **123**, 429–444.

[25] Belavin, A. A., Polyakov, A. M., Schwartz, A. S. and Tyupkin, Y. S. 1975. Pseudoparticle solutions of the Yang–Mills equations. *Phys. Lett.*, B **59**, 85–87.

[26] Bender, C. M. 1978. Perturbation Theory in large order. *Adv. Math.*, **30**, 250–267.

[27] Bender, C. M. and Caswell, W. E. 1978. Asymptotic graph counting techniques in ψ^{2N} field theory. *J. Math. Phys.*, **19**, 2579–2586.

[28] Bender, C. M. and Orszag, S. A. 1999. *Advanced Mathematical Methods for Scientists and Engineers*. Springer-Verlag.

[29] Bender, C. M. and Wu, T. T. 1969. Anharmonic oscillator. *Phys. Rev.*, **184**, 1231–1260.

[30] Bender, C. M. and Wu, T. T. 1973. Anharmonic oscillator. 2: a study of perturbation theory in large order. *Phys. Rev.*, D **7**, 1620–1636.

[31] Bender, C. M. and Wu, T. T. 1976. Statistical analysis of Feynman diagrams. *Phys. Rev. Lett.*, **37**, 117–120.

[32] Beneke, M. 1999. Renormalons. *Phys. Rep.*, **317**, 1–142.

[33] Berg, B. and Lüscher, M. 1979. Computation of quantum fluctuations around multi-instanton fields from exact Green's functions: the $\mathbb{CP}^{N-1}$ case. *Commun. Math. Phys.*, **69**, 57–80.

[34] Bernard, C. W. 1979. Gauge zero modes, instanton determinants, and QCD calculations. *Phys. Rev.*, D **19**, 3013–3019.

[35] Bessis, D., Itzykson, C. and Zuber, J. B. 1980. Quantum field theory techniques in graphical enumeration. *Adv. Appl. Math.*, **1**, 109–157.

[36] Bogomolny, E. B. and Fateev, V. A. 1977. Large order calculations in gauge theories. *Phys. Lett.*, B **71**, 93–96.

[37] Brézin, E. and Wadia, S. (eds.) 1991. *The Large N Expansion in Quantum Field Theory and Statistical Physics*. World Scientific.

[38] Brézin, E., Le Guillou, J. C. and Zinn-Justin, J. 1977. Perturbation theory at large order. 1. The ψ^{2N} interaction. *Phys. Rev.*, D **15**, 1544–1557.

[39] Brézin, E., Le Guillou, J. C. and Zinn-Justin, J. 1977. Perturbation theory at large order. 2. Role of the vacuum instability. *Phys. Rev.*, D **15**, 1558–1564.

[40] Brézin, E., Itzykson, C., Parisi, G. and Zuber, J. B. 1978. Planar diagrams. *Commun. Math. Phys.*, **59**, 35–51.

[41] Brower, R. C., Spence, W. L. and Weis, J. H. 1979. Bound states and asymptotic limits for QCD in two-dimensions. *Phys. Rev.*, D **19**, 3024–3049.

[42] Caliceti, E., Graffi, S. and Maioli, M. 1980. Perturbation theory of odd anharmonic oscillators. *Commun. Math. Phys.*, **75**, 51–66.

[43] Caliceti, E., Meyer-Hermann, M., Ribeca, P., Surzhykov, A. and Jentschura, U. D. 2007. From useful algorithms for slowly convergent series to physical predictions based on divergent perturbative expansions. *Phys. Rep.*, **446**, 1–96.

[44] Callan, C. G. and Coleman, S. R. 1977. The fate of the false vacuum. 2. First quantum corrections. *Phys. Rev.*, D **16**, 1762–1768.

[45] Callan, C. G., Dashen, R. F. and Gross, D. J. 1976. The structure of the gauge theory vacuum. *Phys. Lett.*, B **63**, 334–340.

[46] Chadha, S., Di Vecchia, P., D'Adda, A. and Nicodemi, F. 1977. Zeta function regularization of the quantum fluctuations around the Yang–Mills pseudoparticle. *Phys. Lett.*, B **72**, 103–108.

[47] Christos, G. A. 1984. Chiral symmetry and the $U(1)$ problem. *Phys. Rep.*, **116**, 251–336.

[48] Cicuta, G. M. 1982. Topological expansion for $SO(N)$ and $Sp(2n)$ gauge theories. *Lett. Nuovo Cimento*, **35** 87–92.

[49] Coleman, S. R. 1977. The fate of the false vacuum. 1. Semiclassical theory. *Phys. Rev.*, D **15**, 2929–2936.

[50] Coleman, S. 1985. *Aspects of Symmetry*. Cambridge University Press.

[51] Coleman, S. R. and De Luccia, F. 1980. Gravitational effects on and of vacuum decay. *Phys. Rev.*, D **21**, 3305–3315.

[52] Coleman, S. R., Glaser, V. and Martin, A. 1978. Action minima among solutions to a class of Euclidean scalar field equations. *Commun. Math. Phys.*, **58**, 211–221.

[53] Collins, J. C. and Soper, D. E. 1978. Large order expansion in perturbation theory. *Ann. Phys.*, **112**, 209–234.

[54] Cooper, F. and Freedman, B. 1983. Aspects of supersymmetric quantum mechanics. *Ann. Phys.*, **146**, 262–288.

[55] Cooper, F., Khare, A. and Sukhatme, U. 1995. Supersymmetry and quantum mechanics. *Phys. Rep.*, **251**, 267–385.

[56] Costin, O. 2009. *Asymptotics and Borel Summability*. Chapman-Hall.

[57] Cvitanovic, P. 1976. Group theory for Feynman diagrams in non-Abelian gauge theories. *Phys. Rev.*, D **14**, 1536–1553.

[58] Cvitanovic, P. 2008. *Group Theory: Birdtracks, Lie's and Exceptional groups*. Princeton University Press.

[59] Cvitanovic, P. *et al.* 2011. *Chaos: Classical and Quantum*. Gone with the Wind Press. http://chaosbook.org/.

[60] D'Adda, A., Di Vecchia, P. and Luscher, M. 1978. A $1/N$ expandable series of nonlinear sigma models with instantons. *Nucl. Phys.*, B **146**, 63–76.

[61] D'Adda, A., Di Vecchia, P. and Luscher, M. 1979. Confinement and chiral symmetry breaking in $\mathbb{CP}^{n-1}$ models with quarks. *Nucl. Phys.*, B **152**, 125–144.

[62] Daniel, M. and Viallet, C. M. 1980. The geometrical setting of gauge theories of the Yang–Mills type. *Rev. Mod. Phys.*, **52**, 175–197.

[63] Dashen, R. F., Hasslacher, B. and Neveu, A. 1974. Nonperturbative methods and extended hadron models in field theory. 1. Semiclassical functional methods. *Phys. Rev.*, D **10**, 4114–4129.

[64] David, F. 1991. Phases of the large N matrix model and non-perturbative effects in 2-D gravity. *Nucl. Phys.*, B **348**, 507–524.

[65] David, F. 1993. Non-perturbative effects in matrix models and vacua of two-dimensional gravity. *Phys. Lett.*, B **302**, 403–410.

[66] Delabaere, E., Dillinger, H. and Pham, F. 1997. Exact semiclassical expansions for one-dimensional quantum oscillators. *J. Math. Phys.*, **38**, 6126–6184.

[67] Del Debbio, L., Giusti, L. and Pica, C. 2005. Topological susceptibility in the $SU(3)$ gauge theory. *Phys. Rev. Lett.*, **94**, 032003.

[68] Di Francesco, P. 2006. 2D quantum gravity, matrix models and graph combina-torics. In: *Applications of Random Matrices in Physics*, E. Brézin *et al.* (eds.), pp. 33–88. Springer-Verlag.

[69] Di Francesco, P., Ginsparg, P. H. and Zinn-Justin, J. 1995. 2-D gravity and random matrices. *Phys. Rep.*, **254**, 1–133.

[70] Dingle, R. B. 1973. *Asymptotic Expansions: their Derivation and Interpretation*. Academic Press.

[71] Di Vecchia, P. 1979. An effective Lagrangian with no $U(1)$ problem in $\mathbb{CP}^{n-1}$ models and QCD. *Phys. Lett.*, B **85**, 357–360.

[72] Di Vecchia, P. and Veneziano, G. 1980. Chiral dynamics in the large n limit. *Nucl. Phys.*, B **171**, 253–272.

[73] Donaldson, S. K. and Kronheimer, P. B. 1990. *The Geometry of Four-Manifolds*. Oxford University Press.

[74] Donoghue, J. F., Golwich, E. and Holstein, B. R. 1994. *Dynamics of the Standard Model*. Cambridge University Press.

[75] Dorey, N., Hollowood, T. J., Khoze, V. V. and Mattis, M. P. 2002. The calculus of many instantons. *Phys. Rep.*, **371**, 231–459.

[76] Dorigoni, D. 2014. An introduction to resurgence, trans-series and alien calculus. arXiv:1411.3585 [hep-th].

[77] Dunne, G. V. 2002. Perturbative–nonperturbative connection in quantum mechanics and field theory. In: *Continuous Advances in QCD*, K. A. Olive *et al.* (eds.), pp. 478–505. World Scientific.

[78] Dunne, G. V. 2008. Functional determinants in quantum field theory. *J. Phys. A: Math. Theor.*, **41**, 304006.

[79] Dunne, G. V. and Min, H. 2005. Beyond the thin-wall approximation: precise numerical computation of prefactors in false vacuum decay. *Phys. Rev.*, D **72**, 125004.

[80] Dunne, G. V. and Unsal, M. 2012. Resurgence and trans-series in quantum field theory: the $\mathbb{CP}^{N-1}$ model. *JHEP*, **1211**, 170.

[81] Dunne, G. V. and Unsal, M. 2014. Uniform WKB, multi-instantons, and resurgent trans-series. *Phys. Rev.*, D **89**, 105009.

[82] Dyson, F. J. 1952. Divergence of perturbation theory in quantum electrodynamics. *Phys. Rev.*, **85**, 631–632.

[83] Einhorn, M. B. and Wudka, J. 2003. On the Vafa–Witten theorem on spontaneous breaking of parity. *Phys. Rev.*, D **67**, 045004.

[84] Eynard, B. 2004. Topological expansion for the 1-Hermitian matrix model correla-tion functions. *JHEP*, **0411**, 031.

[85] Eynard B. and Orantin, N. 2007. Invariants of algebraic curves and topological expansion. *Commun. Num. Theor. Phys.*, **1**, 347–452.

[86] Feynman, R. P. 1998. *Statistical Mechanics*. Westview Press.

[87] Forrester, P. J. 2010. *Log-Gases and Random Matrices*. Princeton University Press.

[88] Forrester, P. J. and Warnaar, S. O. 2008. The importance of the Selberg integral. *Bull. Am. Math. Soc. (N.S.)*, **45**, 489–534.

[89] Frishman, Y. and Sonnenschein, J. 2010. *Non-Perturbative Field Theory*. Cambridge University Press.

[90] Fujikawa, K. 1980. Path integral for gauge theories with fermions. *Phys. Rev.*, D **21**, 2848–2858.

[91] Fujikawa, K. and Suzuki, H. 2004. *Path Integrals and Quantum Anomalies*. Oxford University Press.

[92] Gasser, J. and Leutwyler, H. 1984. Chiral perturbation theory to one loop. *Ann. Phys.*, **158**, 142–210.

[93] Gibbons, G. W. and Hawking, S. W. 1977. Action integrals and partition functions in quantum gravity. *Phys. Rev.*, D **15**, 2752–2756.

[94] Ginsparg, P. H. and Moore, G. W. 1993. Lectures on 2-D gravity and 2-D string theory. hep-th/9304011.

[95] Ginsparg, P. H. and Zinn-Justin, J. 1990. 2-d gravity + 1-d matter. *Phys. Lett.*, B **240**, 333–340.

[96] Giusti, L., Rossi, G. C. and Veneziano, G. 2002. The $U_A(1)$ problem on the lattice with Ginsparg–Wilson fermions. *Nucl. Phys.*, B **628**, 234–252.

[97] Giusti, L., Rossi, G. C. and Testa, M. 2004. Topological susceptibility in full QCD with Ginsparg–Wilson fermions. *Phys. Lett.*, B **587**, 157–166.

[98] Giusti, L., Petrarca, S. and Taglienti, B. 2007. Theta dependence of the vacuum energy in the $SU(3)$ gauge theory from the lattice. *Phys. Rev.*, D **76**, 094510.

[99] Gopakumar, R. 1996. The master field revisited. *Nucl. Phys. Proc. Suppl.*, **45B**, 244–250.

[100] Graffi, S., Grecchi, V. and Simon, B. 1970. Borel summability: application to the anharmonic oscillator. *Phys. Lett.*, B **32**, 631–634.

[101] Gross, D. J. and Matytsin, A. 1994. Instanton induced large N phase transitions in two-dimensional and four-dimensional QCD. *Nucl. Phys.*, B **429**, 50–74.

[102] Gross, D. J. and Witten, E. 1980. Possible third order phase transition in the large N lattice gauge theory. *Phys. Rev.*, D **21**, 446–453.

[103] Gross, D. J., Pisarski, R. D. and Yaffe, L. G. 1981. QCD and instantons at finite temperature. *Rev. Mod. Phys.*, **53**, 43–80.

[104] Gross, D. J., Perry, M. J. and Yaffe, L. G. 1982. Instability of flat space at finite temperature. *Phys. Rev.*, D **25**, 330–355.

[105] Grunberg, G. 1994. Perturbation theory and condensates. *Phys. Lett.*, B **325**, 441–448.

[106] Herbst, I. W. and Simon, B. 1978. Some remarkable examples in eigenvalue perturbation theory. *Phys. Lett.*, B **78**, 304–306.

[107] Herrera-Siklody, P., Latorre, J. I., Pascual, P. and Taron, J. 1997. Chiral effective Lagrangian in the large $N(c)$ limit: the nonet case. *Nucl. Phys.*, B **497**, 345–386.

[108] Huang, S., Negele, J. W. and Polonyi, J. 1988. Meson structure in QCD in two-dimensions. *Nucl. Phys.*, B **307**, 669–704.

[109] Jack, I. and Osborn, H. 1984. Background field calculations in curved space-time. 1. General formalism and application to scalar fields. *Nucl. Phys.*, B **234**, 331–364.

[110] Jackiw, R. 1977. Quantum meaning of classical field theory. *Rev. Mod. Phys.*, **49**, 681–706.

[111] Jackiw, R. 1985. Topological investigations of quantized gauge theories. In: *Current Algebra and Anomalies*, S. B. Treiman, R. Jackiw, B. Zumino and E. Witten (eds.), pp. 240–360. World Scientific.

[112] Jackiw, R. and Rebbi, C. 1976. Vacuum periodicity in a Yang–Mills quantum theory. *Phys. Rev. Lett.*, **37**, 172–175.

[113] Jafarizadeh, M. A. and Fakhri, H. 1997. Calculation of the determinant of shape invariant operators. *Phys. Lett.*, A **230**, 157–163.

[114] Jentschura, U. D. and Zinn-Justin, J. 2011. Multi-instantons and exact results. IV: Path integral formalism. *Ann. Phys.*, **326**, 2186–2242.

[115] Jentschura, U. D., Surzhykov, A. and Zinn-Justin, J. 2010. Multi-instantons and exact results. III: Unification of even and odd anharmonic oscillators. *Ann. Phys.*, **325**, 1135–1172.

[116] Jurkiewicz, J. and Zalewski, K. 1983. Vacuum structure of the $U(N \to$ infinity) gauge theory on a two-dimensional lattice for a broad class of variant actions. *Nucl. Phys.*, B **220**, 167–184.

[117] Kalashnikova, Y. S. and Nefediev, A. V. 2002. Two-dimensional QCD in the Coulomb gauge. *Phys. Usp.*, **45**, 347–368.

[118] Kalashnikova, Y. S., Nefediev, A. V. and Volodin, A. V. 2000. Hamiltonian approach to the bound state problem in QCD_2. *Phys. Atom. Nucl.*, **63**, 1623–1628.

[119] Kaul, R. K. and Rajaraman, R. 1983. Soliton energies in supersymmetric theories. *Phys. Lett.*, B **131**, 357–361.

[120] Konishi, K. and Paffuti, G. 2009. *Quantum Mechanics. A New Introduction*. Oxford University Press.

[121] Koplik, J., Neveu, A. and Nussinov, S. 1977. Some aspects of the planar perturbation series. *Nucl. Phys.*, B **123**, 109–131.

[122] Le Guillou, J. C. and Zinn-Justin, J. (eds.) 1990. *Large Order Behavior of Perturbation Theory*. North-Holland.

[123] Lenz, F., Thies, M., Yazaki, K. and Levit, S. 1991. Hamiltonian formulation of two-dimensional gauge theories on the light cone. *Ann. Phys.*, **208**, 1–89.

[124] Li, M., Wilets, M. and Birse, M. C. 1987. QCD In two-dimensions in the axial gauge. *J. Phys.*, G **13**, 915–923.

[125] Lipatov, L. N. 1977. Divergence of the perturbation theory series and the quasiclassical theory. *Sov. Phys. JETP*, **45**, 216–223.

[126] Lucini, B. and Panero, M. 2013. $SU(N)$ gauge theories at large N. *Phys. Rep.*, **526**, 93–163.

[127] Lüscher, M. 1982. Dimensional regularization in the presence of large background fields. *Ann. Phys.*, **142**, 359–392.

[128] Lüscher, M. 1982. A semiclassical formula for the topological susceptibility in a finite space-time volume. *Nucl. Phys.*, B **205**, 483–503.

[129] Lüscher, M. 2004. Topological effects in QCD and the problem of short-distance singularities. *Phys. Lett.*, B **593**, 296–301.

[130] Lüscher, M. 2010. Properties and uses of the Wilson flow in lattice QCD. *JHEP*, **1008**, 071.

[131] Lüscher, M. and Palombi, F. 2010. Universality of the topological susceptibility in the $SU(3)$ gauge theory. *JHEP*, **1009**, 110.

[132] Majumdar, S. N. and Schehr, G. 2014. Top eigenvalue of a random matrix: large deviations and third order phase transitions. *J. Stat. Mech.*, P01012.

[133] Majumdar, S. N. and Vergassola, M. 2009. Large deviations of the maximum eigenvalue for Wishart and Gaussian random matrices. *Phys. Rev. Lett.*, **102**, 060601.

[134] Manohar, A. V. 1998. Large N QCD. arXiv:hep-ph/9802419.

[135] Mariño, M. 2004. Les Houches lectures on matrix models and topological strings. hep-th/0410165.

[136] Mariño, M. 2008. Nonperturbative effects and nonperturbative definitions in matrix models and topological strings. *JHEP*, **0812**, 114.

[137] Mariño, M. 2014. Lectures on non-perturbative effects in large N gauge theories, matrix models and strings. *Fortschr. Phys.*, **62**, 455–540.

[138] Mariño, M. and Putrov, P. 2009. Multi-instantons in large N matrix quantum mechanics. arXiv:0911.3076 [hep-th].

[139] Mariño, M., Schiappa, R. and Weiss, M. 2008. Non-perturbative effects and the large-order behavior of matrix models and topological strings. *Commun. Num. Theor. Phys.*, **2**, 349–419.

[140] McKane, A. J. and Tarlie, M. B. 1995. Regularisation of functional determinants using boundary perturbations. *J. Phys.*, A **28**, 6931–6942.

[141] Meggiolaro, E. 1998. The topological susceptibility of QCD: from Minkowskian to Euclidean theory. *Phys. Rev.*, D **58**, 085002.

[142] Mehta, M. L. 2004. *Random Matrices*. Elsevier.

[143] Miller, P. 2006. *Applied Asymptotic Analysis*. American Mathematical Society.

[144] Münster, G. 1982. The $1/N$ expansion and instantons in $\mathbb{CP}^{N-1}$ models on a sphere. *Phys. Lett.*, B **118**, 380–384.

[145] Münster, G. 1983. A study of $\mathbb{CP}^{N-1}$ models on the sphere within the $1/N$ expansion. *Nucl. Phys.*, B **218**, 1–31.

[146] Negele, J. W. 1982. The mean-field theory of nuclear structure and dynamics. *Rev. Mod. Phys.*, **54**, 913–1015.

[147] Negele, J. W. and Orland, H. 1998. *Quantum Many-Particle Systems*. Westview Press.

[148] Nepomechie, R. I. 1985. Calculating heat kernels. *Phys. Rev.*, D **31**, 3291–3292.

[149] Neuberger, H. 1980. Instantons as a bridgehead at N = infinity. *Phys. Lett.*, B **94**, 199–202.

[150] Neuberger, H. 1981. Nonperturbative contributions in models with a nonanalytic behavior at infinite N. *Nucl. Phys.*, B **179**, 253–282.

[151] Osborn, H. 1981. Semiclassical functional integrals for selfdual gauge fields. *Ann. Phys.*, **135**, 373–415.

[152] Parisi, G. 1978. Singularities of the borel transform in renormalizable theories. *Phys. Lett.*, B **76**, 65–66.

[153] Perelomov, A. M. 1987. Chiral models: geometrical aspects. *Phys. Rep.*, **146**, 135–213.

[154] Peskin, M. E. and Schroeder, D. V. 1995. *An Introduction to Quantum Field Theory*. Addison-Wesley.

[155] Polyakov, A. M. 1977. Quark confinement and topology of gauge groups. *Nucl. Phys.*, B **120**, 429–458.

[156] Polyakov, A. M. 1987. *Gauge Fields and Strings*. Harwood Academic Publishers.

[157] Rajaraman, R. 1982. *Solitons and Instantons*. North-Holland.

[158] Ramond, P. 2001. *Field Theory. A Modern Primer*, second edition. Westview Press.

[159] Salomonson, P. and van Holten, J. W. 1982. Fermionic coordinates and supersymmetry in quantum mechanics. *Nucl. Phys.*, B **196**, 509–531.

[160] Schafer, T. and Shuryak, E. V. 1998. Instantons in QCD. *Rev. Mod. Phys.*, **70**, 323–426.

[161] Schwab, P. 1982. Semiclassical approximation for the topological susceptibility in $\mathbb{CP}^{N-1}$ models on a sphere. *Phys. Lett.*, B **118**, 373–379.

[162] Schwab, P. 1983. Two instanton contribution to the topological susceptibility in $\mathbb{CP}^{N-1}$ models on a sphere. *Phys. Lett.*, B **126**, 241–246.

[163] Schwarz, A. S. 1979. Instantons and fermions in the field of instanton. *Commun. Math. Phys.*, **64**, 233–268.

[164] Seara, T. M. and Sauzin, D. 2003. Ressumació de Borel i teoria de la ressurgència. *Bull. Soc. Catlana Mat.*, **18**, 131–153.

[165] Seiler, E. 2002. Some more remarks on the Witten–Veneziano formula for the eta-prime mass. *Phys. Lett.*, B **525**, 355–359.

[166] Shenker, S. H. 1992. The strength of nonperturbative effects in string theory. In: *Random Surfaces and Quantum Gravity*, O. Álvarez, E. Marinari and P. Windey (eds.), pp. 191–200. Plenum Press.

[167] Shifman, M. 2012. *Advanced Topics in Quantum Field Theory*. Cambridge University Press,

[168] Shore, G. M. 1979. Dimensional regularization and instantons. *Ann. Phys.*, **122**, 321–372.

[169] Simon, B. 1982. Large orders and summability of eigenvalue perturbation theory: a mathematical overview. *Int. J. Quant. Chem.*, **21**, 3–25.

[170] Stone, M. 1977. Semiclassical methods for unstable states. *Phys. Lett.*, B **67**, 186–188.

[171] Stone, M. and Reeve, J. 1978. Late terms in the asymptotic expansion for the energy levels of a periodic potential. *Phys. Rev.*, D **18**, 4746–4751.

[172] Takhtajan, L. 2008. *Quantum Mechanics for Mathematicians*. American Mathematical Society.

[173] 't Hooft, G. 1974. A planar diagram theory for strong interactions. *Nucl. Phys.*, B **72**, 461–473.

[174] 't Hooft, G. 1974. A two-dimensional model for mesons. *Nucl. Phys.*, B **75**, 461–470.

[175] 't Hooft, G. 1976. Computation of the quantum effects due to a four-dimensional pseudoparticle. *Phys. Rev.*, D **14**, 3432–3450.

[176] Tong, D. 2005. TASI lectures on solitons: instantons, monopoles, vortices and kinks. hep-th/0509216.

[177] Vafa, C. and Witten, E. 1984. Parity conservation in QCD. *Phys. Rev. Lett.*, **53**, 535–536.

[178] Vandoren, S. and van Nieuwenhuizen, P. 2008. Lectures on instantons. arXiv:0802.1862 [hep-th].

[179] Vassilevich, D. V. 2003. Heat kernel expansion: user's manual. *Phys. Rep.*, **388**, 279–360.

[180] Veneziano, G. 1979. $U(1)$ without instantons. *Nucl. Phys.*, B **159**, 213–224.

[181] Vicari, E. 1999. The Euclidean two point correlation function of the topological charge density. *Nucl. Phys.*, B **554**, 301–312.

[182] Vicari, E. and Panagopoulos, H. 2009. Theta dependence of $SU(N)$ gauge theories in the presence of a topological term. *Phys. Rep.*, **470**, 93–150.

[183] Wadia, S. R. 1979. A study of $U(N)$ lattice gauge theory in 2-dimensions. EFI-79/44-CHICAGO, arXiv:1212.2906 [hep-th].

[184] Wadia, S. R. 1980. N = infinity phase transition in a class of exactly soluble model lattice gauge theories. *Phys. Lett.*, B **93**, 403–410.

[185] Weinberg, S. 1975. The $U(1)$ problem. *Phys. Rev.*, D **11**, 3583–3593.

[186] Weinberg, S. 1996. *The Quantum Theory of Fields*. Volume II: *Modern Applications*. Cambridge University Press.

[187] Witten, E. 1979. Instantons, the quark model, and the $1/N$ expansion. *Nucl. Phys.*, B **149**, 285–320.

[188] Witten, E. 1979. Current algebra theorems for the $U(1)$ goldstone boson. *Nucl. Phys.*, B **156**, 269–283.

[189] Witten, E. 1979. Baryons in the $1/N$ expansion. *Nucl. Phys.*, B **160**, 57–115.

[190] Witten, E. 1980. The $1/N$ expansion in atomic and particle physics. In: *Recent Developments in Gauge Theories*, G. 't Hooft *et al.* (eds.), pp. 403–419. Plenum Press.

[191] Witten, E. 1980. Quarks, atoms, and the $1/N$ expansion. *Phys. Today*, **33**, 38–43.

[192] Witten, E. 1980. Large N chiral dynamics. *Ann. Phys.*, **128**, 363–375.

[193] Witten, E. 1981. Dynamical breaking of supersymmetry. *Nucl. Phys.*, B **188**, 513–554.

[194] Witten, E. 1982. Instability of the Kaluza–Klein vacuum. *Nucl. Phys.*, B **195**, 481–492.

[195] Witten, E. 1998. Theta dependence in the large N limit of four-dimensional gauge theories. *Phys. Rev. Lett.*, **81**, 2862–2865.

[196] Yaris, R., Bendler, J., Lovett, R., Bender, C. M. and Fedders, P. A. 1978. Resonance calculations for arbitrary potentials. *Phys. Rev.*, A **18**, 1816–1825.

[197] Ynduráin, F. J. 2006. *The Theory of Quark and Gluon Interactions*. Springer-Verlag.

[198] Zinn-Justin, J. 1983. Multi-instanton contributions in quantum mechanics. 2. *Nucl. Phys.*, B **218**, 333–348.

[199] Zinn-Justin, J. 2002. *Quantum Field Theory and Critical Phenomena*. Oxford University Press.

[200] Zinn-Justin, J. and Jentschura, U. D. 2004. Multi-instantons and exact results I: conjectures, WKB expansions, and instanton interactions. *Ann. Phys.*, **313**, 197–267.

[201] Zinn-Justin, J. and Jentschura, U. D. 2004. Multi-instantons and exact results II: specific cases, higher-order effects, and numerical calculations. *Ann. Phys.*, **313**, 269–325.

Author Index

Subject Index

anomaly, 183, 184, 187, 236, 237
anti-instanton, 124, 128, 163, 164

baryon, 232
beta function, 111, 170, 198, 217
Bethe–Salpeter equation, 293
Borel–Padé resummation, 84, 97
Borel resummation, 82–85, 97, 99
    lateral, 85, 86, 148
Borel transform, 81–83, 87–89, 97, 148
bounce, 17

Chern–Simons form, 116
chiral symmetry, 171, 172, 175, 177, 179–181
chiral symmetry breaking (χSB), 172, 173, 237
collective coordinate, 22–24, 47, 66, 127, 139, 143,
    305

decay rate, 11, 35, 62, 66, 67, 78, 98
density of eigenvalues, 250, 252, 253, 255, 256, 259,
    261, 263, 314
determinant
    functional, 21, 23, 26, 28, 30, 32, 34, 36–38, 46, 49,
      51, 166, 168, 169
    spectral, 49
dilute instanton approximation, 47
dimensional regularization, 320
discontinuity, 14, 16, 86, 87, 89, 91, 253–256, 260
double-well potential, 38, 40, 42, 49, 54, 57, 102, 104,
    148, 163
dynamically generated scale, 111, 198

energy
    ground state, 3, 4, 7, 12, 25, 35, 36, 39, 58, 59, 78,
      92, 97, 266
equation of motion (EOM), 3, 16–18, 22, 59, 63, 124,
    161
Euler angles, 327

fluctuation, 19, 21, 24, 60, 134, 136–140, 163, 164,
    166, 199, 212, 299, 306
four-manifold, 130, 132

Gell-Mann–Okubo mass formula, 179
glueball, 230–232
Goldstone boson, 173, 175, 176, 180, 183
Gross–Witten–Wadia (GWW) model, 262, 273, 311,
    325
Gross–Witten–Wadia (GWW) phase transition, 263,
    313

Hartree–Fock approximation, 232, 278, 296, 298
heat kernel, 143, 185, 319, 330, 331, 334–336, 340,
    341, 343, 348
homotopy group, 118, 120
Hubbard–Stratonovich transformation, 298, 317

index, 164, 188
instanton, 3, 17, 63, 161, 163, 164, 201, 203, 207,
    208, 210
    complex, 99, 100, 102
    gas, 45, 47, 48
    in YM theory, 124, 127, 132, 148
    large N, 301, 309–311, 315, 316, 319, 324

large N phase transition, 307

Majumdar–Vergassola right deviation function, 315
Marcenko–Pastur law, 314
master field, 234, 252
matrix model
    Gaussian, 244, 245, 247, 248, 255
    quartic, 256, 266, 310
    Wishart, 313, 315
meson, 173, 175, 176, 178, 179, 230–232, 274, 286,
    292, 296
mode
    negative, 24, 25, 34, 40, 66, 72